DIVERSITY AND DYNAMICS IN FOREST ECOSYSTEMS

DIVERSITY AND DYNAMICS IN FOREST ECOSYSTEMS

Edited by
Munesh Kumar, PhD
Nazir A. Pala, PhD
Jahangeer A. Bhat, PhD

AΛP | APPLE ACADEMIC PRESS

First edition published 2022

Apple Academic Press Inc.
1265 Goldenrod Circle, NE,
Palm Bay, FL 32905 USA

4164 Lakeshore Road, Burlington,
ON, L7L 1A4 Canada

CRC Press
6000 Broken Sound Parkway NW,
Suite 300, Boca Raton, FL 33487-2742 USA

2 Park Square, Milton Park,
Abingdon, Oxon, OX14 4RN UK

Library and Archives Canada Cataloguing in Publication

Title: Diversity and dynamics in forest ecosystems / edited by Munesh Kumar, PhD, Nazir A. Pala, PhD, Jahangeer A. Bhat, PhD.

Names: Kumar, Munesh (Lecturer in forestry), editor. | Pala, Nazir A., editor. | Bhat, Jahangeer A., editor.

Description: First edition. | Includes bibliographical references and index.

Identifiers: Canadiana (print) 20210279966 | Canadiana (ebook) 20210280034 | ISBN 9781771889797 (hardcover) | ISBN 9781774638743 (softcover) | ISBN 9781003145318 (ebook)

Subjects: LCSH: Forest ecology. | LCSH: Forests and forestry.

Classification: LCC QH541.5.F6 D56 2022 | DDC 577.3—dc23

Library of Congress Cataloging-in-Publication Data

Names: Kumar, Munesh (Lecturer in forestry), editor. | Pala, Nazir A., editor. | Bhat, Jahangeer A., editor.

Title: Diversity and dynamics in forest ecosystems / edited by Munesh Kumar, PhD, Nazir A. Pala, PhD, Jahangeer A. Bhat, PhD.

Description: First edition. | Palm Bay : Apple Academic Press, [2022] | Includes bibliographical references and index. | Summary: "Providing a wealth of in-depth knowledge of forest ecosystems, this new volume explores a collection of important topics on forest community dynamics. It looks at the diversity of forest ecosystems and explores such aspects as forest products in enhancing local livelihoods and community participation, forage production, forest conservation and sustainable management, regeneration patterns, seed handling, and more. Chapters in Diversity and Dynamics in Forest Ecosystems present new research on forest products, livelihood generation mechanisms of forest-dependent communities, utilization patterns of untapped resources from forests, and the structure of different ecosystems from the tropical to the temperate landscape. This book also features different drivers of community dynamics, such as the role of seed handling in forests, the influence of altitudinal variations, and protected and community-conserved forests on the forest diversity. Chapters also consider the role of nontimber forest products and their significance in livelihood diversification for tribal communities and forage crop genetic resources, and forest resource extraction by forest fringe dwellers. Also explored are aspects of soil organic carbon in agroforestry systems and integrated approaches of sustainable agroforestry development in diverse forest ecosystems. This edition also examines the vegetation structure and regeneration aspects of timberline zone, including diversity of herbaceous flora along the altitudinal gradient. The abundance of in-depth knowledge of the diversity and dynamics of forest ecosystems in this volume will be valuable in conservation and management of forests, which play an important role in the world environment. Forests are presently facing multiple disturbances, and this volume will help forestry professionals and others formulate further strategies to mitigate global climate change and other challenges"-- Provided by publisher.

Identifiers: LCCN 2021036505 (print) | LCCN 2021036506 (ebook) | ISBN 9781771889797 (hardback) | ISBN 9781774638743 (paperback) | ISBN 9781003145318 (ebook)

Subjects: LCSH: Forest ecology. | Forest dynamics. | Forest biodiversity.

Classification: LCC QH541.5.F6 D57 2022 (print) | LCC QH541.5.F6 (ebook) | DDC 577.3--dc23

LC record available at https://lccn.loc.gov/2021036505

LC ebook record available at https://lccn.loc.gov/2021036506

ISBN: 978-1-77188-979-7 (hbk)
ISBN: 978-1-77463-874-3 (pbk)
ISBN: 978-1-00314-531-8 (ebk)

About the Editors

Munesh Kumar, PhD

Associate Professor, Department of Forestry and Natural Resources, H.N.B. Garhwal University, Srinagar Garhwal, Uttarakhand, India

Munesh Kumar, PhD, is Associate Professor in the Department of Forestry and Natural Resources, H.N.B. Garhwal University, Srinagar Garhwal, Uttarakhand, India. He also served at Mizoram University as Assistant Professor in the Department of Forestry. His area of research interest is Forest Ecology, Agroforestry, Ethnobotany. He obtained his MSc and the PhD degree in Forestry from H.N.B. Garhwal University. He has more than 15 years of teaching and research experience. He has published more than 120 research papers in international and national journals.

Nazir A. Pala, PhD

Assistant Professor cum Scientist, Division of Silviculture and Agroforestry, Faculty of Forestry, SKUAST- Kashmir, India

Nazir A. Pala, PhD, is working as Assistant Professor cum Scientist in the Division of Silviculture and Agroforestry, Faculty of Forestry, SKUAST-Kashmir. Dr. Pala has completed his schooling from Jammu and Kashmir BOSE and BSc, MSc and PhD in Forestry from HNB Garhwal University, Srinagar Garhwal, Uttarakhand. Dr. Pala has nine years of research and teaching experience at the UG and PG level. Dr. Pala has so far co-authored two books and published more than 100 research papers in both national and internal reputed journals. Dr. Pala is actively involved in several national-level research projects at the capacity of PI/Co-PI and apart from guiding students at the PG level.

Jahangeer A. Bhat, PhD

Head, Department of Forestry,
College of Agriculture, Fisheries and Forestry,
Fiji National University, Republic of Fiji Islands

Jahangeer A. Bhat, PhD, is Head of the Department of Forestry at the College of Agriculture, Fisheries and Forestry at Fiji National University, Republic of Fiji Islands. Dr. Bhat is a counselor, mentor, and coordinator for forestry academic programs. He has been instrumental in developing HE and TVET streams of forestry and allied programs and worked closely in the area of accreditation with the Fiji Higher Education Commission and forestry stakeholders. Before joining Fiji National University, he worked at H.N.B. Garhwal University and has 11 years of research and eight years of teaching experience, with a publication record of more than 50, which includes research articles, review papers, conference papers, and books with national and international repute. Dr. Bhat is reviewing research articles for a number of scientific journals and has handled research projects in his capacity as PI and Co-PI. His major interests lie in emerging issues in forestry, including conservation of biodiversity, traditional knowledge of plants, and sustainable management of forest resources, with his main focus of research on vegetation ecology, ethnobotany, and evaluation of ecosystem services, forest plant biodiversity, climate change, and sociocultural issues in forestry. Dr. Jahnageer A. Bhat is currently associated with the College of Horticulture and Forestry, Rani Lakshmi Bai Central Agricultural University, Jhansi (U.P.), India.

Contents

Contributors

Suheel Ahmad
Regional Research Station, ICAR-Indian Grassland and Fodder Research Institute, CITH Campus, Rangreth, Srinagar–190007, Jammu and Kashmir, India, E-mail: suhail114@gmail.com

Khushnooda Anjum
Faculty of Forestry, Sher-e-Kashmir University of Agricultural Sciences and Technology of Kashmir, Benhama, Ganderbal–191201, Jammu and Kashmir, India

J. A. Baba
KVK/ETC Malangpora, Pulwama, Sher-e-Kashmir University of Agriculture Sciences and Technology of Kashmir, Jammu and Kashmir, India

Muneesa Banday
Faculty of Forestry, Benhama, Ganderbal, Sher-e-Kashmir University of Agricultural Sciences and Technology of Kashmir, Jammu and Kashmir, India, E-mail: 13forestry08@gmail.com

S. S. Bargali
Department of Botany, DSB Campus, Kumaun University, Nainital – 263001, Uttarakhand, India/ Department of Environmental science, H.N.B. Garhwal University, Srinagar Garhwal, Uttarakhand, India

G. M. Bhat
Faculty of Forestry, Sher-e-Kashmir University of Agricultural Sciences and Technology of Kashmir, Benhama, Ganderbal–191201, Jammu and Kashmir, India

Jahangeer A. Bhat
Department of Forestry, College of Agriculture, Fisheries and Forestry, Koronivia, PO Box–1544, Nausori, Fiji National University, Republic of Fiji Islands, Fax: +679 340 0275 / Presently at College of Horticulture and Forestry, Rani Lakshmi Bai Central Agricultural University, Jhansi – 284003 (UP), India, E-mail: jahan191@gmail

Sheeraz Saleem Bhat
Regional Research Station, ICAR-Indian Grassland and Fodder Research Institute, CITH Campus, Rangreth, Srinagar–190007, Jammu and Kashmir, India, E-mail: shrzbhat@gmail.com

A. B. Bhatt
Ecology Laboratory, Department of Botany and Microbiology, HNB Garhwal University Srinagar, Garhwal, Uttarakhand–246174, India

Arvind Bijalwan
College of Forestry, VCSG Uttarakhand University of Horticulture and Forestry, Ranichauri–249199, Tehri Garhwal, Uttarakhand, India, E-mail: arvindbijalwan276@gmail.com

David Lopez Cornelio
School of Natural Resources and Applied Sciences, Solomon Islands National University, P.O Box R113, Honiara, E-mail: david.cornelio@sinu.edu.sb

M. Vassanda Coumar
Indian Institute of Soil Science, Bhopal, Madhya Pradesh, India

Manzoor A. Dar
Doon (PG) College of Agriculture, Science and Technology, Camp Road, Central Hope Town–248197, Dehradun, Uttarakhand, India, E-mail: manzoorhmd5@gmail.com

Sudhansu Sekhar Dash
Botanical Survey of India, CGO Complex, Salt Lake City, Kolkata–700064, West Bengal, India, E-mail: ssdash2002@gmail.com

Manmohan Jagatram Dobriyal
College of Horticulture and Forestry, Rani Lakshmi Bai Central Agricultural University, Jhansi–284003, Uttar Pradesh, India

A. A. Gatoo
Faculty of Forestry, Sher-e-Kashmir University of Agricultural Sciences and Technology of Kashmir, Benhama, Ganderbal–191201, Jammu and Kashmir, India

Peerzada Ishtiyak
Faculty of Forestry, Benhama, Ganderbal, Sher-e-Kashmir University of Agricultural Sciences and Technology of Kashmir, Jammu and Kashmir, India

Ajaz Ul Islam
Faculty of Forestry, Benhama, Ganderbal, Sher-e-Kashmir University of Agriculture Sciences and Technology of Kashmir, Jammu and Kashmir, India

M. A. Islam
Faculty of Forestry, Sher-e-Kashmir University of Agricultural Sciences and Technology of Kashmir, Benhama, Ganderbal–191201, Jammu and Kashmir, India, E-mail: ajaztata@gmail.com

Munesh Kumar
Department of Forestry and Natural Resources, HNB Garhwal University, (A Central University), Srinagar, Garhwal, Uttarakhand, India, E-mail: muneshmzu@yahoo.com

Vikas Kumar
High Altitude Biology, CSIR-Institute of Himalayan Bioresource Technology, Palampur–176061, Himachal Pradesh, India

R. K. Maikhuri
G.B. Pant Institute of Himalayan Environment and Development, Garhwal Unit, Post Box No.–92, Srinagar, Garhwal, Uttarakhand, India / Department of Environmental science, H.N.B. Garhwal University, Srinagar Garhwal, Uttarakhand, India

Ajay Maletha
Amity Institute of Forestry and Wildlife, Amity University, Sector–125 Noida, Uttar Pradesh, India, E-mail: maletha.jay@gmail.com

A. R. Malik
Faculty of Forestry, Sher-e-Kashmir University of Agriculture Sciences and Technology of Kashmir, Benhama, Ganderbal–191201, Jammu and Kashmir, India; High Mountain Arid Agriculture Research Institute, Krishi Vigyan Kendra, Leh, Ladakh, Jammu and Kashmir, India, E-mail: malikrashid2@gmail.com

Abdul Raouf Malik
Sher-e Kashmir University of Agricultural Sciences and Technology of Kashmir, Jammu and Kashmir, India

Zubair A. Malik
Ecology Laboratory, Department of Botany and Microbiology, HNB Garhwal University Srinagar, Garhwal, Uttarakhand–246174, India; Department of Biology, Government HSS, Harduturoo, Anantnag, Jammu and Kashmir–192201, India, E-mail: malikmzubair081@gmail.com

Nazim Hamid Mir
Regional Research Station, ICAR-Indian Grassland and Fodder Research Institute, CITH Campus, Rangreth, Srinagar–190007, Jammu and Kashmir, India

J. A. Mugloo
Faculty of Forestry, Benhama, Ganderbal, Sher-e-Kashmir University of Agriculture Sciences and Technology of Kashmir, Jammu and Kashmir, India

Shah Murtaza
Faculty of Forestry, Sher-e-Kashmir University of Agricultural Sciences and Technology of Kashmir, Benhama, Ganderbal–191201, Jammu and Kashmir, India

D. Namgyal
Faculty of Forestry, Benhama, Ganderbal, Sher-e-Kashmir University of Agriculture Sciences and Technology of Kashmir, Jammu and Kashmir, India

Saswat Nayak
Department of Forestry, School of Earth Sciences and Natural Resource Management, Mizoram University, Aizawl–796004, Mizoram, India; College of Forestry, Orissa University of Agriculture and Technology, Bhubaneswar–751003, Odisha, India

A. K. Negi
Department of Forestry and Natural Resources, HNB Garhwal University (A Central University) Srinagar, Uttarakhand–246174, India

Nazir A. Pala
Faculty of Forestry, Sher-e-Kashmir University of Agricultural Sciences and Technology of Kashmir, Benhama, Ganderbal–191201, Jammu and Kashmir, India, E-mail: nazirpaul@gmail.com

Samiran Panday
Department of Botany, Budge Budge College, 7-DBC Road, Budge Budge, South 24 Parganas–700137, West Bengal, India

P. K. Chandrasekhara Pillai
Ex-Head, Silviculture Department and Scientist-in-Charge, Kerala Forest Seed Centre, Kerala Forest Research Institute, Peechi, Thrissur–680 653, Kerala, India, E-mail: pkcpillai@gmail.com

Susheel Kumar Raina
Regional Research Station, ICAR-National Bureau of Plant Genetic Resources, CITH Campus, Rangreth, Srinagar–191132, Jammu and Kashmir, India

Rameez Raja
Faculty of Forestry, Benhama, Ganderbal, Sher-e-Kashmir University of Agricultural Sciences and Technology of Kashmir, Jammu and Kashmir, India

Kulasekaran Ramesh
Indian Institute of Oilseeds Research, Hyderabad, Telangana, India

Megna Rashid
Faculty of Forestry, Sher-e-Kashmir University of Agricultural Sciences and Technology of Kashmir, Benhama, Ganderbal–191201, Jammu and Kashmir, India

M. M. Rather
Faculty of Forestry, Sher-e-Kashmir University of Agricultural Sciences and Technology of Kashmir, Benhama, Ganderbal–191201, Jammu and Kashmir, India

Dinesh Singh Rawat
Department of Botany and Microbiology, HNB Garhwal University, Srinagar, Garhwal–246174, Uttarakhand, India

Uttam Kumar Sahoo
Department of Forestry, School of Earth Sciences and Natural Resource Management,
Mizoram University, Aizawl–796004, Mizoram, India, E-mails: uksahoo_2003@rediffmail.com;
uttams64@rediffmail.com

Paramjit Singh
Department of Botany, Central University of Punjab, Bathinda–151001, Punjab, India

Bipin Kumar Sinha
Botanical Survey of India, CGO Complex, Salt Lake City, Kolkata–700064, West Bengal, India

P. A. Sofi
Faculty of Forestry, Sher-e-Kashmir University of Agricultural Sciences and Technology of Kashmir,
Benhama, Ganderbal–191201, Jammu and Kashmir, India

Sheikh Mohammad Sultan
Regional Research Station, ICAR-National Bureau of Plant Genetic Resources, CITH Campus,
Rangreth, Srinagar–191132, Jammu and Kashmir, India

Tarun Kumar Thakur
Department of Environmental Science, Indira Gandhi National Tribal University (IGNTU),
Amarkantak, Madhya Pradesh, India

N. P. Todaria
Department of Forestry and Natural Resources, HNB Garhwal University, (A Central University),
Srinagar, Garhwal, Uttarakhand, India

Anup Prakash Upadhyay
Indian Institute of Forest Management, Bhopal, Madhya Pradesh, India

Pooja Verma
Indian Institute of Forest Management, Bhopal, Madhya Pradesh, India

A. A. Wani
Faculty of Forestry, Sher-e-Kashmir University of Agricultural Sciences and Technology of Kashmir,
Benhama, Ganderbal–191201, Jammu and Kashmir, India

Mudasir Youssouf
Centre for Environmental Science and Technology, Central University of Punjab, Bathinda,
Punjab–151001, India

Abbreviations

ACU	average cattle unit
ACZs	agroclimatic zones
AH	agri-horticulture
AS	agri-silviculture
ASH	agri-silvi-horticulture
ASP-I	assessment period I
CBD	convention of biological diversity
CBH	circumference at breast height
CCA	canonical correspondence analysis
CCAs	community-conserved areas
CD	concentration of dominance
CITES	Convention of International Trade in Endangered Species
DRDO	Defense Research and Development Organization
FCGRs	forage crop genetic resources
GA_3	gibberellic acid
GBH	girth at breast height
GHG	greenhouse gas
GIS	geographic information system
H_2O_2	hydrogen peroxide
IAA	indole acetic acid
IHR	Indian Himalayan region
IV	indicator value
IVI	important value index
JFM	joint forest management
KWLS	Kedarnath Wildlife Sanctuary
KWS	Kamlang Wildlife Sanctuary
MAI	mean annual increment
MAPs	medicinal and aromatic plant species
MC	moisture content
MPTs	multipurpose trees
NDBR	Nanda Devi Biosphere Reserve
NNP	Namdapha National Park
NTFPs	non-timber forest products
OSR	observed species richness

PAN	protected area network
PGRs	plant genetic resources
PSH	pastoral-silvi-horticulture
RET	rare, endangered, and threatened
SOC	soil organic carbon
SP	silvi-pastoral
SPSS	statistical package for social sciences
SR	species richness
TBC	total basal cover
TIV	total importance value
TTZ	topographical tetrazolium
TWINSPAN	two-way indicator species analysis

Foreword

It gives me imminence pleasure to know that three of my former students presently working at different Universities have come together with this very explanatory edition in the form of a book entitled "Diversity and Dynamics of Forest Ecosystems." This book edition contains different chapters covering various aspects of the forest ecosystem and its dynamics. The forest ecosystem is a dynamic one, and the biodiversity it contains provides various ecosystem services to mankind. Biodiversity is believed to be closely related to ecosystem functions, and the relationship between biodiversity and ecosystem functions is a central topic of ecological research. The rapid increase in the global population with the advancement in lifestyle, modernization, and rising aspiration of people has increased the pressure on forest resources multi-folds. Therefore, there is an urgent need to understand the forest ecosystem and its dynamics in terms of conservation, utilization, and sustainable management of resources in conjunction with combating biodiversity loss. This edition has compiled all the important and necessary contents for future conservation strategies. I am glad to see the coverage and contents with exhaustive information and would like to congratulate authors for their contribution and editors of this book for its compilation and editing. I also congratulate the publisher, CRC press, for accepting this edition for publication. I am sure that this edition will be of great assistance with the updated and latest information to the targeted readers, particularly students, scholars, and academicians.

—**Prof. N. P. Todaria**
Former Dean, School of Agriculture and Allied Science,
Former Head, Department of Forestry and Natural Resources,
HNB Garhwal University, Srinagar, Uttarakhand, India

Preface

We are pleased to bring out this edited book titled, *Diversity and Dynamics in Forest Ecosystems*. The inference of forest community dynamics and diversity of ecosystems touches new dimensions as desirable for sustainable forest management and conservation in today's world. The understanding of forest structure, diversity, community participation, conservation, and sustainable management of forest resources is of utmost importance to formulate further strategies for global climate change and mitigation. Therefore, it becomes imperative to study the aspects of forestry in terms of community dynamics, structure, and diversity at the different ecosystem level to framing up of policies for sustainable development.

This edition is comprised of fourteen chapters in which efforts were given to bring most of the research works pertaining to forest products, livelihood generation mechanism of forest-dependent communities, utilization pattern of untapped resources from forests to the structure of different ecosystems from tropical to the temperate landscape. This book is also featuring the different drivers of community dynamics, such as the role of seed handling in forests, influence of altitudinal variations, protected and community conserved forests on forest diversity. The role of non-timber forest products (NTFPs) and their significance in livelihood diversification for tribal communities and forage crop genetic resources (FCGRs), livelihood generation, and forest resource extraction by the forest fringe dwellers are discussed. The aspects like soil organic carbon (SOC) in agroforestry systems and integrated approach of sustainable agroforestry development in cold arid deserts and the lowland tropical forests of Eastern Himalaya in relation to the environment and tree attributes have been brought up in this book. This edition also encompasses the vegetation structure and regeneration aspects of the timberline zone, including the diversity of herbaceous flora along the altitudinal gradient.

Finally, the editors keenly hope that this edition will disseminate the vital aspects of forest community dynamics and its diversity in different ecosystems. This edition will be helpful for researchers, academicians, and scholars interested in this field of forestry.

We would like to convey our deep appreciation and acknowledgment to all the contributors and well-wishers. The team from AAP–CRC Press also deserves all the praise for their best efforts for their suggestions and publishing this edited book.

—*Editors*

Introduction

Forests provide many services for mankind, such as food, medicine, timber, freshwater, fresh air, and have esthetic values. Understanding of the dynamics is very significant so that forests can be maintained by knowing their biodiversity and natural habitat forests. Biodiversity changes and community dynamics are determined by various factors. Therefore, it is indispensable to understand forest dynamics for effective forest management and conservation drives.

Generally, the factors accountable for community dynamics and changes in diversity are natural and anthropogenic disturbances, and the frequency and intensity of these disturbances lead to changes in community dynamics and biodiversity. However, small-scale changes create and maintain diversity and heterogeneity in a community.

The present edited book provides information on diversity and dynamics in forest ecosystems. This book presents fourteen chapters, and each provides different information about forest resource utilization patterns, their dynamics, and diversity.

Chapter 1 discusses the economic contribution of non-timber forest products (NTFPs) of *Madhuca latifolia* (Mahua) in Odisha and evaluates the importance of this tree for the household economy of tribes of Odisha. Mahua species provide subsistence to the forest fringe dwellers during lean seasons and is being used for a variety of uses for household consumption. However, Mahua products have not been very rosy because of not being a freely tradable non-timber forest product.

Chapter 2 discusses grass/legume intercropping for forage production, and orchard floor management is discussed for its scope and opportunities of fodder crops, which has by and large remained untapped for fodder development. The utilization of these orchards can give a big boost to livestock development. Proper orchard floor management is vital to the health and productivity of fruit trees, with management practices impacting tree growth, fruit yield, and fruit quality. Current recommended orchard floor management practices consist of maintaining a vegetation-free tree row and a grass cover crop in the alleyway. The management of these systems can present a challenge regarding the selection of the proper grass and legume species as well as the maintenance of the optimum balance between the two species in the grass-legume stand.

The results of Chapter 3 indicate that farmers are showing interest in adopting agroforestry than monocropping, which has a significant impact on SOC content, and that can lead to increased carbon sequestration potential and improved soil quality.

The authors of Chapter 4 have conducted a study in the Namdapha National Park (NNP), eastern Himalaya, and found that climber abundance, species richness (SR), and diversity dwindle with an increase in canopy cover and slope angles; with a decrease in altitude and tree species richness.

Chapter 5 discusses vegetation analysis and regeneration pattern of dominant tree species in the timberline zone of the Nanda Devi Biosphere Reserve (NDBR) aimed to generate baseline information on the ecological characteristics of timberline ecotone in NDBR, especially the extent and changes in the certain tree species and their response and shift in multiple directions under the changing climatic scenarios. The findings of the study would be highly useful in designing strategies and action plans for sustainable management and conservation of timberline ecotone of the region.

Chapter 6 provides herbaceous diversity along the altitudinal gradient in a protected area to understand the function and structure of the ecosystem and diversity decreasing with the increasing altitude.

The content of Chapter 7 has forage crop genetic resources (FCGRs) of North-Western Himalayas: an underutilized treasure gives a glimpse of the huge gap between the demand and supply of both green and dry fodder, which affects the livestock productivity badly. Authors propose that we need to devise sustainable protocols for sustainably managing, conserving, and maintaining the diversity of FCGRs through promoting *in-situ* conservation. Broadening of genetic base and improvement programs of FCGRs are required, which can be achieved through germplasm exchange at national and international levels. In addition to this, an international treaty on their conservation and germplasm exchange is needed for sustainable fodder production and livestock productivity in the region.

Chapter 8 highlights the role of livelihood security and forest resource extraction by forest fringe communities in the Indian Himalayan Region (IHR). Increased household size and a low level of literate population greatly influence the quantity of forest products gathered. Agroforestry plantation is recognized as the most competent land-use option for ensured sustainable development, increased productivity of land, eased environmental stress, and enhanced livelihoods.

Chapter 9 discusses an integrated approach of sustainable agroforestry development in the cold arid deserts region of Ladakh, which has been suggested for modification in the system to provide sustainable livelihood and environment security. These developed models have the potential to increase the productivity of available farm resources without undermining ecology and environmental sanctity. The developed models have the potential for new avenues of employment generation, particularly through the integration of dairy farming, goat farming, and herbal farming in the system. These models will help in social, economic, and environmental development.

Chapter 10 shows the role of traditional practices in forest conservation has focused on the involvement of local people and their knowledge/experience of resource values and management options for forest conservation.

Chapter 11 provides a new strategy of the Solomon Islands government to restructure and further develop the forestry sector with the aim to quantify the growth performance of four exotic timber species commonly planted in the country. It contributes to the optimization of tree plantations by enhancing their productivity and predicting with more accuracy the economic benefits.

Chapter 12 discuss the diversity and regeneration of tree species in the Western Himalayas and reported that tree species of lower altitude showed lognormal curves while those of middle and higher altitudes showed geometric curves. Authors have also discussed that some tree species exhibited discontinuous regeneration due to the absence of some of their diameter classes, and such species would be in trouble in the future and may result in a change in the composition of their respective communities.

Chapter 13 provides an important aspect of success for any afforestation and reforestation program. It provides a glimpse of the handling of forestry seeds scientifically. The authors focused on the quality of seeds ensure through the collection from genetically superior stands/trees and scientific handling practices.

Chapter 14 suggests the home garden forest resources are the key option for socioeconomic development, poverty reduction, and livelihood security; hence, the policy must be directed towards livelihood diversification through sustainable production/collection, extraction, and commercialization of these resources. Knowledge of different stages of seed handling procedures will be beneficial for the production of superior planting stock; it will also help to achieve high survival rates.

—*Editors*

CHAPTER 1

Role of Non-Timber Forest Products from *Madhuca latifolia* in Enhancing Local Livelihoods and Household Dependency in Odisha

SASWAT NAYAK[1,2] and UTTAM KUMAR SAHOO[1]

[1]Department of Forestry, School of Earth Sciences and Natural Resource Management, Mizoram University, Aizawl–796004, Mizoram, India, E-mails: uksahoo_2003@rediffmail.com; uttams64@rediffmail.com (U. K. Sahoo)

[2]College of Forestry, Orissa University of Agriculture and Technology, Bhubaneswar–751003, Odisha, India

ABSTRACT

This study evaluates the importance of Mahua (*Madhuca latifolia* Macb.) and contribution of mahua flower and fruits to the household economy of tribes of Odisha, based on field surveys carried in 10 agroecological zones. The number of mahua trees that the households owned varied from 4.5 (in Nabarangpur) to 9 (in Angul). The annual income from 'NTFPs other than Mahua' was found highest (Rs. 5412) in Koraput and lowest (Rs. 1072) in Bolangir. The annual income from flowers and seed oil of Mahua was found highest (Rs. 22306) in Bolangir and lowest (Rs. 7631) in Baleswar. The contribution of Mahua products to the total income was maximum in Bolangir (95.49%) and minimum in Nuapada (60.23%). Similarly, the contribution of Mahua products to the total income generated from all sources was found maximum (21.94%) in Sundargarh and minimum (14.12%) in Baleswar. Mahua was found providing critical subsistence to the forest fringe dwellers during lean seasons and is being used for a variety of uses for not only household consumption but also for sell, however, the earning from various Mahua products has not been very satisfactory in Odisha as it is not a freely tradable non-timber forest product despite the fact that it has good prospective to make a better living to rural households.

1.1   INTRODUCTION

Since ages, the forests and their associated products have been sustaining livelihoods (Momo et al., 2006; Bwalya, 2013; Mukul et al., 2016) particularly for the forest-depended communities which live in creeping poverty (Shackleton, 2004; Shackleton et al., 2007, 2011; Timko et al., 2010; Kabubo-Mariara, 2013). It is estimated that globally about 20 crore native communities are completely relying on forest for their subsistence (Langat, 2015) and there are about 10.95 to 17.45 crore of human population depend on forest to a varying degree. Non-timber forest products (NTFPs) contribute significantly to socioeconomic status of rural population in developing countries (Heubach et al., 2011; Uberhuang et al., 2012; Akani and State, 2013; Prabhakaran et al., 2013; Dolni and Chatterjee, 2014; Dagm et al., 2016; Suleiman et al., 2017) and in India there are about one fifth (3000 species) of the total floral species (15000 species) yields NTFPs. Another report mentioned that nearly 50 crore people residing in and around the forest primarily gather NTFPs for their subsistence (Alexander et al., 2002). The role of NTFPs in village economy (Malhotra et al., 1991; Mukhopadhy, 2009; Malhotra and Bhattacharya, 2010) and socioeconomic factors contributing to forest dependency (Kamang et al., 2008; Garekae et al., 2017; Htun et al., 2017) have been studied. These reports suggest that the forests contribute significantly to the food supply of the tribal population as well as their livelihood system (Babulo et al., 2008; Bhattacharya and Hayat, 2009; Reddy and Chakravorty, 1999; Ghosal, 2011; Shit and Pati, 2012; Pandey et al., 2016) and thus becomes an intrinsic part of the social life of the tribal and other communities living near the forests (Falcon and Arnold, 1991; Malhotra et al., 1991). The preference of extraction of some NTFPs, especially the fuelwood species are based on forest dwellers' perception (Sahoo et al., 2014).

Madhuca latifolia (Mahua) is one of the few trees that have a special status among NTFPs in Central and Eastern India as it is associated with the tribal livelihood systems in various ways. It is a multipurpose tree species contributing the requirement of food, fodder, and fuel. Besides that, it is also an important source of seasonal income. Almost all parts of this poor man's fruit tree (as it is popularly called) are of immense use. From birth to death, many ceremonies of the tribes find the use of Mahua. Mahua oil is being applied on the newborn child after cutting the umbilical cord, during a wedding the bride and groom are made to hold the stick of the mahua tree, and besides, mahua drinks are served in tribal marriages. The newly developed tender shoots of the tree are used as vegetable for culinary. The newly developed green branches are used as miswak by the tribal. In tribal weddings, the central pillar of the marriage mandap is made up of branches of the tree as they considered it as beneficent. The raw fruit is cooked as a vegetable is a rich source of energy as it contains nearly 68% sugar. It has got 6.3% of protein. Apart from that, it has also got a significant amount of vitamin B and C (Ramadan et al., 2016). Large seeds present in the fruits contain 20 to 50% fatty oil is extracted, which is mainly used for cooking, also as hair oil and therapeutic massages. The sap produced from the mahua tree is used as a gum which is applied on nets and hung on trees for catching small game birds. The leaves are stitched into bowls used for steaming food, eating plates, to drink mahua liquor and rice wine. The most economic part of the tree is flowers which are dried and used to make the liquor.

During famine-like situation when there is a scarcity of rice and grains, the dried flowers are mixed with jaggery and eaten as sweet dishes like laddus and also cooked as food due to its high vitamins, minerals, and iron content. The fruit pulp is used for alcoholic fermentation due to its high sugar content. Fatty oil extracted from seeds is also used for the manufacture of soap and the oilseed cake as manure. The smoke generated from the Mahua cake used as an insecticide and believed to drive away snakes. Tribals use Mahua cake as a fish poison as well as in the treatment of snakebites (Bhatia, 1970; Lakshman, 1983; Vinothkumar et al., 2018). Mahua oil is used in medicine as a moisturizing agent, curing of skin disorders, rheumatism, headache, laxative, and piles. Seeds also produce 22% water-soluble gum and Husks for preparation of active carbon (Mishra et al., 2009; Ganvir and Dwivedi, 2012).

This tree is a way of life for the tribal and is a cultural identity under which villager meetings are often held. While other trees are cut, this tree is left untouched due to its economic importance and religious beliefs, thereby creating awareness to conserve the species (Ahirwar, 2015). In Odisha, there are several forest-dwelling tribes depending on the forests for their daily livelihoods and for income generation activities. To meet the daily household needs, the tribes collect Mahua seeds and flowers for their own consumption and generate income by selling it. However, it is imperative to clearly understand how Mahua has been contributing to the socio-economic domains and rural livelihoods of these communities and besides, to understand the linkages between various socio-economic factors affecting the level of dependency on Mahua with particular reference to flower and seeds in different districts/agroecological zones of Odisha. Therefore; the objectives of this chapter was to analyze:

1. How important is Mahua's access and other NTFPs for the livelihoods of communities?
2. Whether the poorest communities more or less dependent on NTFPs than other groups?
3. What is the extent to which Mahua flowers and seeds have been contributing to the economies of the households?
4. How do access to NTFPs and their impact on livelihood vary across the survey sample, e.g., a district or across an agroclimatic zone? In addition, how do land ownership, tenure, and user rights affect the livelihoods of communities?

1.2 MATERIALS AND METHODS

1.2.1 STUDY AREA

Odisha is located between 17° 49′ N to 22° 34′ North latitudes and 81° 27′ E to 87° 29′ East longitude on the eastern coasts of India. It is encompassed on the Northeast by West Bengal, North by Jharkhand, West by Chhattisgarh, South by Andhra Pradesh, and East by the Bay of Bengal. It has a geographical area of 1,55,707 sq. km (4.74% of India's total area) and has a coastline of about 529 km. The study was carried out in Sundargarh, Mayurbhanj, Baleswar, Nayagarh, Kandhamal, Nabarangpur, Koraput, Malkangiri, Nuapada, Bolangir, and Angul district under all ten agroclimatic zones (ACZs) of Odisha. The detailed demographic profile of the district and studied villages is presented in Tables 1.1 and 1.2.

1.2.2 METHOD OF INVESTIGATION

The steps followed in the present study are selection of the area, collection of specific records of the relevant factor, designing the sampling technique, finalizing the period of investigation, preparation of the interview schedule, rapport building with respondents, collection of data, processing, and analysis of data. In this method of study, 40 respondents were selected randomly from villages under each agroclimatic zone. The selected villages were situated from the outskirt to 8 km of the natural forest areas with sufficient mahua trees on field bunds, which are dependent on mahua trees.

TABLE 1.1 Demographic Profile of the Districts of Study Area

District	No. of Blocks	No. of Gram Panchayat	No. of Village	No. of Household	Population	Area (ha)	SC (%)	ST (%)	Rural Population	Sex Ratio	Density (Persons km−2)
Sundargarh	17	262	1762	476142	2093437	9712	9.16	50.75	1355340	973	216
Mayurbhanj	26	382	3950	583670	2519738	10418	7.33	58.72	2326842	1006	242
Baleswar	12	289	2932	532281	2320529	3806	20.62	11.88	2067236	957	610
Nayagarh	08	180	1692	227927	962789	3390	14.17	6.1	883051	915	248
Kandhamal	12	153	2587	171120	733110	8021	15.76	53.58	660831	1037	91
Nabarangpur	10	169	891	272537	1220946	5291	14.53	55.79	1133321	1019	231
Koraput	14	226	2042	336200	1379647	8807	14.25	50.56	1153478	1032	157
Malkangiri	7	108	1055	136882	613192	5791	22.55	57.83	536664	1020	106
Nuapada	05	109	668	151761	610382	3852	13.46	33.8	576328	1021	158
Bolangir	14	285	1783	413833	1648997	6575	17.88	21.05	1451616	987	251
Angul	08	209	1871	296168	1273821	6375	18.81	14.1	1067275	943	200

Source: Census of India (2011), Odisha, Series-22, Part XII-B.

TABLE 1.2 Demographic Profile of Villages under Different Agroclimatic Zones (ACZs) of Odisha

Agroclimatic Zone	District	Block	Village	Households No.	Population	Scheduled Caste (%)	Scheduled Tribe (%)	Total Land (Ha)
ACZ-1	Sundargarh	Hemgiri	Khodbahal	141	616	14.77	27.92	656
		Hemgiri	Kuchedega	204	979	6.74	37.39	666
		Hemgiri	Ramalata	68	327	72.17	0.00	125
ACZ-2	Mayurbhanj	Bangiriopshi	Chandbil	77	383	0.00	84.33	187
		Bisoi	Sanabalichua	81	374	0.27	87.17	346
ACZ-3	Baleswar	Nilagiri	Tulasidhia	509	2289	21.32	18.79	451
		Nilagiri	Gengutasahi	21	134	0.00	88.81	33
ACZ-4	Nayagarh	Daspalla	Banigochha	248	1149	7.05	14.27	288
		Daspalla	Neliguda	199	862	33.41	0.46	197
		Daspalla	Padapanda	50	232	0.00	26.72	34
ACZ-5	Kandhamala	Kotagarh	Srirampur	298	1588	33.19	9.38	369
		K Nuagaon	Kanjamendi	95	401	1.25	27.68	104
ACZ-6	Nabarangpur	Jharigan	Dhamnaguda	328	1705	10.67	37.65	355
		Dabugaon	Checheriguda	224	1075	21.67	34.42	308
		Dabugaon	Badaliguda	194	940	8.30	63.30	238
ACZ-7	Malkangiri	Mathili	Katapalli	113	622	3.22	79.26	490
	Koraput	Baipariguda	Haladikund	196	979	2.86	41.06	336
ACZ-8	Nuapada	Khariar	Bhaludungri	203	896	18.75	0.00	167
		Nuapada	Tanawat	442	2147	8.90	41.97	832
		Khariar	Kotipadar	318	1410	8.01	0.00	189

TABLE 1.2 *(Continued)*

Agroclimatic Zone	District	Block	Village	Households No.	Population	Scheduled Caste (%)	Scheduled Tribe (%)	Total Land (Ha)
ACZ-9	Bolangir	Patnagarh	Hudapalli	36	151	6.62	49.67	308
		Patnagarh	Uluba	573	2361	5.68	27.19	351
		Khaprakhol	Nandupala	233	519	17.34	4.24	570
ACZ-10	Angul	Kishorenagar	Kasturibahal	187	800	17.75	5.75	233
		Kishorenagar	Baniadohali	83	563	0.00	46.71	238
		Kishorenagar	Raibahal	56	298	8.39	56.71	203
		Kishorenagar	Jhampuli	144	729	29.90	4.39	251

Source: Census of India (2011), Odisha, Series-22, Part XII-B.

1.2.3   ASSESSMENT

In accordance with the objectives of the study and considering the limitations of the research with respect to time, manpower and other facilities, the assessment was carried out in 27 villages of eleven districts under ten ACZs. The details of villages selected in different districts are given in Table 1.2. The distribution of *Madhuca latifolia* in Odisha is extensive covering all ACZs in both natural forest and traditional agroforestry systems contributing to the rural population, there was a strong need to study the impact of mahua tree produces on their livelihood. The dependency on mahua tree for the socio-economic condition of these people was studied through scientific surveys, considering their views, constraints, supply chain, and expert suggestion. The selection of the study area, however, was based on the following main consideration:

1. The study site had plenty of availability of *Madhuca latifolia* as a major livelihood option.
2. The sites were the representatives of different ACZs of Odisha.
3. Co-operation from the collector's to be high and therefore, reliable data should be collected.

1.2.4   PREPARATION OF SURVEY SCHEDULE

In conformity with the set objectives of the study, a set of preliminary survey schedules was designed for the collection of data for the study. The survey schedule carefully designed in such a way that all the factors associated with the economics of various components could be ascertained. Simple questions regarding their basic factors were incorporated in the schedule. The draft schedule was pre-tested by interviewing some sample collectors and traders of selected area by the researcher himself and with discussion with facilitators' forest officials, members of the respective villages. Some parts of the draft schedule were modified on the basis of the actual experiences gained during the pre-testing. The final schedule was prepared in a simple manner maintaining logical sequences and necessary adjustments.

1.2.5   PERIOD OF DATA COLLECTION

The researcher explained the purpose of the study to the respondents to get valid and authentic information. The help of village committee chairman, field-level facilitators, and also Forest Department officials were taken for appointments with the interviewer, the researcher conducted the interview schedule personally to the respondents. Compatibility was established with the respondents through spontaneous discussion regarding the objectives of the interview. Co-operation was attained from respondents during data collection from April 2017 to June 2017 in a phased manner.

1.2.6 CHARACTERISTICS OF STUDY AND DEVELOPMENT OF THE RESEARCH INSTRUMENT

The characteristics included in the study were: (i) family size of respondents, (ii) landholding of respondents, (iii) employment in various sectors, (iv) presence of *Madhuca latifolia* in fields and homesteads, (v) collection of NTFPs including mahua flowers and seeds and their availability, (vi) processing, home consumption and selling of mahua flowers and seeds, (vii) storage methodology adopted for mahua flowers and seeds, (viii) total annual income from different sources, (xi) comparative economics and (x) constraints in mahua flowers and seed collection and its trade. Family size of a respondent was determined in terms of the total number of members in the family of each respondent. The family member included respondent himself, spouse, sons, daughters, and other dependents. The scoring was made by the actual member of the family, expressed by respondents. If a respondent had four members in his family, his score was given as 4. Farm size was measured in terms of acres by collecting the information from the farmers. Some farmers expressed the landholding in terms of acre or guntha then these data were converted to acres and recorded. The nature of employment in different sectors such as NTFP collection, agriculture, farm labor, and other sectors were recorded and represented in the table for analysis. The availability of different NTFPs, including mahua flowers and seeds in the localities, amount of bearing in tree, collection quantity, and their season of collection were asked and recorded. The indigenous methods followed in the processing of mahua flowers and seeds, the quantity consumed in-home, and the quantity sold with their rates were recorded, calculated, and represented in the table. The selling of NTFPs in different supply chain was identified. Different storage methodology practiced was recorded at primary collector as well as a licensed trader's level. The total earnings of all family members of respondents from agriculture, business, service, NTFP collection including mahua flowers and seeds and other sources as contained in items of interview schedule. The amount is recorded separately and the total is calculated and expressed. The researcher analyzed the total income and the relative contribution of income from mahua flowers and mahua seeds, other NTFPs to total income. Problems faced in mahua flowers and seeds collection were recorded by asking the respondents as prepared in the interview schedule. The respondents were also asked to give their opinion towards the problems along with their extent of confrontation in the collection and marketing of mahua flowers.

1.2.7   DATA COLLECTION, PROCESSING, AND ANALYSIS

The study was conducted for field-level primary data, and the researcher himself collected data required for the study. The survey data were gathered by three main methods. These are (i) direct observation, (ii) interviewing respondents (iii) record kept by respondents. Data were collected through field visits in the study area and personal interview with the sample mahua flowers collectors and traders. Interviews were carried on in respondent's house as per their convenience. They provided information based on their past experiences and memory. In order to minimize the response error, cross-questions were asked in simple Odia. In some areas where the respondents don't understand Odia the help of local tribal interpreter were taken. After completion of each interview, each interview schedule was checked and noted properly. Compilation, tabulation, and analysis of all field survey data was done in accordance with the objective of study. During this process, local units were converted into standard units, and all the responses in the interview schedule are given numerical coded value whenever necessary. To expedite tabulation, the responses to the question in the interview schedules were transferred to a master sheet. The respondents were then classified into several categories for describing the different characteristics and their constraint they are facing. Descriptive analyzes such as range, average, percentage, increasing or decreasing order were used whenever possible.

1.3   RESULTS

1.3.1   SOCIOECONOMIC STATUS OF MAHUA DEPENDENT FOREST DWELLERS

Analyzing the family members of the respondents of village's dependent on mahua under different ACZs, it was found that an average number of members per household ranges between 4.7–6.6 in all ACZs. The maximum average number of members per household (6.6) was found in agroclimatic zone-7 and 9 and the minimum average number of members per household (4.7) in agro-climatic zone-4 and 6 (Table 1.3). Agroclimatic zone wise trend showed that average numbers of males per household ranges from 1.8–2.3 with maximum (2.3) in agroclimatic zone-9 and minimum (1.8) in ACZs- 3, 4, 6 and 8. The average number of females per household ranges from 1.5–2.3 with maximum (2.3) in agroclimatic zone-10 and minimum (1.5) in ACZs-6. Similarly, among the respondents, Small family size was found maximum (70%) in agroclimatic zone-1 and minimum (7.5%) in agroclimatic zone-7. Medium family size was found maximum (87.5%) in agroclimatic zone-7 and minimum (20%) in agro-climatic zone-1, whereas, large family size was found maximum (17.5%) in agroclimatic zone-10 and minimum (2.5%) in agroclimatic zone-8 (Table 1.3).

1.3.2 AGRICULTURAL LAND HOLDINGS

The average agricultural land/household among the respondents of villages' dependent on mahua under different ACZs range between 1.8–7.9 acre in all ACZs. The maximum agricultural land per household (7.9) was found in agroclimatic zone-10 and minimum average number of members per household (1.8) in agroclimatic zone-6 (Table 1.4 and Figure 1.4). Among the respondents, agroclimatic zone wise trend showed that marginal landholding was found maximum (40%) in agroclimatic zone-7 and minimum (5%) in agroclimatic zone-9, whereas, no marginal landholding was found in ACZs-4 and 10. Small landholding was found maximum (62.5%) in agroclimatic zone-3 and minimum (10%) in agroclimatic zone-10. Small-medium landholding was found maximum (50%) in agroclimatic zone-1 and minimum (7.5%) in agroclimatic zone-7. Medium landholding was found maximum (55%) in agroclimatic zone-4 and minimum (5%) in agroclimatic zone-5, whereas, no medium landholding was found in ACZs-3 and 6. Large landholding was found maximum (15%) in agroclimatic zone-9 and minimum (5%) in agroclimatic zone-5 and 7, whereas, no large landholding was found in ACZs-2, 3, and 6 (Table 1.4 and Figure 1.5).

1.3.3 COLLECTION AND CONTRIBUTION OF MADHUCA LATIFOLIA TO THE ANNUAL INCOME

The average number of mahua trees per household among the respondents of villages under different ACZs ranges between 4.5–9.0 across all ACZs.

TABLE 1.3 Family Size of Sample Rural Population (%) Dependent on *Madhuca latifolia*

Agroclimatic Zone	Average No. of Members/Family	Average no. of Males/Family	Average No. of Females/Family	Average No. of Children/Family	Family Size Small (Up to 4)	Family Size Medium (Up to 5–8)	Family Size Large (> 8)
ACZ-1	6.2	2.2	2.2	1.8	70.00	20.00	10.00
ACZ-2	5.6	2.0	1.8	1.8	27.50	60.00	12.50
ACZ-3	5.7	1.8	2.0	1.9	17.50	75.00	7.50
ACZ-4	4.7	1.8	1.7	1.2	50.00	40.00	10.00
ACZ-5	5.6	1.9	1.7	2.0	25.00	65.00	10.00
ACZ-6	4.7	1.8	1.5	1.4	60.00	35.00	5.00
ACZ-7	6.6	2.2	2.0	2.4	7.50	87.50	5.00
ACZ-8	5.5	1.8	1.9	1.8	22.50	75.00	2.50
ACZ-9	6.6	2.3	2.0	2.3	12.50	75.00	12.50
ACZ-10	6.5	2.2	2.3	2.0	30.00	52.50	17.50
Mean	**5.77**	**2.00**	**1.91**	**1.86**	**32.25**	**58.50**	**9.25**

TABLE 1.4 Agricultural Land Holdings (Acre) of Sample Rural Population (%) Dependent on *Madhuca latifolia*

Agroclimatic Zone	Average Agricultural Land/Household	Major Crops Cultivated	Marginal (up to 0.5 ha)	Small (> 0.5–1.0 ha)	Small Medium (> 1.0–2.0 ha)	Medium (> 2.0–4.0 ha)	Large (> 4.0 ha)
ACZ-1	3.9	Paddy, vegetables	10.0	20.0	50.0	10.0	10.0
ACZ-2	2.6	Paddy, Groundnut	32.5	40.0	15.0	12.5	0.0
ACZ-3	1.9	Paddy, vegetables	10.0	62.5	27.5	0.0	0.0
ACZ-4	6.0	Paddy, Maize, Brinjal, Green gram, Black gram	0.0	17.5	17.5	55.0	10.0
ACZ-5	2.3	Paddy, Vegetables	30.0	40.0	20.0	5.0	5.0
ACZ-6	1.8	Paddy. Maize	35.0	45.0	20.0	0.0	0.0
ACZ-7	2.7	Paddy, Millet	40.0	40.0	7.5	7.5	5.0
ACZ-8	3.4	Paddy, Green gram, Black gram	7.5	42.5	32.5	15	2.5
ACZ-9	6.3	Paddy, Maize (Sweet corn), Black gram, Groundnut	5.0	17.5	35.0	27.5	15.0
ACZ-10	7.9	Paddy, vegetables	0.0	10.0	35.0	42.5	12.5
Mean	**3.88**	**Paddy**	**17.0**	**33.5**	**26**	**17.5**	**6.0**

The maximum No. of trees per household (9.0) was found in agroclimatic zone-10 and minimum average number of members per household (4.5) in agroclimatic zone-6 (Table 1.5). Analyzing the respondents of village's dependent on mahua the average annual income from agricultural produce varies from Rs. 10622–Rs. 76636 among all ACZs with maximum in agroclimatic zone-9 and minimum in agroclimatic zone-2. Annual income from wages found highest (Rs. 35452) in agroclimatic zone-7 and lowest (Rs. 6500) in agroclimatic zone-4. Annual Income from 'NTFPs other than Mahua' was found highest (Rs. 5412) in agroclimatic zone-7 and lowest (Rs. 1072) in agroclimatic zone-9. Annual income from flowers and seed oil of Mahua was found highest (Rs. 22306) in agroclimatic zone-9 and lowest (Rs. 7631) in agroclimatic zone-3 (Table 1.6). Analyzing the contribution of Mahua products to the total income generated from NTFPs, it was found that maximum contribution was found in the case of agroclimatic zone-9 (95.49%) and minimum contribution in the case of agroclimatic zone-8 (60.23%). Similarly, the contribution of Mahua products to the total income generated from all sources was found maximum (21.94%) in agroclimatic zone-1 and minimum (14.12%) in agroclimatic zone-3 (Table 1.6).

1.3.4   COLLECTION OF MADHUCA LATIFOLIA FLOWERS AND SEEDS

Analyzing the respondents views from all ACZs on basis of flowering two different categories of trees are present one is early flowering type and other late flowering type although gap in flowering is not wide. It was found that in agroclimatic zone 5, 6 and 7 such type of late-flowering *Madhuca latifolia* trees exists. The showering season of *Madhuca latifloia* flowers was from March-April in all parts of Odisha. The corollas, after getting matured after 30 days of initiation of flowers, changed its color to creamy white started falling on the ground after getting molted on the tree. Normally showering took place in the early part of the day between 4 O'clock–11 O'clock. The primary collectors clean the ground to avoid unwanted materials mixed with the flowers during collection. The collection starts early morning and continues even up to noon. During collection whole family are involved to get maximum collection, which lasts for three to four weeks. The whole collection period can be marked into three phases starting, middle, and end of which the middle phase had most collection when most of the matured flowers showered. It was revealed by the respondents that better sunshine favors more showering of flowers. It was also found that older trees produce smaller flowers than trees with younger age. Flowers of *Madhuca latifolia*

TABLE 1.5 Status, Collection, and Utilization of *Madhuca latifolia* and its Produces in Sample Villages of Different Agroclimatic Zones (ACZs) of Odisha

Agroclimatic Zone	No. of Trees/HH	Period of Collection of Mahua Flower	Collection of Mahua Flower/HH (QTLs)	Mahua Flowers Selling Price (@Rs./Kg)	Period of Collection of Mahua Fruits/Seeds	Collection of Mahua Seeds/HH (Kg)	Oil Extracted from Seed/HH (Liter)
ACZ-1	8.9	March 3rd week–April 3rd week	4.36	29.00	June 1st week–June 4th week	71.00	28.80
ACZ-2	7.9	March 3rd week–April 3rd week	3.64	23.66	June 1st week–June 4th week	59.92	24.17
ACZ-3	7.3	March 3rd week–April 3rd week	2.58	25.20	June 1st week–June 4th week	31.26	13.50
ACZ-4	6.0	March 3rd week–April 3rd week	3.33	24.17	June 1st week–June 4th week	93.33	32.92
ACZ-5	5.9	March 3rd week–April 4th week	2.30	24.05	June 1st week–July 1st week	56.16	24.42
ACZ-6	4.5	March 3rd week–April 4th week	2.74	23.74	June 1st week–July 1st week	39.04	15.35
ACZ-7	7.3	March 3rd week–April 4th week	3.76	23.17	June 1st week–July 1st week	38.23	14.70
ACZ-8	5.4	March 3rd week–April 3rd week	2.76	24.28	June 1st week–July 1st week	41.96	18.43
ACZ-9	6.6	March 3rd week–April 3rd week	7.07	22.60	June 1st week–July 1st week	327.4	129.12
ACZ-10	9.0	March 3rd week–April 3rd week	4.11	25.00	June 1st week–July 1st week	71.47	26.59
Mean	**6.88**	March 3rd week–April 3rd week	**3.67**	**24.49**	June 1st week–June 4th week	**82.98**	**32.80**

TABLE 1.6 Contribution of *Madhuca latifolia* to the Total Income/Household/Annum

Agrocli-matic Zone	Income from Agri-cultural Produce (Rs.)	Income from Wages (Rs.)	Income from NTFPs Other Than Mahua (Rs.)	Income from Mahua Flower (Rs.)	Market Value Mahua Seed Oil (Rs.)	Income from Mahua Tree Products (Rs.)	Income from NTFP's (Rs.)	Total Income (Rs.)	Contribution of Mahua Products to Income from NTFPs (%)	Contribution of Mahua Products to Total Income (%)
	A	B	C	D	E	F(D + E)	G(C + F)	H(A + B + G)		
ACZ-1	16070	28980	4540	12497	1440	13937	18477	63527	75.43	21.94
ACZ-2	10622	28400	1130	9090	1188	10278	11408	50430	90.09	20.38
ACZ-3	14378	27333	4716	6956	675	7631	12347	54059	61.80	14.12
ACZ-4	49320	6500	1850	8166	2062	10229	12079	67899	84.68	15.07
ACZ-5	18500	18356	4230	7650	845	8525	12725	49581	66.99	17.19
ACZ-6	18548	22133	4737	7025	767	7793	12530	53212	62.19	14.65
ACZ-7	13608	35452	5412	8688	735	9423	14517	63579	64.91	14.82
ACZ-8	21810	13384	5151	6882	921	7803	12955	47194	60.23	16.53
ACZ-9	76636	15968	1072	16818	5488	22306	23378	115982	95.41	19.23
ACZ-10	51857	16329	1247	10294	1329	11623	12870	81057	90.31	14.34

were collected during the period from March 3rd week–April 3rd week in all ACZs except agroclimatic zone-5, 6 and 7 where the collection was done during the period March 3rd week-April 4th week (Table 1.5). The average collection of mahua flowers per household among the respondents of villages under different ACZs ranges between 2.30 QTLs-7.07 QTLs across all ACZs. The maximum collection of mahua flowers per household (7.07 QTLs) was found in agroclimatic zone-9 and minimum collection of mahua flowers per household (2.30 QTLs) in agroclimatic zone-5 (Table 1.5 and Figure 1.6). Seeds of *Madhuca latifolia* were collected for extraction of oil during the period from June 1st week–June 4th week in all ACZs-1, 2 and 3, whereas, the collection was done during the period June 1st week–July 1st week in rest of the ACZs. The average collection of mahua seeds per household among the respondents of villages under different ACZs ranges between 31.26 kg–327.4 kg across all ACZs. The maximum collection of mahua seeds per household (327.4 kg) was found in agroclimatic zone-9 and minimum collection of mahua flowers per household (31.26 kg) in agroclimatic zone-3 (Table 1.5).

1.3.5 PROCESSING OF MADHUCA LATIFOLIA FLOWERS AND SEEDS

The flowers collected were taken to home and spread on opened grounds for sun drying and made upside-down 2 times a day to ensure uniform drying. Sun drying is done continuously for 3–4 days till the flowers turn reddish in color with a reduction in weight up to 65–75%. As per respondents view cloudy weather affects the quality of flowers making it dark in color which ultimately reduces its market price. The selling rate of dried mahua flowers at the primary collector level varies from place to place and ranges between Rs. 22.60–Rs. 29.00/kg overall ACZs of Odisha. The maximum selling rate (Rs. 29.00) at the primary collector level found in agroclimatic zone-1 whereas the minimum selling rate (Rs. 22.60) in agroclimatic zone-9 (Table 1.5 and Figure 1.7). The fruits collected during June-July which was then de-pulped to get the seeds from it. The seeds were dried for 1 day and then the thin seed cover was removed to expose the kernel. The kernels were made into two haves and then sun-dried on cleaned open home yards for 4–5 days till it turns into reddish color. During this drying period, the reduction in weight of seed is about 15–20%. The seeds are prone to quick decaying and turn black in color due to fungus attack for which it is dried before marketing. The seeds were then taken by the respondents to local mechanical oil extractors (Ghani's) for extraction of oil from dried kernels. Most of the oils were utilized for home consumption for cooking purposes, whereas some sold in the local markets. The average utilization of mahua seed oil per household among the respondents of villages under different ACZs ranges between 13.50 liter-129.12 liter across all ACZs. The maximum utilization of mahua seed oil per household (129.12 liters) was found in agroclimatic zone-9 and minimum utilization of mahua seed oil per household (13.50 liter) in agroclimatic zone-3 (Table 1.5).

1.3.6   MARKETING OF MADHUCA LATIFOLIA FLOWERS

The dried mahua flowers collected by the primary collectors during harvesting season passed through intermediate traders and stored, which were then sold in the lean season when the prices were higher. During March 2000, the Government of Odisha adopted a new NTFP resolution in which 67 NTFPs, including mahua flowers, which were earlier under control of state forest department were included under the control of gram panchayats. After analyzing the respondents, it was found that the traders who procure the mahua flower from villages of a panchayat must register themselves for license with that panchayat by paying a nominal fee. The schematic pathway of marketing of mahua flowers in Odisha has been elicited (Figure 1.1). From the information provided by respondents including, primary collectors, grass-root level middleman and licensed traders. The primary collectors collect mahua flowers from tree from their own land or from nearby forestlands which was dried for 3–4 days till its color changes to red. As per the traders view, the dried flowers are categorized into three classes depending upon the color of flowers, i.e., red, yellow, and black. The red flower is well dried and contains less moisture content (MC); yellow flowers are having more MC and less dried than red-colored flowers whereas the black colored flowers are somewhat degraded due to higher MCs. The procurement price of red-colored flower is highest followed by yellow-colored flower and minimum for black colored flower. The license holders collect dried mahua flowers from primary collectors directly, from grass root level middlemen (collects dried flower from primary collectors) and from local weekly markets (Haats) where primary collectors sell. The average procurement price from primary collector was @Rs. 24.49/kg during 2016–2017, however, in some remote villages the procurers exchange mahua flowers with un-iodized salts in the volume ratio of flower: salt: 1: 2.5. The license holder traders sundried the procured mahua flowers for another 2–3 days after spreading it over-drying yards, after which it was kept in the stored house. After the rainy season and onset of winter on October onwards the stored dried mahua flowers were then sold. The dried flowers selling rate found to be fluctuated and were sold @Rs. 60/ kg in 2016–2017 and Rs. 35/kg in 2017–2018. These products were taken further by other traders, wholesalers who sell them to retailers, distilleries, and other users. Some of the flowers were utilized within the state, whereas a considerable portion are illegally transported to bordering states like West Bengal, Andhra Pradesh and Chhattisgarh. After analyzing the data collected from the traders, it was found that an average number of licensed traders per district was about 18 numbers with average storage capacity of dried mahua flowers of each trader about 2000 Qtl. With nearly 22 districts involved in the trade of Mahua flowers of 7,92,000 Quintals occurs in the state contributing about Rs. 194 Crore to the primary collectors of the state.

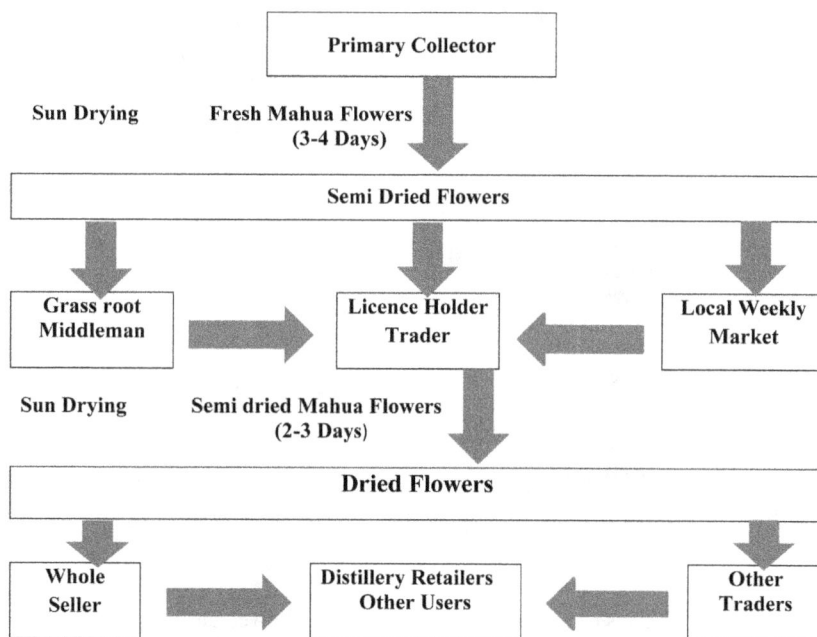

FIGURE 1.1 Collection, processing, and marketing of *Madhuca latifolia* flowers.

1.3.7 CONSTRAINTS, VIEWS, AND SUGGESTIONS FOR BETTER MANAGEMENT MAHUA TREE AND MAHUA PRODUCTS

The rural population residing near the natural forest of Odisha is dependent on various NTFPs of which mahua flowers and seeds contribute a substantial portion of their household income. The present investigation found that there are some constraints around the sustainable use of mahua flowers and seeds for better and secure livelihood. Among the respondents, 57% (Figure 1.2) perceived that the fluctuating rates of mahua flowers are discouraging the rural population for collection of flowers. About 15% of them did thought that the collection of mahua flowers and seeds are mostly restricted to agricultural field and forests outskirts because of fear of attack by wild animals, infection from wild plants in natural forests, etc. About 12% expressed that lack of other users of mahua flowers besides liquor industries affect the procuring price of mahua flowers. Nearly 7% told that better drying could have increased the rate of collection and reduced degradation of flowers. About 5% of respondents reported that better accessibility to forest patches will enhance the collection from natural forests. 4% correspondents reported fewer yields of some trees declines the production in the flowers particularly in less flowering years. The present use of mahua flower for various purposes is shown in Figure 1.3.

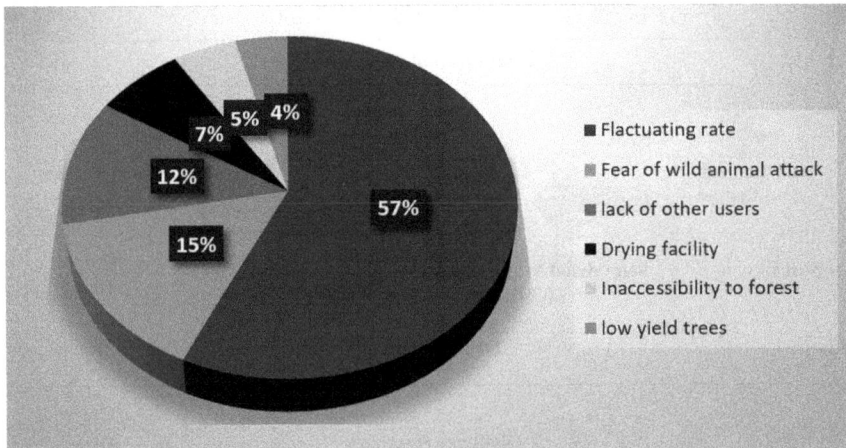

FIGURE 1.2 Constraints of primary collectors of mahua flower and seed.

1.4 DISCUSSION

1.4.1 *ACCESS OF MAHUA AND OTHER NTFPS TO THE FOREST FRINGE COMMUNITIES/DWELLERS*

In the studied area, it was found that every mahua tree has an owner, usually an Adivasi family. The ownership of each tree is recorded with the forest department and is passed from one generation to the next. The community conserved this species and secured tenure rights of each tree mostly outside the forest area (near the village boundary) and in agricultural fields. The production of flower and fruits increases with age and girth of the trees. Our findings suggest that a single tree can produce up to 200 kg of flowers a year, and the older trees are more fecund or fertile, producing abundant quantities of flowers. Mahua trees bloom between the latter half of March and April and shed its flower in the morning every day, over a period of 15–20 days. Some reports suggest that customary puja is performed under the Mahua tree in some villages sometime after holi and prior to the collection of mahua flower. The purchase price of flowers varies during the year. Besides, mahua other NTFPs are also being collected from the forest area. Most of these NTFPs are non-nationalized and therefore they are fully access to the forest fringe communities/Adivasis in the state. A wide variety of other NTFPs that are available (Table 1.7) in the study area around the year further evince their role in the livelihood of tribes of Odisha.

Fresh Mahua Flower												
Present Utilisation Status	Future Scope for Utilisation											
Sun Drying	Puree/pulp/ whole flower						Juice					
Uses	Ladoo	Citric acid	Concentrated	→	Bakeries	→	Buscuit	←	Bakeries	←	Concentrated	Jelly
Food	Sauce	Medicinal				→	Cake	←				
Cattle food	Jam	uses			Beverages	→	Sharbat	←	Beverages	←		
Liquor	Pickle					→	Wine	←				
				→	Confectionary	→	Candy	←	Confectionary	←		

FIGURE 1.3   Utilization of *Madhuca latifolia* flowers.

1.4.2   DEPENDENCE ON MAHUA ACROSS THE PEOPLE OF DIFFERENT SOCIO-ECONOMIC STRATA

Mahua tree nevertheless is the most sacred tree among the tribes of Odisha. The different parts of the plants are used as NTFPs. The dried flowers are roasted, made into balls or laddus, and are often used as snack; besides, the dried flowers are roasted with sesame, char, cashew, and other wild seeds and used for household consumption. The households, irrespective of their landholding size and/economic status, have been depending on this species for various use from cultural to meeting economic or family household needs. All the households in the study sites consume the flower as vegetables and during lean season, mahua provides critical subsistence mostly to the land less and low landholding families. Besides, Mahua contributes significantly to the livelihood of all households who collect it

TABLE 1.7 List of Non-Timber Forests (Other Than Mahua Flower and Seeds), Their Season of Availability/Collection/Harvesting, Parts Used and Household Consumption/Sell in the Study Area

Name of the NTFPs	Species	Family	Months of Availability	Consumption/Sale	Parts Used
Khaira	*Acacia catechu* (L.f.) Willd	Leguminosae	Jan-Feb, Dec	Sale	heartwood
Chirata/Kalmegh	*Andrographis paniculata* (Burm.f.) Wall. Nees.	Acanthaceae	Aug-Oct	Sale	The aerial portion of the plant
Bael fruit	*Agele mormelos* (L.) Correa	Rutaceae	Feb-Mar	Sale	Fruit
Sitafal	*Annona squamosa* Linn.	Annonaceae	Sep-Nov	Both	Seed, flower
Neem seed	*Azadirachta indica* A. Juss	Meliaceae	May	Sale	Seed
Axle wood	*Anogeissus latifolia (Roxb. ex DC.) Wall. ex Beddome.*	Combretaceae	Jun-Sept	Sale	Leaves for tanning and gums/resin
Jackfruit	*Artocarpus heterophyllus Lam.*	Moraceae	Mar-Jun	Sale	Wild edible fruit
Satavar	*Asparagus racemosa* Willd	Asparagaceae	Mar, Dec	Sale	Root for medicine
Patal plate	*Bahunia roxburghiana* Voigt	Fabaceae	Mar-Apr	Sale	Leaf, seed
Chironji	*Buchnania lanzan* Spreng	Anacardiaceae	Apr-May	Sale	Seeds
Palash fruit	*Butea monosperma* (Lam.) Taub.	Fabaceae	Feb	Sale	Fruits/seeds
Kumbhi	*Careya arborea* Roxb	Lecythidaceae	Nov-Dec, Jan-Feb	Sale	Bark and leaves
Satinwood	*Chloroxylon swietenia* (Roxb.) DC	Rutaceae	Throughout the year	Sale	Coagulation plant Agricultural tools
Garari	*Cleistanthus collinus* (Roxb.) Benth.ex. Hook.f.	Euphorbiaceae	May	Sale	Fertilizer
Bidi patta	*Dispyros melanoxylon*	Ebenaceae	May	Sale	Edible fruits
Karanja seeds	*Pongamia pinnata*	Fabaceae	Feb-Jun	Sale	Seeds
Kendu fruits	*Dispyros melanoxylon*	Ebenaceae	Apr-Jun	Self-consumption	Fruits

TABLE 1.7 (Continued)

Name of the NTFPs	Species	Family	Months of Availability	Consumption/Sale	Parts Used
Amla	Emblica officinalis	Euphorbiaceae	May, Sep-Nov	Both	Fruits
Bamboo	Dendrocalamus strictus (Roxb.) Nees	Poaceae	Mar-Jun	Both	Agricultural tools Thatching
Indian coral tree	Erythrina suberosa Roxb.	Fabaceae	Throughout the year	Self-consumption	Soft wood
Indian boxwood	Gardenia latifolia Ait.	Rubiaceae	Feb-April	Both	Fruits and seeds
Dhaman	Grewia tilifolia Vahl.	Boraginaceae	September	Both	Fruits and flower
Anjan	Hardwica binnata roxb.	Detarioideae	Oct-Feb	Self	Leaves for fodder
Kurchi	Holiarhena antidyzenterica (Linn.) Wall. Synonym H. pubescens (Buch. Ham.) Wall. Ex G. Don.	Apocynaceae	Feb-Jun	Self	Fruits and flower
Nevaari	Ixora arborea Roxb. Ex Sm	Rubiaceae.	Root and leaves throughout the year	Both	Leaves, flower, root and fruit
Crape Murtle	Lagerstoemia parviflora Roxb.	Lythraceae	Throughout the year	Both	Bark and gum
Indian Ash tree	Lannea coromandelica (Houtt.) Merr	Anacardiaceae	April-June	Self	Leaves for food
Menda	Litsea glutinosa	Lauraceae	Jan, Mar, Dec	Both	Bark and gum
Honey	Madhuca Latifolia (Roxb.) J. F. Macbr	Sapotaceae	Apr-May, Oct	Both	Flower, fruits, seeds
Mango	Mangifera indica	Anacardiaceae	May-Jul	Both	Fruits
Night Jasmine	Nyctanthes arbortristis L.	Oleaceae	Throughout the year	Both	Flower
Khajur leaves	Phoenix sylvestris	Arecaceae	Jan-Feb, Nov-Dec	Both	Leaves
Indian kino tree	Pterocarpus marsupium Roxb.	Fabaceae	March	Self	Fodder plant

TABLE 1.7 *(Continued)*

Name of the NTFPs	Species	Family	Months of Availability	Consumption/Sale	Parts Used
Kusum	*Schleichera oleosa* (Lour.) Oken	Sapindaceae	Mar-April	Both	Flower
Bhelwa	*Semecarpus anacardium* L.	Anacardiaceae	Jan-Feb	Sell	Wild edible fruit, fertilizer
Sal Seed	*Shorea robusta*	Dipterocarpaceae	Apr-Jun	Sell	Seeds and leaves
Jamun	*Syzigium cumuni*	Myrtaceae	May-Jul	Both	Fruits
Imli	*Tamarindus indica*	Fabaceae	May	Both	Fruits, firewood
Indian redwood	*Soymida febrifuga* (Roxb.) Juss	Meliaceae	Feb-Jun	Both	Bark, fruits, flowers
Bahera	*Terminalia bellerica* (Gaertn.) Roxb.	Combretaceae	Feb-Mar	Sale	Fruits
Harar	*Terminalia chebula* Retz.	Combretaceae	Jan-Mar, Dec	Sale	Fruits
Marda	*Terminalia tomentosa* (Roxb.) Wight and Arn.	Combretaceae	Apr-May	Sale	Fruits
Dhawi/Dhobo Flower	*Woodfolia fructose*	Lythraceae	Feb	Sale	Flower for churnas and ghritas for various diseases
Ber	*Zizyphus mauritiana*	Rhamnaceae	Jan-Feb, May, Nov-Dec	Both	Wild edible fruits
Fuelwood	Several species	—	Throughout the year	Both	Firewood

for self-consumption and for sale. Studies in Bihar, Madhya Pradesh and Andhra Pradesh has shown that NTFP contributing to household income ranged from 10–55% and about 80% of forest dwellers depending on forests for 25 to 50% of their food requirements (Pandey et al., 2016). Mahua played important financial empowerment to the women through NTFPs collected and sold from forest fringe. Several other studies have shown the high dependence of forest dwellers on a wide variety of NTFPs. For example, Saha, and Sundriyal (2011) and Lalremruata et al. (2006) reported 19.32% of the total household income for different communities comes from NTFPs in the humid tropics of northeast India, Sahoo et al. (2010a–c) assessed the degree of NTFPs dependency by the tribes of Mizoram which reveal that NTFPs collection is a source of income for the forest dwellers in the state. There have been some studies which clearly distinguish NTFP dependency among the landless and landowners. Poverty and fuelwood usage were the factors for landowners while rice insufficiency, off-farm income and fuelwood usage mostly affect the NTFPs dependency for landless people (Soe and Yeo-Cheng, 2019).

1.4.3 EXTENT OF MAHUA AND OTHER NTFPS CONTRIBUTION TO THE ECONOMIES OF THE HOUSEHOLDS

The occupation pattern of the studied sites showed that paddy is the major crop. Maize, millet, groundnut, and vegetables too are also grown in a varying degree by the tribes/forest fringe dwellers. However, the hilly terrain does not favor better paddy growth, and it was observed the households having limited acres of land using for paddy cultivation with very low productivity. The tribals are mostly engaged in agricultural operations during the month of June to October as most of the farming operations are done in this period. During the agricultural stress period and non-farming months, the dependence on NTFPs increases, and during this period, more collection of forest produce takes place. Studies carried out elsewhere and in India suggest that the rate of extraction of NTFPs is linked to the degree of agricultural stress (Thomas et al., 2011; Delacote, 2007). Several studies have also revealed that NTFPs/ forest dependence decreases with increase off-farm or non-forest income (Hedge and Enters, 2000; Bahuguna, 2000). Higher agricultural productivity and agriculture income results in less extraction of forest resources (Guntilake, 1988; Momo et al., 2006; Rayamajhi et al., 2012). The income contribution of mahua to the total NTFPs income per household is substantial (60% in

ACZ-8 to >90% in ACZ-9) which confer the substantial role of mahua in the livelihood of the forest dwellers compared to other NTFPs. The income from mahua varies from Rs. 7631 to 22306 per household; this requires 15–20 days of hard work during the flowering season. In a season each household collects about 2.30–7.07 quintal of mahua flower per household, which contributes to 14.12% of the total family income (in ACZ-3) and 21.94% of the total family income in ACZ-1 (Table 1.6).

1.4.4   VARIATIONS IN NTFPS AVAILABILITY AND LIVELIHOOD IMPLICATION ACROSS THE ACZS IN CLIMATE CHANGE SCENARIO

The constraint analysis revealed that the availability of various NTFPs and particularly mahua scenario in Odisha is no rosier. The current climate change coupled with other developmental activities in the state in the past few years is adding to the curtailment of forest resource availability for livelihoods of tribes. Declining productivity of forest produce such as food, fuel, medicinal, and herbs seem to dispossess the rural poor from a supplementary source of income, food, and healthcare (Basu, 2009), thus concentrated efforts to understand and assess how NTFPs can contribute to mitigation and adaption of climate change, but also by looking at if and how NTFPs are impacted by changing climates (Kirilenko and Sedjo, 2007). Upon interaction with the tribes, they revealed that there is a significant change in the phenology of mahua with respect to flowering and fruiting. Changes in flower and fruiting period, i.e., mid-March to mid-February, affects the change in local forest ecology. This also affects the pollination pattern and reduced honey collection by up to 90% in the Wayanad district of Kerala due to climate change (Manoj, 2011; WWF, 2011). Similar changes in mahua flowering and fruiting are reported in Mandla district of Madhya Pradesh (Sushant, 2013). The study argues that during mid-February, the agricultural season is about to end and farmers are preoccupied with harvesting their produce. Early-onset of mahua flowering during the harvest period leaves less time for the community to collect mahua. Studies in Madhya Pradesh (Malhotra and Bhattacharya, 2010) have revealed that drought reduced agricultural production, which affects the tribals demand, and they extract NTFPs to fulfill households need. A similar report was also reported from the tribal belt of Odisha, where paddy to fulfil the low paddy production, NTFPs can be an alternative for food, nutritional, and income security of the population in future.

1.4.5   *LAND OWNERSHIP, TENURE, AND USER RIGHTS AFFECTING THE LIVELIHOODS*

Many NTFPs in the state of Odisha operate in open or in semi-open access systems of resource tenure, resulting in exploitation of NTFPs. This happens mostly during the lean seasons when the poor households with little income on agriculture have no choice other than going to the forests and extracting the NTFPs to the extent possible to meet the basic needs. Women are more depended on subsistence forest income and their involvement in NTFP collection is more notable than the men in the study area. Mahua is a non-nationalized non-timber forest product; however, in Odisha, it is kept under the control of Gram Panchayats as per the NTFP policy of 2000. It is still not a freely tradable item because of its classification as an intoxicant by the Orissa Excise Act, 1915, and Odisha government imposes 4% VAT on the flowers and 14% VAT on the finished products which affect the livelihood of the primary producers to a great extent. The neighboring states like Bihar and Chhattisgarh have lower taxes and duties on mahua, and therefore, these states have advantages over Odisha for boarder trading.

According to the licensed traders they were allowed to collect and sell mahua flowers within 30 km radius and beyond that it requires transit permit which prevents better marketing potentials for the produces within the state (Figures 1.4–1.7).

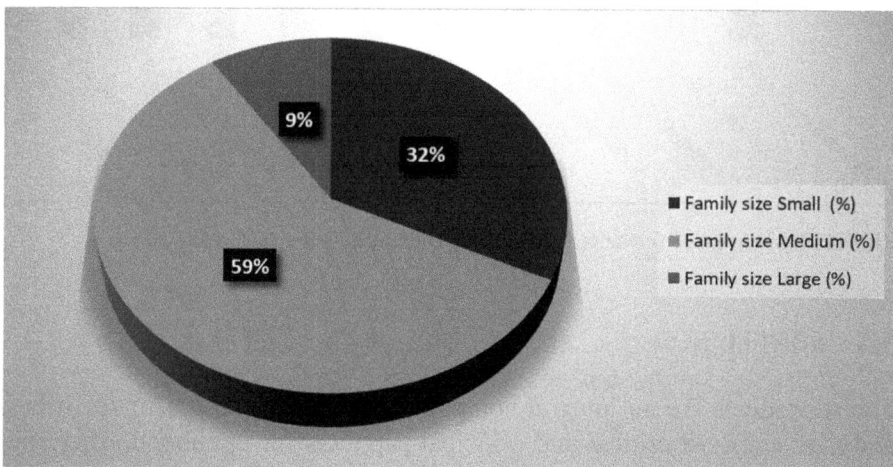

FIGURE 1.4   Percentage of family size of respondents.

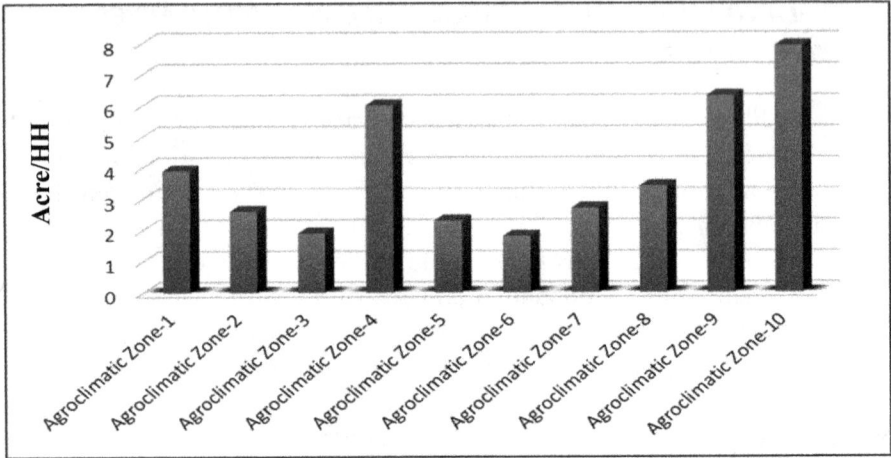

FIGURE 1.5   Average landholding (acre/household) of respondents under different agroclimatic zones.

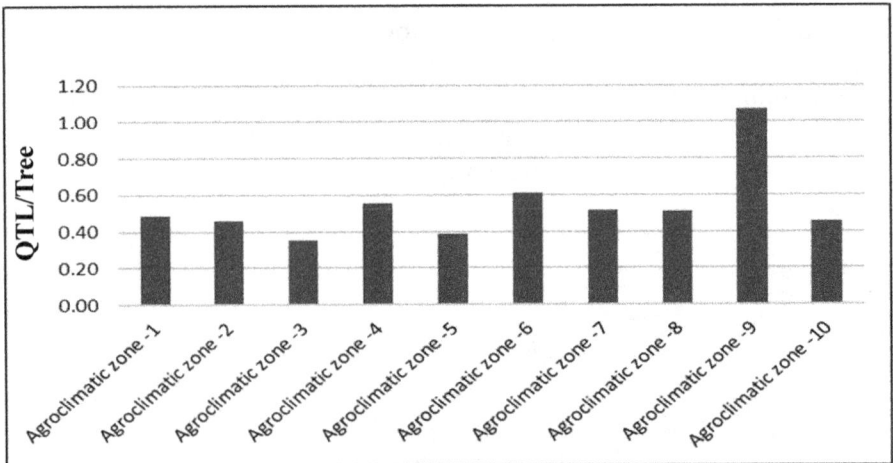

FIGURE 1.6   Yield of flowers (quintals)/tree under different agroclimatic zones.

1.5   CONCLUSION

Forest products are an integral component of the livelihoods of the tribes, and they are most critical and principal suppliers for livelihood. There are various ways in which forest can contribute to the tribal livelihoods such as providing household needs on timber, fuelwood, wild foods, medicinal plants, forest-based agriculture, food security (as safety net during agricultural stress), providing dietary supplements, income generation, income from forest-based labor, etc. However, there is limited marketing and use of the mahua products. Widening the scope of mahua flowers marketing, diversifying the use of mahua flowers and seeds, creating important infrastructures for processing of mahua flowers, capacity building of mahua flower and seed collectors and planting of high yielding mahua trees are some of the suggestive measures which may be undertaken to boost the rural economy especially for the forest dwellers in the state of Odisha.

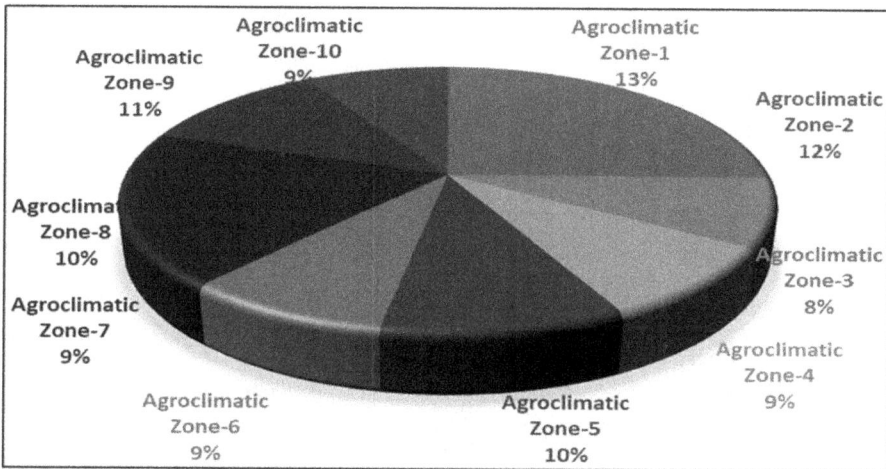

FIGURE 1.7 Contribution (%) of mahua flowers and seeds to total income of respondents under different agroclimatic zones.

KEYWORDS

- agroclimatic zones
- livelihood
- *Madhuca latifolia*
- mahua tree
- non-timber forest products
- rural economy

REFERENCES

Ahirwar, R. K., (2015). Indigenous knowledge of traditional magico-religious beliefs plants of district Anuppur, Madhya Pradesh, India. *American Journal of Ethnomedicine, 2*(2), 103–109.

Akanni, K. A., & State, O., (2013). Economic benefits of non-timber forest products among rural communities in Nigeria. *Environ. Nat. Resour. Res.* 3(4), 19–26.

Alexander, S. J., McLain, R. J., & Blanter, K. A., (2001). Socio-economic research on non-timber forest products in the Pacific Northwest. *Journal of Sustainable Forestry, 13*, 95–105.

Babulo, B., et al., (2008). The economic contribution of forest resource use to rural livelihoods in Tigray. *Forest Policy and Economics, 11*, 109–117.

Bahuguna, V. K., (2000). Forests in the economy of the rural poor: An estimation of the dependence level. *Ambio., 29*, 126–129 (Cameroon).

Basu, J. P., (2009). Adaptation, non-timber forest products and rural livelihood: An empirical study in West Bengal, India. *Earth and Environmental Science, 6*, 3–8.

Bhatia, H. L., (1970). Use of Mahua oil cake in fishery management. *Indian Farming, 20*, 39–40.

Bhattacharya, P., & Hayat, S. F., (2009). Sustainable NTFP management for livelihood and income generation of tribal communities: A case from Madhya Pradesh, India. In: Uma, S. R., Hiremath, A. J., Joseph, G. C., & Rai, N. D., (eds.), *Non Timber Forest Products: Conservation Management and Policy in the Tropics* (pp. 21–34). ATTREE & University of Agriculture Science, Bangalore.

Bwalya, S. M., (2013). Household dependence on forest income in rural Zambia. *Zambia Social Journal, 2*(1), 67–86.

Census of India, (2011). *District Census Handbook: Village and Town Wise Primary Census Abstract (PCA), Series-22, Part XII-B*. Directorate of census operations Odisha.

Dagm, F., Wubalem, T., & Abdella, G., (2016). Economic contribution to local livelihoods and households dependency on dry land forest products in hammer District, Southeastern Ethiopia. *International Journal of Forestry Research*, 11–12.

Delacote, P., (2007). Agricultural expansion, forest products as safety nets, and deforestation. *Environment and Development Economics, 12*, 235–249.

Dolni, G., & Chatterjee, S., (2014). The importance of non-timber forest products in tribal livelihood: A case study of Santal community in Purulia district, West Bengal. *Indian Journal of Geography and Environment, 13*, 110–120.

Falconer, J., & Arnold, J. E. M., (1991). *Household Food Security and Forestry: An Analysis of Socioeconomic Issues* (p. 154). FAO, Rome.

Ganvir, V. N., & Dwivedi, A. P., (2012). Preparation of adsorbent from mahua oil seeds cakes and its characterization. *International Journal of Advanced Engineering Research and Studies, 1*(III), 270–271.

Garekae, H., Thakadu, O. T., & Lepetu, J., (2017). Socio-economic factors influencing household forest dependency in Chobe Enclave, Botswana. *Ecological Processes, 6*, 40. doi: 10.1186/s13717-017-0107-3.

Ghosal, S., (2011). Importance of non-timber forest products in native household economy, full length research paper. *Journal of Geography and Regional Planning, 4*(3), 159–168.

Gunatilake, H. M., (1998). The role of rural development in protecting tropical rainforests: Evidence from Sri Lanka. *Journal of Environmental Management, 53*, 273–292.

Hedge, R., & Enters, T., (2000). Forest products and household economy: A case study from Mudumalai Wildlife Sanctuary, Southern India. *Environmental Conservation, 27,* 250–259.

Heubach, K., Wittig, R., Nuppenau, E. A., & Hahn, K., (2011). The economic importance of non-timber forest products (NTFPs) for livelihood maintenance of rural west African communities: A case study from northern Benin. *Ecological Economics, 70,* 1991–2001.

Htun, T. T., Wen, Y., & Ko, K. A. C., (2017). Assessment of forest resources dependency for local livelihood around protected areas: A case study in Popa Mountain Park, Central Myanmar. *International Journal of Sciences, 6*(1), 34–43.

Kabubo-Mariara, J., (2013). Forest-poverty nexus: Exploring the contribution of forests to rural livelihoods in Kenya. *Nat Res Forum, 37*(3), 177–188.

Kamanga, P., Vedeld, P., & Sjaastad, E., (2008). Forest incomes and rural livelihoods in Chiradzulu District, Malawi. *Ecological Economics, 68,* 613–624.

Kirilenko, A. P., & Sedjo, R. A., (2007). Climate change impacts on forestry. *Proceedings of the National Academy of Sciences, 104,* 19697–19702.

Lakshman, A. K., (1983). Mahua oil cake in fish culture. *Environment and Ecology, 1,* 163–167.

Lalremruata, J., Sahoo, U. K., & Lalramnghinglova, H., (2006). Inventory on non-timber forest products of Mizoram in North-East India. *Journal of Non-Timber Forest Products, 14*(3), 173–180.

Langat, D. K., (2015). *Role of Forest Resources to Local Livelihoods: The Case of East Mau Forest Ecosystem.* Kenya Kenya Forestry Research Institute.

Malhotra, K. C., & Bhatacharya, P., (2010). *Forest and Livelihood* (p. 246). Publ. CES, Hyderabad.

Malhotra, K. C., Deb, D., Dutta, M., Vasulu, T. S., Yadav, G., & Adhikari, M., (1991). *Role of Non Timber Forest Produce in Village Economy: A Household Survey in Jamboni Range.* Mimeographed, Indian Institute of Biosocial Research and Development, Calcutta.

Mamo, G., Sjaastad, E., & Vedeld, P., (2006). Economic dependence on forest resources: A case from Dendi District, Ethiopia. *Forest Policy and Economics, 9,* 916–927.

Manoj, E. M., (2011). *Honey Collection-Climate Change Affecting the Livelihood of Tribal People of Wayanad, Kerala.* The Hindu.

Mishra, S., Prakash, D., & Gottipati, R., (2009). Characterization and utilization of Mahua oil cake: A new adsorbent for removal of congo red dye from aqueous phase. *Electronic Journal of Environmental, Agricultural and Food Chemistry, 8*(6), 425–436.

Mukhopadhyay, D., (2009). Impact of climate change on forest ecosystem and forest fire in India. *Earth and Environmental Science, 6*(3), 20–27.

Mukul, S. A., Rashid, A. Z. M. M., Uddin, M. B., & Khan, N. A., (2016). Role of non-timber forest products in sustaining forest-based livelihoods and rural households' resilience capacity in and around protected area: A Bangladesh study. *J Environment Planning and Management, 59*(4), 628–642.

Pandey, P., Tripathi, Y. C., & Kumar, A., (2016). Non timber forest products (NTFPs) for sustained livelihood: challenges and strategies. *Research Journal of Forestry, 10,* 1–7.

Prabhakaran, R., Kumar, T. S., & Rao, M. V., (2013). Role of non-timber forest products in the livelihood of Malayali tribe of Chitteri hills of Southern Eastern Ghats, Tamil Nadu, India. *Journal of Applied Pharmaceutical Science, 3*(5), 56–60.

Ramadan, M. F., Mohdaly, A. A. A., Asiri, A. M. A., Tadros, M., & Niemeyer, B., (2016). Functional characteristics, nutritional value and industrial applications of *Madhuca longifolia* seeds: An overview. *Journal of Food Science and Technology, 53*(5), 2149–2157.

Rayamajhi, S., Smith-hall, C., & Helles, F., (2012). Empirical evidence of the economic importance of Central Himalayan forests to rural households. *Forest Policy and Economics*, *20*, 25–35.

Reddy, S. R. C., & Chakravarty, S. P., (1999). Forest dependence and income distribution in a subsistence economy: Evidence from India. *World Development*, *27*, 1141–1149.

Saha, D., & Sundriyal, R. C., (2011). Utilization of non-timber forest products in humid tropics: Implications for management and livelihood. *Forest Policy and Economics*, *14*, 28–40.

Sahoo, U. K., Jeeceelee, L., Lalremruati, J. H., Lalremruata, J., Lalliankhuma, C., & Lalramnghinglova, H., (2010). Role of NTFPs in the livelihood of communities in and around Dampa Tiger Reserve in North-East India. *The Bioscan*, *1*(II), 779–790.

Sahoo, U. K., Lalremruata, J., & Lalramnghinglova, H., (2014). Assessment of fuelwood based on community preference and wood constituent properties of tree species in Mizoram, north-east India. *Forests, Trees, and Livelihood*, *23*(4). doi: 10.1080/14728028.2014.943684.

Sahoo, U. K., Lalremruata, J., Jeeceelee, L., Lalremruati, J. H., Lalliankhuma, C., & Lalramnghinglova, H., (2010). Folk utilization of non-timber forest products (NTFPs) by the tribal communities in and around Dampa Tiger Reserve in Mizoram. *The Bioscan*, *1*(II), 721–729.

Sahoo, U. K., Lalremruata, J., Lalramnghinglova, H., Lalremruati, J. H., & Lalliankhuma, C., (2010). Livelihood generation through non-timber forest products by rural poor in and around Dampa tiger reserve in Mizoram *Journal of Non-Timber Forest Products*, *17*(2), 147–161.

Shackleton, C. M., (2004). *Assessment of the Livelihoods Importance of Forestry, Forests and Forest Products in South Africa (Manuscript)*. Grahamstown: Rhodes University.

Shackleton, C. M., Shackleton, S. E., Buiten, E., & Bird, N., (2007). The importance of dry woodlands and forests in rural livelihoods and poverty alleviation in South Africa. *Forest Policy Economics*, *9*(5), 558–577.

Shackleton, S., Delang, C. O., & Angelsen, A., (2011). From subsistence to safety nets and cash income: Exploring the diverse values of non-timber forest products for livelihoods and poverty alleviation. In: Shackleton, S., Shackleton, C., & Shanley, P., (eds.), *Non-Timber Forest Products in the Global Context* (p. 289). Springer: Berlin/Heidelberg, Germany. ISBN: 9783642179822.

Shit, P. K., & Pati, C. K., (2012). Non-timber forest products for livelihood security of tribal communities: A case study in Paschim Medinipur District, West Bengal. *Journal of Human Ecology*, *40*(2), 149–156.

Soe, K. T., & Yea-Chang, Y., (2019). Livelihood dependency on non-timber forest products: Implications for REDD+. *Forests*, *10*, 427. doi: 10.3390/f10050427.

Suleiman, M. S., Wasonga, V. O., Mbau, J. S., Suleiman, A., & Elhadi, Y. A., (2017). Non-timber forest products and their contribution to household's income around Falgore game reserve in Kano, Nigeria. *Ecological Processes*, *6*(23), 1–14.

Sushant, (2013). Impact of climate change in eastern Madhya Pradesh, India. *Tropical Conservation Science-Special*, *6*(3), 338–364.

Thomas, M., Sahu, P., Shrivastava, A., & Hussain, Z., (2011). Biodiversity and livelihood options of people in Chambal ravine of Morena District, Madhya Pradesh, India. *Journal of Tropical Forestry*, *27*, 40–56.

Timko, J. A., Waeber, P., & Kozak, R. A., (2010). The socio-economic contribution of non-timber forest products to rural livelihoods in Sub-Saharan Africa: Knowledge gaps and new directions. *International Forest Review*, *12*, 284–294.

Uberhuaga, P., Smith-Hall, C., & Helles, F., (2012). Forest income and dependency in lowland Bolivia. *Environmental Development and Sustainability, B*, 3–23.

Vonothkumar, R., Dey, R., & Srivinasan, M., (2018). Piscicide effects of mahua oil cake from finfish culture system. *International Journal of Trends in Scientific Research and Development*, 2(5), 1205–1211.

WWF, (2011). *Climate Change in India: A Case Study of Orissa*. http://zeenews.india.com/myearth2011/orissa.aspx (accessed on 4 December 2020).

CHAPTER 2

Grass/Legume Intercropping for Forage Production and Orchard Floor Management in Jammu and Kashmir

SUHEEL AHMAD,[1] NAZIM HAMID MIR,[1] SHEERAZ SALEEM BHAT,[1] and ABDUL RAOUF MALIK[2]

[1]ICAR-Indian Grassland and Fodder Research Institute, Regional Research Station, Srinagar–190007, Jammu and Kashmir, India, E-mail: suhail114@gmail.com (S. Ahmad)

[2]Sher-e Kashmir University of Agricultural Sciences and Technology of Kashmir, Jammu and Kashmir, India

ABSTRACT

Hortipastoral systems refer to a fruit tree-based agroforestry system which combines fruit trees and forage crops and has been recognized as a sustainable land management system owing to its diversified output and several environmental benefits. The resource base available with the farming community is limited, and with the rise in population and no further scope for extending area under cultivation in the state, the per capita availability of land is declining. The land limits has made difficult to the farmers to fulfill land requirement for cultivation of fodder and forage. As per the estimates of the horticulture department of Jammu and Kashmir, more than 325,000 hectares is under fruit orchards. The scope exists that the introduction of fodder crops as inter-crop in orchards, which has by and large remained untapped for fodder development. The utilization of these orchards can give a big boost to livestock development. Orchard floor management practices have been found not only important for the overall health and productivity of fruit trees but also indispensable for improving soil quality, ensuring weed suppression, and orchard access.

Grassing down the alleyways with perennial grasses and legumes and maintaining a vegetation-free tree row has been recommended orchard floor management practice. Hortipastoral systems, therefore, is one of the ideal strategy of land use management that combines forage crops, preferably perennial, and fruit trees into an integrated production system to get maximum provisioning and environmental benefits. The yield and quality of forage crops in hortipastoral systems can be enhanced by introducing shade-tolerant grass-legume mixtures. However, management of such an integrated system can be challenging because of complexity in design and management. There is a need for long-term investigation involving socio-economic and ecological principles, to find the right mix of trees and forage crops for maximum benefits.

2.1   INTRODUCTION

Agroforestry systems, through the integration of crops and trees, have emerged as a dynamic and sustainable land use option (Kidd and Pimental, 1992; Nair, 1988) that have been based on socio-economic and ecological principles for enhanced social, economic, and environmental benefits (Mead, 2004). Agroforestry has been considered as an alternative to subsistence and intensive arable farming (Griffith, 2000; Caravaca et al., 2002) as it provides valuable ecosystem services including improvement in soil fertility, protection of watersheds, mitigation/adaptation of climate change, biodiversity conservation and rehabilitation of degraded lands (Garrity, 2004; Idol et al., 2011). Although there has been an expansion and intensification of research in smallholder agroforestry systems (Leakey et al., 2012), very less information is available on the use of forage crops (perennial grasses and legumes) as an efficient and sustainable floor management practice in fruit orchards.

Temperate horticulture is of immense importance for the economic development in hilly regions. In the fruit map of India, Jammu, and Kashmir accounts for 48% of total temperate fruit production followed by Himachal Pradesh and Uttarakhand. Among these states, Jammu, and Kashmir has emerged as leading apple producing state in the country and during last decade the area under crop has increased 16 times, production by about sixty times and productivity by about five times (Bhat et al., 2013). Although, yield of commercially important apple varieties is highest in the country (11.29 t/ha) in Kashmir yet it compares poorly to the yields obtained by advanced countries which is 60 t/ha (Anonymous, 2009). Most of the area occupied by apple in this region is under rainfed condition. Having a slow growth initially or during formative years, the interspaces not only go without productive use but also become vulnerable to several obnoxious weeds, like, *Anthemis cotula, Carthamus lanatus, Artemisia scoparia, Conium maculatum*, etc. Thus intercrops of compatible along with main crop help to minimize weed in initial growth and further increase the yield. These weeds also hamper various orchard operations. The utilization of these orchards as a niche area for forage resource augmentation can give a big boost to livestock development. Intercropping of perennial forage grasses and/or legumes with fruit crops is a sustainable option for high forage and fruit production, besides providing other environmental benefits. Due to increased population of both humans and livestock, poor productivity of grassland resources and deficit in forage supply and farmer's inability to spare their cultivated land for forage production, horti-pastoral systems (orchards+ pasture+ livestock) wherein the interspaces between fruit tree species are utilized for cultivation of grasses and grass-legume mixtures, seems to be a very promising intervention with respect to profitability and orchard floor management. These systems have been found to exhibit sustained or improved fertility by way of increased biomass, organic matter addition, aiding in the availability of other important nutrients, increasing the beneficial soil micro-fauna and flora, reduced run-off and increased infiltration (Atucha et al., 2011; Ahmad et al., 2018).

2.2   GRASSES AND/OR LEGUMES FOR ORCHARD FLOOR MANAGEMENT

There exist a lot of opportunities to improve overall orchard productivity and sustainability through exploitation and management of orchard floor (Granatstein and Sanchez, 2009). Orchard floor management practices generally range from managing existing resident vegetation (weeds/sod) by frequent cuttings, planting a cover crop between tree rows or through clean cultivation. Good orchard floor management practices are indispensable for the productivity of fruit trees and improvement of soil quality and health. Grassing down the alleyways with perennial grasses and legumes and maintaining a vegetation-free tree row has been a recommended orchard floor management practice (Parker et al., 1993; Dabney et al., 2001; Merwin, 2004). Grass-clover mixtures, when grown in the interspaces of apple trees, have been found to increase shoot growth and yield (Kuhn and Pederson, 2009).

Welker and Glenn (1989) noted that peach (*Prunus persica* L.) trees competing with proximate sod covers were stunted during the early years of a 6-year study but adapted to grass competition over time and eventually became more yield-efficient than trees in herbicide-treated weed-free rows. Hoagland et al. (2008) elucidated that a cover crop consisting of mixed legumes in an apple orchard improved soil fertility and availability of nutrients, especially nitrogen. Yao et al. (2009) reported that apple roots grew deeper and survived longer beneath sod than pre-herb treatments. Further, Ramos et al. (2010) reported that cover crops integrated in almond orchards enhanced soil quality by increasing soil organic carbon (SOC), stability of soil aggregates, and improving soil biological properties. Clean cultivation or cultivation of annual crops decreases soil organic matter and exposes the soil to erosion (Caravaca et al., 2002). Hence, it is advocated to plant perennial forage species in the alleyways of fruit orchards, which will not only provide nutritious fodder but shall also ensure protection of soil. Compared with mono-crop systems, forage grass/legume mixtures increase herbage yield and produce forage with a more balanced nutrition for livestock (Barnett and Posler, 1983; Giambalvo et al., 2011; Geirus et al., 2012). Orchard floor management practices involving grasses and legumes (hortipastoral systems) are important in fruit production to maintain soil tilth and fertility, reduce weed invasion, aid in moderation of soil temperature and moisture extremes, provide a habitat for beneficial insects and decrease soil erosion (Atucha et al., 2011; Ahmad et al., 2018). However, the establishment and management of hortipastoral systems present a challenge in identifying a proper mix of grass/legume mixtures that are compatible with the fruit trees (Kyriazopoulos et al., 2013).

2.2.1 *FORAGE YIELD AND QUALITY IN HORTIPASTORAL SYSTEMS*

Forage yield and quality in hortipastoral systems can be improved by introducing appropriate grass/legume mix with suitable agronomic practices. Misri (1988) conducted a study on the green forage yield of various combinations of grasses and legumes in an apple orchard and reported that ryegrass + red clover and *Dactylis* + red clover combination recorded green forage yield of 48.0 and 42.0 t ha^{-1}, respectively. Singh (1995) reported that rye and orchard grass were found to be the best grass species and clovers and lucerne the best perennial legumes for introduction in apple orchards. Makaya and Gangoo (1995) reported that the average green fodder yield of *Dactylis glomerata*, red, and white clover was 22.03, 24.96 and 24.58

t/ha, respectively as compared to the natural vegetation (14.64 t/ha) while working on the forage yield of grasses and legumes in an almond orchard. The traditional agroforestry systems identified in Kashmir Himalaya include; boundary plantations, agri-silviculture (AS) on sloping lands, AS in plains, horti-silviculture, horti-silvi-pasture, horti-silvi-agriculture, silvopasture, hortipasture, and home gardens (Mughal and Bhattacharya, 2002; Dar et al., 2018). However, in these identified agroforestry systems, there is less representation of all the components, and in most systems, tree component is less, and utilization of space is not efficient. There is a need to refurbish these systems for improved socio-economic and environmental benefits. Ram et al. (2005), while working on the performance of *Ziziphus mauritiana* based Horti-pasture system found that intercropping of guinea grass with Caribbeanstylo resulted in significantly higher dry forage and crude protein yields compared to guinea grass + Caribbean stylo + dinanath grass + Caribbean stylo and natural pasture. Ram et al. (2006) conducted a study on annona-based hortipasture system and found that total crude protein yield was significantly increased in intercropping of *Stylosanthus hamata* + buffelgrass as compared to *Stylosanthus scabra*.

Ram and Parihar (2008) reported that intercropping of dhawalu grass (*Chrysopogon fulvus*) with *Stylosanthus hamata* in 1:1 row ratio resulted in significantly higher total green forage (25.83 t/ha) and dry matter yield (6.73 t/ha) as compared to sole stand of legume and grass. Grass-legume intercropping, besides increasing the biomass production, provides better quality forage, rich in protein and carbohydrates, which would be helpful to improve the milk production and animal health. Legumes, being rich in proteins, enhance forage quality and also add nitrogen to the soil through biological nitrogen fixation (Ram, 2008; Ram and Parihar, 2008). Several legumes, like red and white clover, sainfoin, alfalfa, berseem, shaftal, etc., have been used in temperate regions to improve pasture production and soil N status. White clover is a short-lived perennial stoloniferous legume with a prostrate growth habit. As the name itself, the flowers are white-colored and are clustered into heads (Baker and Williams, 1987; Frame et al., 1998). Red clover is a cool-season perennial forage legume with fusiform root and short stock having many stems that arise from basal leaves. It has an upright growth habit and a strong deep taproot from which finer roots arise and provides high quality forage (Thomas, 2003). Fescues have become one of the most popular and valuable grasses for use in agroforestry and soil conservation systems in temperate regions of the world. Tall fescue (*Festuca arundinaceae*) is a temperate perennial bunchgrass adapted to a wide range of soils (Stephenson and Posler, 1988) owing to its certain characteristics like, deep taproot, drought hardiness, good summer growth and its compatibility with legumes (Lowe and Bowdler. 1995; Reed, 1996; Charlton and Stewart, 2006). *Dactylis glomerata* L. (commonly known as orchard grass) is a widespread temperate perennial forage grass, adapted especially to shaded conditions (Lin et al., 1999; Devkota et al., 2009). Due to the high forage quality, persistence, and shade tolerance, orchard grass is a comparatively well-suited grass species in temperate agroforestry systems (Koukoura and Kyriazopoulos, 2007; Peri et al., 2007; Kyriazopoulos et al., 2013).

2.2.2 OTHER BENEFITS OF GRASS-LEGUME COVER CROPS (AHMAD ET AL., 2020)

- Reduced soil disturbance which boosts soil fertility and improves soil structure and moisture conservation.
- Improve soil health by promoting beneficial soil organisms.
- Legumes help in biological nitrogen fixation, thereby making it available to the plants.
- Help in stabilizing soil by reducing erosion and improving water infiltration Suppress weeds by covering bare areas.
- Improves orchard access during wet weather.
- Soil carbon sequestration by increasing organic matter in the soil.
- Improvement of recreation and esthetic values owing to greener landscape.
- Improvement of water quality owing to reduced erosion, leaching losses and improved water-holding capacity.
- Diversified output in the form of fruits, fodder, livestock products, fuelwood, etc.

2.3 SCOPE AND POTENTIAL OF HORTIPASTORAL SYSTEMS IN JAMMU AND KASHMIR

Jammu and Kashmir is the northern most region of India, which shares international borders with China, Pakistan, and Afghanistan. The Union Territory of Jammu and Kashmir is divided into two distinct geographical, cultural, and linguistic regions that include sub-tropical and sub-temperate Jammu and temperate Kashmir valley (Wani and Wani, 2007). The climate of the region is determined by altitudinal gradient, the elevation increasing from 330 m in Jammu to about 3305 m in Kashmir. With the increase in the elevation, the rainfall decreases from 1052 mm in Jammu to 662 mm in Srinagar with a mean annual temperature of 24.5°C in Jammu and 13.3°C in Kashmir (Wani et al., 2014). The length of the crop-growing season also decreases as we proceed from south to north. In Jammu, the crop can be grown around the year while in Kashmir valley, double cropping is possible (Ahmad and Verma, 2011). The resource base available with the farming community is limited. With the rise in the population and no further scope for extending area under cultivation, the per capita availability of land is declining further.

The limited land creates difficulty to the farmers to fulfill fodder and forage cultivation (Anonymous, 2012; Wani et al., 2014). In comparison to the requirement of about 139.13 and 58.53 lakh metric tons of green and dry fodder, respectively, the availability is about 64 and 35 lakh metric tons (Anonymous, 2013). Consequently, there is a huge gap between demand and supply of fodder, especially during winter months. The major challenge is to maintain the gap of forage production and requirement for sustaining the productivity of livestock in the region (Ahmad et al., 2016). Livestock rearing plays a significant role in the economy of the people living in hilly regions with dependence largely on grasslands and pastures. In Jammu and Kashmir, according to 18th livestock census (2007) there were 3.45 million cattle, 1.05 million buffalo, 3.68 million sheep, and 2.07 million goats. In the last decade the total cattle population has remained almost stagnant but annual growth rate in crossbred cattle was 4.49% p.a. Buffalo has also registered a growth rate of 2.94% p.a. during 1997–2007, whereas in sheep and goats it was 1.5 and 1.05% p.a. respectively (Table 2.1). Balance sheet for the entire state of Jammu and Kashmir to feed such a burgeoning population of livestock is presented in Table 2.2.

The density of livestock per sq. km of area of Jammu and Kashmir was 98 animals against 90 animals for the 16th livestock census. The number of livestock per 1000 of population as per 2003 census was 926 animals while as at all India level the number of livestock per 1000 population works out to be only 457 animals. Average livestock per household, as per census 2001, works out to six animals per household for Jammu and Kashmir as compared to about three animals per household at All India level.

The region has huge potential for the quality temperate horticultural crops because it has created a niche of apple, pears, and dry fruits, i.e., almond, and walnut production. In temperate region fruits like, apple, walnut, and almond has ranked first, second, and third, respectively both in their area contributing of 64.05, 27.31 and 8.65% respectively and production. Including this the region has a variety of agro-climatic, i.e., sub-tropical, sub temperate, temperate, and cold arid and each region has specific fruit in the year (Tables 2.3–2.5).

TABLE 2.1   Livestock Population and Their Growth Rate in Jammu and Kashmir

Year Population	Crossbred Cattle	Total Cattle	Buffalo	Sheep	Goat
1987	4,90,919	2,765,699	5,95,257	2,493,454	1,396,025
1997	1,083,000	3,175,000	7,87,000	3,170,000	1,864,000
2003	1,320,000	3,083,842	1,039,461	3,410,676	2,054,923
2007	1,680,724	3,446,150	1,051,490	3,680,232	2,068,653
Growth Rate (Percent p.a.)					
1987–1997	0.79	0.14	0.28	0.24	0.29
1997–2003	3.35	−0.48	4.75	1.23	1.64
2003–2007	6.23	2.82	0.29	1.92	0.17
1997–2007	4.49	0.82	2.94	1.50	1.05

TABLE 2.2   Fodder Balance Sheet on Dry Matter Basis ('000 tons)

Region	Availability	Requirement	Deficit
Jammu	5545.58	8188.00	2642.42 (32.27%)
Kashmir	1866.49	3635.00	1768.51 (48.65%)
Ladakh	108.57	740.00	631.43 (85.32%)
Jammu and Kashmir	7420	12563.00	5142.36 (40.93%)

Figures in parenthesis represent percentage.

Source: Wani et al. (2014).

The continuous effort is made to enhance forage production through alternate land-use systems. Use of forage crops (temperate perennial grass/legume mixtures) in the interspaces of fruit orchards offers a huge potential as a sustainable orchard floor management approach, combined with grazing by sheep for the improvement of physical, chemical, and biological properties of orchard soils and augmenting forage resource availability (Sharma, 2004; Kumar and Choubey, 2008; Khan and Kumar, 2009; Ramos et al., 2011; Wani et al., 2014; Ahmad et al., 2018). The utilization of these orchards can give a big boost to livestock development in the region (Table 2.3).

TABLE 2.3   Area under Fruits in Jammu and Kashmir (Area in 000 Hectares)

Year	Apple	Pear	Apricot	Cherry	Other Fresh	Walnut	Almond	Other Dry	Total Fruits
2004–2005	107.93	10.54	4.93	2.55	41.62	74.89	15.43	0.42	258.31
2005–2006	111.88	11.00	5.16	2.59	43.60	77.22	15.55	0.41	267.41
2006–2007	119.04	11.25	5.43	2.75	46.24	81.39	16.37	0.62	283.09
2007–2008	127.80	12.10	4.78	3.14	48.32	82.05	16.40	0.55	295.14
2008–2009	133.10	12.35	4.92	3.30	49.65	84.56	17.18	0.56	305.62
2009–2010	138.19	12.55	5.00	3.41	50.57	87.28	17.54	0.60	315.14
2010–2011	141.71	12.53	5.85	3.46	53.50	89.78	17.65	0.58	325.06
2011–2012	154.72	13.21	6.05	3.48	54.11	83.61	16.41	11.19	342.78

Source: Anon (2012–2013); Malik and Choure (2014).

TABLE 2.4   Suitable Fruit Crops for Different Zones

Zone	Districts	Fruit Crop
Temperate zone	Anantnag, Pulwama, Shopian, Budgam, Baramulla, Srinagar, Ganderbal, Kulgam, Kupwara, and Budgam	Apple, Almond, Pear, Cherry, Walnut, and Apricot
Intermediate zone	Doda, Rajouri, Poonch, and parts of Udhampur	Peach, Plum, Apricot, Olive, and pomegranate
Sub-tropical zone	Jammu, Kathua, Samba, and parts of Udhampur	Mango, Citrus, Ber, Aonla
Cold arid zone	Leh and Kargil	Apricot, Apple

Source: Ahmad and Verma (2011).

2.4   CONCLUSION

There is a tremendous scope for expansion of livestock oriented activities in the entire north-western Himalayan region, particularly in Jammu and Kashmir because of local market and significant quantities of the demand of livestock produce is met through import from neighboring states of Haryana, Punjab, and Rajasthan. In view of fodder shortage and limitation to expand the area under fodder cultivation, the effective utilization of interspaces of fruit orchards offer a unique opportunity to mitigate the fodder shortages up to a greater extent. Presently research has been mainly conducted on cultivation of green fodder in irrigated areas, but focus has to be given to the integration of fruits and forages. There is scanty information on fruit tree and pasture association, especially in temperate agroforestry systems, hence further research aimed at exploring new species mix is required to optimize production, diversification, and environmental sustainability (Jose et al., 2004). The synergies and trade-offs of fruit tree-pasture combinations should be exploited by evaluating different plant species under a given soil and climatic condition. Establishment and upscaling of hortipastoral models should be replicated at farmers' fields. There is a need for development of recommendations/advisories for managing animal, forage crops, and fruit trees. This will require greater knowledge regarding three-way interactions between livestock, pasture, and fruit trees. To encourage more farmers/ orchardists and educated youth in the animal husbandry and horticulture sector, hortipastoral models need to be developed along with the introduction of fodder conservation techniques. This approach would enhance the supply of nutritious fodder, thereby ensuring sustainable livestock production in the region.

TABLE 2.5 Suitable Grasses and Legumes in Various Agroclimatic Zones of Jammu and Kashmir

Zone	Grasses	Legumes
Temperate zone	*Dactylis glomerata, Festuca arundinacea, Lolium perenne, Phleum pratense, Bromus unioloides, Phalaris spp., Poa pratensis, Lolium multiflorum, Agrostis spp. Avena sativa*	*Trifolium pratense, T. repens, Onobrychis viciifolia, Medicago sativa, Trifolium alexandrinum, Lotus corniculatus, Coronilla varia, Trifolium resupinatum*
Intermediate zone	*Dactylis glomerata, Festuca arundinacea, Lolium perenne, Dicanthium annulatum, Chloris gayana, Chrysopogon fulvus, Heteropogon contortus, Setaria spp., Avena sativa*	*Trifolium alexandrinum, Onobrychis viciifolia, Stylosanthus hamata, Macroptelium atropupreum,*
Sub-tropical zone	*Dicanthium annulatum, Chloris gayana, Chrysopogon fulvus, Heteropogon contortus, Cenchrus ciliaris, C. setigerous, Paspalum notatum, Avena sativa*	*Trifolium alexandrinum, Stylosanthus hamata, Stylosanthus scabra,*
Cold arid zone	*Festuca arundinacea, Avena sativa, Phalaris spp., Dactylis glomerata, Elymus spp.*	*Medicago sativa, Medicago falcata, Lotus corniculatus, Astragalus spp., Caragana spp., Melilotus officinalis, Cicer microphyllum*

Source: Ahmad et al. (2016).

KEYWORDS

- **agroforestry**
- *Dactylis glomerata* **L.**
- **grasses and legumes**
- **hortipasture**
- **importance value index**
- **orchard floor**

PLATE 2.1   An apple-based hortipasture system at ICAR-CITH experimental farm.

PLATE 2.2   Almond-based hortipastoral system at RRS-IGFRI, Srinagar.

PLATE 2.3   Haymaking in an apple orchard at Shopian.

PLATE 2.4   Sheep grazing in an apple orchard at Kulgam.

REFERENCES

Ahmad, M. F., & Verma, M. K., (2011). Temperate fruit scenario in Jammu and Kashmir: Status and strategies for enhancing productivity. *Indian Horticulture Journal, 1*(1), 01–09.

Ahmad, S., Bhat, S. S., & Mir, N. H., (2020). Enhancing forage availability and ecosystem services through hortipastoral systems: Challenges and opportunities. In: *Souvenir Cum Abstract Book of International Conference on Advances and Innovations in Agriculture and Allied Sciences* (pp. 25–31). JNU, New Delhi, India.

Ahmad, S., Khan, P. A., Verma, D. K., Mir, N. H., Sharma, A., & Wani, S. A., (2018). Forage production and orchard floor management through grass/legume intercropping in apple-based agroforestry systems. *International Journal of Chemical Studies, 6*(1), 953–958.

Ahmad, S., Singh, J. P., Khan, P. A., & Ali, A., (2016). Pastoralism and strategies for strengthening rangeland resources of Jammu and Kashmir. *Annals of Agri-Bio Research, 21*(1), 49–54.

Anonymous Digest of Statistics, (2012). *Directorate of Economics and Statistics* (p. 700). Planning and Development Department, Government of Jammu and Kashmir.

Anonymous, (2009). *Production and Area Statements for 2008–2009* (pp. 139–154). Department of Horticulture, Government of Jammu and Kashmir.

Anonymous, J. K., (2013). *The Greater Kashmir.* http://jammu.greaterkashmir.com/news/2013/Mar/23/jk-targets-20-lakh-mt-milk-production-this-fiscal-50.asp (accessed on 4 December 2020).

Atucha, A., Merwin, I. A., & Brown, M. G., (2011). Long-term effects of four groundcover management systems in an apple orchard. *Hortscience, 46*(8), 1176–1183.

Baker, M. J., & Williams, W. M., (1987). *White Clover* (p. 534). CAB International, Oxon, UK.

Barnett, F. L., & Posler, G. L., (1983). Performance of cool-season perennial grasses in pure stands and in mixture with legumes. *Agron. J., 75*, 582–586.

Bhat, R., Wani, W. M., Banday, F. A., & Sharma, M. K., (2013). Effect of intercrops on growth, productivity, quality, and relative economic yield of apple cv. Red Delicious. *SKUAST Journal of Research, 15*(1), 35–40.

Caravaca, F., Masciandaro, G., & Ceccanti, B., (2002). Land use in relation to soil chemical and biochemical properties in semi-arid Mediterranean environment. *Soil and Tillage Research, 68*, 23–30.

Charlton, D., & Stewart, A., (2006). *Pasture and Forage Plants for New Zealand* (p. 96). Grassland Association and New Zealand Grassland Trust.

Dabney, S. M., Delgado, J. A., & Reeves, D. W., (2001). Using winter cover crops to improve soil and water quality. *Communications in Soil Science and Plant Analysis, 32,* 1221–1250.

Dar, M., Qaisar, K. N., Ahmad, S., & Wani, A. A., (2018). Inventory and composition of prevalent agroforestry systems of Kashmir Himalaya. *Advances in Research, 14*(1), 1–9.

Devkota, N. R., Kemp, P. D., Hodgson, J., Valentine, I., & Jaya, I. K. D., (2009). Relationship between tree canopy height and the production of pasture species in a silvopastoral system based on alder trees. *Agroforestry Systems, 76*(2), 363–374.

Frame, J., Charlton, J. F. L., & Laidlaw, A. S., (1998). *Temperate Forage Legumes* (p. 327). CAB International, Oxon, UK.

Garrity, D. P., (2004). Agroforestry and the achievement of the millennium development goals. *Agroforestry Systems, 61,* 5–17.

Geirus, M., Kleen, J., Logus, R., & Taube, F., (2012). Forage legume species determine the nutritional quality of binary mixtures with perennial ryegrass in the first production year. *Animal Feed Science and Technology, 172,* 150–161.

Giambalvo, D., Ruisi, P., Di Miceli, G., Frenda, A. S., & Amato, G., (2011). Forage production, N uptake, N_2 fixation, and N recovery of berseem clover grown in pure stand and in mixture with annual ryegrass under different managements. *Plant Soil, 342,* 379–391.

Granatstein, D., & Sanchez, E., (2009). Research knowledge and needs for orchard floor management in organic tree fruit systems. *International Journal of Fruit Science, 9*(3), 257–281.

Griffith, D. M., (2000). Agroforestry: A refuge for tropical biodiversity after fire. *Conservation Biology, 14*(1), 325–326.

Hoagland, L., Carpenter-Boggs, L., Granatstein, D., Mazzola, M., Smith, J., Peryea, F., & Reganold, J. P., (2008). Orchard floor management effects on nitrogen fertility and soil biological activity in a newly established organic apple orchard. *Biology and Fertility of Soils, 45,* 11–18.

Idol, T., Haggar, J., & Cox, L., (2011). Ecosystem services from smallholder forestry and agroforestry in the tropics. In: Campbell, W. B., & Lopes, O. S., (eds.), *Integrating Agriculture, Conservation and Ecotourism: Examples from the Field.* New York: Springer. doi: 10.1007/978-94-007-1309-3_5.

Jose, S., Gillespie, A. R., & Pallardy, S. G., (2004). Interspecific interactions in temperate agroforestry. *Agroforestry Systems, 61*(1), 237–255.

Khan, T. K., & Kumar, S., (2009). *Physical Hortipastoral Models, Their Management and Evaluation* (p. 52). IGFRI Jhansi (India).

Kidd, C. V., & Pimentel, D., (1992). *Integrated Resource Management: Agroforestry for Development* (p. 223). San Diego, Academic Press.

Koukoura, Z., & Kyriazopoulos, A. P., (2007). Adaptation of herbaceous plant species in the understorey of *Pinus brutia. Agroforestry Systems, 70*(1), 11–16.

Kuhn, B. F., & Pedersen, H. L., (2009). Cover crop and mulching effects on yield and fruit quality in un-sprayed organic apple production. *European J. Horticulture Sci., 74,* 247–253.

Kumar, S., & Choubey, B. K., (2008). Performance of aonla (*Emblica officinalis*)-based hortipastoral system in semi-arid region under rainfed situation. *Indian Journal of Agricultural Sciences, 78*(9), 748–751.

Kyriazopoulos, A. P., Abraham, E. M., Parissi, Z. M., Koukoura, Z., & Nastis, A. S., (2013). Forage production and nutritive value of *Dactylis glomerata* and *Trifolium subterraneum* mixtures under different shading treatments. *Grass and Forage Science, 68*(1), 72–82.

Leakey, R. R. B., Weber, J. C., Page, T., Cornelius, J. P., Akinnifesi, F. K., Roshetko, J. M., Tchoundjeu, Z., & Jamnadass, R., (2012). Tree domestication in agroforestry: Progress in the second decade. In: Nair, P. K. R., & Garrity, D. P., (eds.), *The Future of Agroforestry* (pp. 145–173, 541). New York: Springer.

Lin, C. H., McGraw, R. L., George, M. F., & Garrett, H. E., (1999). Shade effects on forage crops with potential in temperate agroforestry practices. *Agroforestry Systems, 44*(2), 109–119.

Lowe, K. F., & Bowdler, T. M., (1995). Growth, persistence, and rust sensitivity of irrigated perennial temperate grasses in the Queensland subtropics. *Australian Journal of Experimental Agriculture, 35*(5), 571–578.

Makaya, A. S., & Gangoo, S. A., (1995). Forage yield of pasture grasses and legumes in Kashmir valley. *Forage Research, 21*(3), 152–154.

Malik, Z. A., & Choure, T., (2014). Horticulture growth trajectory evidences in Jammu and Kashmir (A lesson for Apple Industry in India). *Journal of Business Management and Social Sciences Research, 3*(5), 45–49.

Mead, D. J., (2004). Agroforestry. In: *Forests and Forest Plants, Encyclopedia of Life Science Systems* (Vol. 1, pp. 324–355). Oxford UK: EOLSS Publishers.

Merwin, I. A., (2004). *Groundcover Management Effects on Orchard Production, Nutrition, Soil and Water Quality* (pp. 25–29). New York Fruit Quarterly, Ithaca, NY.

Misri, B., (1988). Forage production in alpine and sub-alpine region of northwest Himalaya. In: *Pasture and Forage Crops Research* (pp. 43–55). Punjab Singh, RMSI, IGFRI, Jhansi.

Mughal, A. H., & Bhattacharya, P. K., (2002). Agroforestry systems practiced in Kashmir Valley of Jammu and Kashmir. *Indian Forester, 128*(8), 846–852.

Nair, P. K. R., (1998). Directions in tropical agroforestry research: Past, present, and future. *Agroforestry Systems, 38*(1), 223–245.

Parker, M. L., Hull, J., & Perry, R. L., (1993). Orchard floor management affects peach rooting. *Horticulture Sci., 118*, 714–718.

Peri, P. L., Lucas, R. J., & Moot, D. J., (2007). Dry matter production, morphology, and nutritive value of *Dactylis glomerata* growing under different light regimes. *Agroforestry Systems, 70*(1), 63–79.

Ram, S. N., & Parihar, S. S., (2008). Growth, yield and quality of mixed pasture as influenced by potash levels. *Indian Journal of Agricultural Research, 42*(3), 228 – 231.

Ram, S. N., (2008). Productivity and quality of pasture as influenced by planting pattern and harvest intervals under semiarid conditions. *Indian Journal of Agricultural Research, 42*(2), 128–131.

Ram, S. N., Kumar, S., & Roy, M. M., (2005). Performance of jujube (*Ziziphus mauritiana*)-based horti-pasture system in relation to pruning intensities and grass-legume associations under rainfed conditions. *Indian Journal of Agronomy, 50*(3), 181–183.

Ram, S. N., Kumar, S., Roy, M. M., & Baig, M. J., (2006). Effect of legumes and fertility levels on buffel grass (*Cenchrus ciliaris*) and Annona *(Annona squamosa)* grown under horti-pasture system. *Indian Journal of Agronomy, 51*(4), 278–282.

Ramos, M. E., Benitez, E., Garcia, P. A., & Robles, A. B., (2010). Cover crops under different managements vs. frequent tillage in almond orchards in semiarid conditions: Effects on soil quality. *Applied Soil Ecology, 44*(1), 6–14.

Ramos, M. E., Robles, A. B., Sanchez-Navarro, A., & Gonzalez-Rebollar, J. L. Soil responses to different management practices in rainfed orchards in semiarid environments. *Soil and Tillage Research, 112*, 85–91.

Reed, K. F. M., (1996). Improving the adaptation of perennial pasture grasses. *New Zealand Journal of Agricultural Research, 39,* 457–464.

Sharma, S. K., (2004). Horti-pastoral based land-use systems for enhancing productivity of degraded lands under rainfed and partially irrigated conditions. *Uganda J. of Agricultural Sciences, 9,* 320–325.

Singh, V., (1995). Technology for forage production in hills of Kumaon. In: Hazra, C. R., & Bimal, M., (eds.), *New Vistas in Forage Production* (pp. 197–202). RMSI, IGFRI, Jhansi.

Stephenson, R. J., & Posler, G. L., (1988). The influence of tall fescue on the germination, seedling growth and yield of bird's foot trefoil. *Grass and Forage Science, 43*(3), 273–278.

Thomas, R. G., (2003). Comparative growth forms of dryland forage legumes. In: Moot, D. J., (ed.), *Legumes for Dryland Pastures* (pp. 19–26). Grassland Research and Practice Series No. 11, New Zealand Grassland Association, Wellington, New Zealand.

Wani, S. A., & Wani, M. H., (2007). *Livestock Crop Production System Analysis for Sustainable Production System in Various Agro-Climatic Zones of J&K.* Lead Paper Presented in 90th Annual Conference of Indian Economic Association.

Wani, S. A., Shaheen, F. A., Wani, M. H., & Saraf, S. A., (2014). Fodder budgeting in Jammu and Kashmir: Status, issues and policy implications. *Indian Journal of Animal Sciences, 84*(1), *54–59.*

Welker, W. V., & Glenn, D. M., (1989). Sod proximity influences the growth and yield of young peach trees. *Journal of the American Society for Horticultural Science, 114*(6), 856–859.

Yao, S. R., Merwin, I. A., & Brown, M. G., (2009). Apple root growth, turnover, and distribution under different orchard groundcover management systems. *HortScience, 44*(1), 168–176.

CHAPTER 3

Tree Diversity and Soil Organic Carbon Status in Agroforestry Systems of Central Province of India

POOJA VERMA,[1] M. VASSANDA COUMAR,[2] ARVIND BIJALWAN,[3]
MANMOHAN JAGATRAM DOBRIYAL,[4] KULASEKARAN RAMESH,[5]
ANUP PRAKASH UPADHYAY,[1] and TARUN KUMAR THAKUR[6]

[1]Indian Institute of Forest Management, Bhopal, Madhya Pradesh, India

[2]Indian Institute of Soil Science, Bhopal, Madhya Pradesh, India

[3]College of Forestry, VCSG Uttarakhand University of Horticulture and
Forestry, Ranichauri–249199, Tehri Garhwal, Uttarakhand, India,
E-mail: arvindbijalwan276@gmail.com

[4]College of Horticulture and Forestry, Rani Lakshmi Bai Central
Agricultural University, Jhansi–284003, Uttar Pradesh, India

[5]Indian Institute of Oilseeds Research, Hyderabad, Telangana, India

[6]Department of Environmental Science, Indira Gandhi National Tribal
University (IGNTU), Amarkantak, Madhya Pradesh, India ·

ABSTRACT

Agroforestry is the most dynamic and sustainable land use practice comprising woody perennial trees in combination with agricultural crops that has enormous benefits for the environment as well as for the farmers. A study was undertaken to assess the diversity of trees and soil organic carbon (SOC) status under agroforestry system in the farmlands of selected villages of Raebareli district, Uttar Pradesh. The main woody perennial tree under agroforestry systems in the study site are *Eucalyptus spp.*, *Mangifera indica*, *Emblica officinalis*, *Dalbergia sissoo*, *Acacia nilotica*, *Madhuca indica* and *Ziziphus mauritiana* having importance value index of 83.54, 39.38, 25.95, 20.70, 18.72, 16.04 and 11.25 respectively. SOC under selected agroforestry trees (*Eucalyptus spp.* and *Emblica officinalis*) was estimated by standardized Walkley and Black method. The results indicated that SOC content decreased with an increase in soil depth. The percent decrease in SOC from the surface (0–15 cm) to subsurface soil (15–30 cm) was 23.7% in sole wheat system as compared to 13.6%, 14.5%, 13.0%, and 14.3% in sole Eucalyptus, sole Aonla, Eucalyptus-wheat, and Aonla-wheat systems, respectively. Further, it was evident from the study that SOC in the surface soil (0–5 cm) under Eucalyptus-wheat system was 9.4% higher as compared to sole forestry system (Eucalyptus), while it was 24.7% higher as compared to sole agriculture system (wheat). Similarly, under Aonla-wheat system, it was 6.7% higher as compared to sole forestry system (Eucalyptus), while it was 28.1% higher as compared to sole agriculture system (wheat). The findings of this study indicate that farmers are showing interest in adopting agroforestry than mono-cropping system, which has a significant impact on SOC content and that can lead to increased carbon sequestration and improved soil quality.

3.1   INTRODUCTION

Agroforestry systems are traditional land-use systems that integrate woody perennial trees/shrubs with agricultural crops/livestock, which are temporally or spatially arranged (Nair et al., 1985). It is being considered as highly profitable, socially compatible, and economically sound land-use systems. Agroforestry is one of the best options for intensification of tree cover outside the forest. The main reason behinds planting trees on their farmlands are additional income, availability of small construction timber and fuelwood. Agroforestry also aids in reducing unemployment and has an immense potential to attain sustainability in agriculture and the environment by enhancing agronomic productivity and mitigating climate change impact. Old and well-established agroforestry systems increase the deep soil organic carbon (SOC) stock (Cardinael et al., 2017). Carbon stored in the trees and the transfer of organic matter from the trees to the soil can increase SOC stocks. Removing atmospheric carbon and storing it in the terrestrial ecosystems is one of the options to compensate greenhouse gas (GHG) emissions (Lal, 2003). Soils play a critical role in the global carbon balance by regulating dynamic bio-geochemical processes and the exchange of GHGs with the atmosphere. Soil C sequestration establishes another realistic opportunity attainable in different agroforestry systems (Lal, 2004). Agroforestry systems also have a huge potential of carbon sequestration to the extent of 10 Mg/ha/year in short rotation period in the plantation of *Eucalyptus* and *Leucaena* (Srinivasarao et al., 2014).

In a country like India, where land-holding size is shrinking and the majority of farming community belongs to the marginal category, combining perennial trees with agriculture crops/livestock is the only way to improve the inclusive farm productivity. In almost every region of the country, traditional agroforestry, commercial agroforestry and tree plantation has been evidenced in several studies. Uttar Pradesh (UP) is one of the largest states of India having a large population of agrarian community. Agroforestry practices in the state differ accordingly to the different agro-climatic zones, site-specific trees and socio-economic status of farmers (Singh, 2014). The farmers of UP has majorly marginal and small landholdings. Henceforth, for increasing the overall profitability and getting the alternate source of income, agroforestry is extensively practiced in the state. Among forest trees *Eucalyptus spp., Populus deltoides, A. nilotica, Madhuca indica, Tectona grandis* and fruits trees like, *Mangifera indica, Emblica officinalis, Psidium guava* can be seen as boundary plantation, scattered in the farmland or in spacing pattern under agri-silviculture (AS) and agri-horticulture (AH) system respectively. Most of the trees favored by the farmers are the multipurpose trees (MPTs) which accomplishes the domestic needs of local people for fuel, fodder, fiber, fruits, etc., (Rai et al., 1995). In Central Uttar Pradesh, traditional as well as commercial agroforestry has been well established and practiced by the farmers.

Raebareli falls under the Central Plain Zone agro-climatic zone, which is characterized by dry sub-humid to semi-arid climate. Distribution of tree depends upon the land holding, requirement of the farmers, and the climate of the region. Raebareli has a developing approach toward agroforestry; characterized by subsistence agriculture region with low crop intensity and irrigation facility. The farmers of the regions are showing curiosity and eagerness towards the agroforestry and farm forestry. Eucalyptus in bund plantation and *E. officinalis* (spacing pattern) is the major trees prevalent under agroforestry in the region. Various studies show the phyto-diversity, traditional knowledge and invasion of alien species of the trees, shrubs, and herbs present in the district (Reddy et al., 2009; Singh and Shukla, 2015), but the tree diversity and SOC studies in the agricultural lands in context with agroforestry is unexplored. In this context, the present study was conducted to assess the diversity of agroforestry trees in the agricultural fields of the farmers. Important agricultural crop grown in the area has also been listed in the chapter. SOC of two agroforestry systems with Eucalyptus and Aonla has also been estimated; as these trees are commercially important in the area for their timber and fruit, respectively. Moreover, framers of the regions are growing many agricultural crops in the association of these trees.

3.2   MATERIALS AND METHODS

3.2.1   STUDY AREA

The study was carried out in randomly selected three villages of Unchahar block of Raebareli District of Uttar Pradesh located in the central part of the state. It forms a part of the Lucknow Division and lies between 25° 49' N to 26° 36' N Latitude and 100° 41' E to 81° 34' E Longitude (Figure 3.1). The climate of the region is drying sub-humid to semi-arid. Majority of the population is involved in agriculture and allied activities. Several fruit and forest trees can be seen in farm forestry and agroforestry systems in the region. Bund based agroforestry-using Eucalyptus species is the most preferred pattern among the farming communities. It is planted along with various annual crops in the farmlands. According to the secondary data collected from the District Agricultural Department of the district, the main Kharif crops grown in the Unchahar block of the district are paddy (*Oryza sativa*), Urd (*Vigna mungo*), Arhar (*Cajanus cajan*), Jowar (*Sorghum bicolor*), Til (*Sesamum indicum*), Bajra (*Pennisetum glaucum*) and Moong (*Vigna radiata*). The main rabi crops are wheat (*Triticum aestivum*), Chana (*Cicer arietinum*), Mustard (*Brassica juncea*), Jau (*Hordeum vulgare*), Masoor (*Lens culinaris*), Alsi (*Linum usitatissimum*), etc.

3.3   METHODS

3.3.1   TREE DIVERSITY ANALYSIS

An attempt is made in the chapter for studying the tree diversity in the farmlands of three villages, namely Barsawan, Kharu, and Umri which have been randomly selected from the block Unchahar of Raebareli district. For investigating the agroforestry status in the villages, 65%, 62.22% and 53.18% of the total household from the village Barsawan, Kharu, and

Umri respectively have been surveyed. Questions regarding land under agroforestry, farm forestry, agriculture have been noted down. Land under agroforestry includes all type of agroforestry, i.e., bund plantation, AS, Horti-silviculture, Agri-silvihorticulture, silvipastoral, and hortipastoral systems. Land under farm forestry comprises all the sole tree farming and fruit tree orchards. Land area under sole agricultural crops and land under other uses includes area under other crops like vegetables, medicinal, etc. Land details of surveyed household have been denoted (Table 3.1) in percentage value.

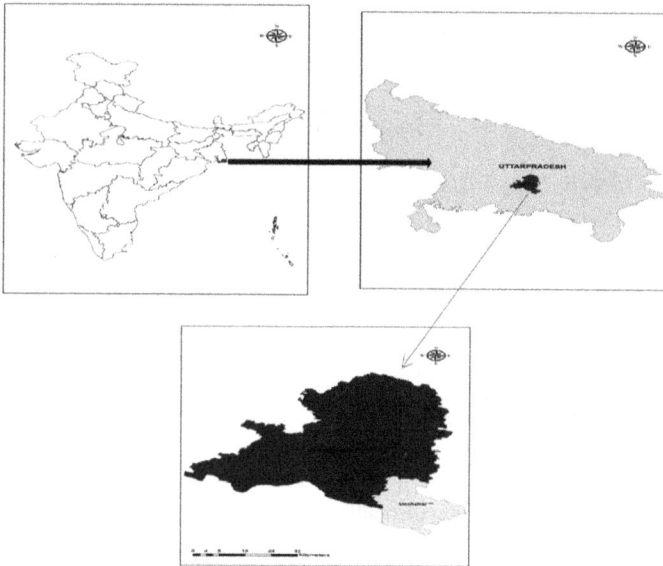

FIGURE 3.1 Location map of the study area.

TABLE 3.1 Village Wise Details of Household Surveyed and Area under Different Cropping Systems (Values in Percentage)

Village Surveyed	Household Surveyed	Land under Agroforestry	Land under Farm Forestry	Land under Agriculture	Other
Barsawan (V1)	65.00	62.65	15.66	19.27	2.40
Kharu (V2)	62.22	54.68	13.54	31.54	0.52
Umri (V3)	53.18	46.27	12.33	35.98	5.39

For phytosociological analysis, ten quadrats of a sample size of 20×20 meter were laid in each selected village for estimating the trees diversity in the agricultural fields. All the woody perennials trees which fall under the quadrats have been documented and counted. Prevalent tree-crop combinations (traditional and planted), i.e., AS, AH, agri-Silvi-horticulture (ASH), and monocropping, mixed cropping, medicinal crops, various cash crops have also been recorded from all the villages. In the existing agroforestry trees, phytosociological studies were carried out to analyze the density, frequency, abundance, and additionally various quantitative characteristics like relative frequency, relative density, relative dominance, and importance value index (IVI). The main purpose of the phytosociological analysis is to estimate the species richness (SR) and diversity which is existing in the selected village study area. The formula used for phytosociological analysis (Phillips, 1959; Curtis and McIntosh, 1950) is given in Figure 3.2.

Density = $\frac{\text{Total number of Individuals in all sampling units}}{\text{Total number of sampling units studied}}$

Frequency = $\frac{\text{Number of sampling units in which species occur}}{\text{Total number of sampling units}} \times 100$

Abundance = $\frac{\text{Total number of individual of a species in all the sampling units}}{\text{Total number of sampling units in which the species occurred}} \times 100$

Relative density = $\frac{\text{Density value of species}}{\text{Sum of density value of all species}} \times 100$

Relative frequency = $\frac{\text{Frequency value of species}}{\text{Sum of frequency value of all species}} \times 100$

Relative dominance = $\frac{\text{Total basal area of the species}}{\text{Total basal area of all species}} \times 100$

IVI (Importance value index) = Relative density + Relative frequency + Relative dominance

---

FIGURE 3.2   Formula used for phytosociological analysis.

3.3.2   SOIL ORGANIC CARBON (SOC) ESTIMATION

Soil samples were collected from the existing agroforestry system (Eucalyptus and Aonla based agroforestry) and crop field (wheat) of selected three villages during Rabi season of 2016–2017 using tube augers. Representative soil samples were collected from five different soil depths (0–5 cm, 5–15 cm, 15–30 cm, 30–60 cm and 60–90 cm) with three replications; a total of 5 sampling points were selected from each location in a zigzag manner and were mixed to prepare a composite sample for analysis. The soil samples were processed, air-dried, and passed through a 0.5 mm sieve for SOC estimation using the standard procedure of Walkley and Black (1934) method.

3.4   RESULTS AND DISCUSSION

3.4.1   *TREE AND CROP DIVERSITY*

Eucalyptus based agroforestry is one of the major agroforestry systems practiced by the farmers in Central Uttar Pradesh. Farmers have shown mixed feelings about eucalyptus plantations. The practice of growing eucalyptus in their farmlands is becoming a common practice in the region (Kumar et al., 2003). Eucalyptus based bund/boundary system and *E. officinalis* based AH is commonly exiting in the selected villages. Both agroforestry systems were more profitable than pure agricultural mono-cropping systems. From the survey, the predominant timber tree species in the sites on the agricultural fields were Eucalyptus, *A. nilotica* (Babool) and *Dalbergia sissoo* (Shisham), while *Mangifera indica* (mango), *E. officinalis* (Amla), *Ziziphus mauritiana* (ber) among the horticultural trees. Eucalyptus is commonly found on the bunds/boundary of the agricultural fields. Sole tree crops/orchards have been observed in the sites were *E. officinalis, M. indica, Eucalyptus, Musa paradisiaca*, etc. Among the fruit trees, *E. officinalis* is an important fruit tree because of its significant contribution to the enhancement of the socio-economic status of farmers through commercial production. The list of fruits and forest trees commonly present in agroforestry systems (in the selected village) are merged and presented in Table 3.2.

3.4.2   *EXISTING AGROFORESTRY SYSTEM IN THE STUDY AREA*

AS (forest tree +agriculture crop), AH (fruit tree+ agricultural crop), and agrisilvihorticulture system (fruit tree+ forest tree + agriculture crop) are the most popular systems retained by large and medium farmers in the region. Silvopastoral system (trees with any fodder crop) is also seen in the study site. Fodder crops common in the region are Sorghum or Jowar/Sudan chari (*Sorghum bicolor*), Bajra (*Pennisetum americanum*) and Barseem (*Trifolium alexandrinum*). Different tree-crop combination and agroforestry systems observed in all the site is depicted in Table 3.3. Traditional agroforestry is still practiced in the study area as it enhances soil fertility and endures to offer much food, fiber, forage, fuelwood, and other resources for construction for the rural people. The tree inventory study identified mango, babul, and mahua as the most common trees found under traditional agroforestry systems in all the study sites. Aonla and Eucalyptus is the another utmost preferred woody perennials under commercial agroforestry. Most of the farmers in the

TABLE 3.2   List of Forest and Fruit Trees Present in the Agroforestry Systems

Scientific Name	Local Name	Family	AF Systems	Parts Used
		Forest Tree Species		
Acacia nilotica	Babooldesi	Fabaceae	AS, ASH	Timber, fuel
Anthocephallus cadamba	Kadamb	Combretaceae	AS	Timber, fuel
Artocarpus heterophyullus	Kathal	Urticaceae	AS, SH	Timber, fuel, Fruit
Azadirachata indica	Neem	Meliaceae	AS	Timber, fuel, medicinal use
Bauhinia variegata	Kachnar	Fabaceae	AS	Fodder, timber, fuel, medicinal use
Butea monosperma	Dhak	Fabaceae	AS	–
Cassia fistula	Amaltas	Cesalpiniaceae	AS	Timber, fuel
Dalbergia sissoo	Shisham	Fabaceae	AS	Timber, fuelwood
Eucalyptus spp.	Eucalyptus	Myrtaceae	AS, SH	Timber, fuel
Ficus bengalensis	Bargad	Moraceae	AS	Timber, fuel, worship
Ficus racemosa	Gular	Moraceae	AS	Timber, fuel
Ficus religiosa	Pipal	Moraceae	AS	Timber, fuel, worship
Moringa oleifera	Sehjan	Moringaceae	AS, ASH	Fuelwood, fruit
Prosopis juliflora	Vilayatibabool	Fabaceae	AS	Timber, fuel
Tectona grandis	Sagwan	Verbenaceae	AS	Timber, fuel
Leucaena leucocephala	Subabool	Fabaceae	AS	Timber, leaves
		Fruit Tree Species		
Aegle marmelos	Bael	Rutaceae	AH	Edible fruit
Carica papaya	Papaya	Caricaceae	AH	Edible fruit, medicinal use
Emblica officinalis	Amla	Phyllanthaceae	AH, AHS	Edible fruit

TABLE 3.2 (*Continued*)

Scientific Name	Local Name	Family	AF Systems	Parts Used
Madhuca indica	Mahua	Sapotaceae	AH	Edible fruit, flowers
Mangifera indica	Aam	Anacardiaceae	AH, AHS	Edible fruit, timber
Morus alba	Shahtoot	Moraceae	AH	Edible fruit, fodder, fuel
Musa paradisiaca	Kela	Musaceae	AH	Edible fruit, worship
Psidium guajava	Amrood	Myrtaceae	AH	Edible fruit
Punica granatum	Anar	Punicaceae	AH, AHS	Edible fruit
Syzygium cumini	Jamun	Myrtaceae	AH	Edible fruit, medicinal use
Tamarindus indica	Imli	Fabaceae	AH	Edible fruit
Ziziphus mauritiana	Ber	Rhamnaceae	AH	Edible fruit

study site have Aonla orchards (spacing of 8 m by 8 m) and inter-row space is efficiently to be utilized for growing agricultural crops. Mixed varieties of Aonla like Chakaiya, Francis, NA7, NA10 has been planted in a single orchard whose planting material has been brought from adjacent district Pratapgarh, which is a leading producer of Aonla in the state. Eucalyptus is the other most common timber tree species planted along the edges, bunds, or middle of the agricultural fields. Eucalyptus is chosen widely in the area for its prominent explicit characteristics, i.e., fast growth, high adaptability to any climatic variation, low maintenance, an alternate source of income and fulfills the demand from wood-based industries. Eucalyptus bund based system is extensively seen in all the three villages and nearby all agricultural crops have been grown with the tree. Apart from Aonla and eucalyptus, farmers/growers are raising banana (*Musa* spp.), guava *Psidium guajava*) and other vegetable crops commercially such as potato (*Solanum tuberosum*), bhindi (*Abelmoschus esculentus*), etc. Medicinal plants such as tulsi (*Ocimum sanctum*), Bhoomi amla (*Phyllanthus niruri*), and Brahmi (*Bacopa monnieri*) are also grown by some of the farmers of Barsawan village.

TABLE 3.3   Various Tree-Crop Combinations and Agroforestry Systems Present in the Study Site

Agroforestry Systems	Tree Component	Crop Components
Agrisilviculture system	Eucalyptus	Wheat, mustard, bajra, jowar
	Eucalyptus	Wheat + mustard
	Shisham	Wheat, mustard, bajra, jowar
	Babul	Wheat, urd, mustard
	Mango	Wheat, mustard
Agrihorticulture system	Mango	Wheat, mustard
	Guava, Mango	Wheat
	Aonla	Wheat, mustard, green gram
	Aonla	Wheat + mustard
Agri-silvihorticulture system	Anar, mango, shisham	Urd, moong
	Eucalyptus, Guava	Wheat, mustard
Silvipastural system	Eucalyptus	Bajra, berseem
	Babul	Sudan chari, berseem
	Shisham	Bajra
Hortipastural system	Aonla	Bajra, berseem
	Mango	Berseem

3.4.3   STRUCTURE AND COMPOSITION OF TREES UNDER AGROFORESTRY

Agrisilviculture system is the most predominant type of agroforestry in the study site. Different tree crop combination observed during the survey has been listed in Table 3.4. Most common agroforestry system persist in the area is Eucalyptus-wheat and eucalyptus-mustard in bund plantations. The data collected from the survey from each plot has been combined and frequency, abundance, density of the trees in the farmlands estimated. Table 3.4 shows the result on the phytosociology of timber and horticultural trees in the study site. From the table we evaluated the highest density (206 trees/ha), frequency (62%) and IVI (83.54) values for Eucalyptus spp. In horticulture trees, the highest density (30.5 trees/ha) and IVI (39.38) values were recorded for *Mangifera indica*. Other trees having IVI values were *E. officinalis* (25.95), *D. sissoo* (20.70), *A. nilotica* (18.72), *M. indica* (16.04), *Z. mauritiana* (11.25), *Prosopis juliflora* (8.60), and *Azadirachta indica* (8.07).

TABLE 3.4   Phytosociological analysis of Trees under Agroforestry System

Trees under Agroforestry	Frequency (%)	Density (tree/ha)	Abundance (tree/ha)	Total Basal Area (m2/30 plots)	IVI
Eucalyptus spp.	62	206	332.25	46393.4	83.54
Mangifera indica	36	30.5	84.5	64516.23	39.38
Emblica officinalis	34	25	73.5	32119.44	25.95
Dalbergia sissoo	16	28	175	27409.49	20.70
Acacia nilotica	40	21	52.5	10640.03	18.72
Madhuca indica	28	12.5	44.5	17258.63	16.04
Ziziphus mauritiana	26	8.5	32.5	7994.62	11.25
Prosopis juliflora	18	7.5	41.5	6668.81	8.60
Azadirachata indica	22	6.5	29.5	3206.89	8.07
Ficus religiosa	4	1	25	17210.91	7.35
Ficus bengalensis	2	0.5	25	18232.22	7.11
Syzygium cumini	16	4.5	28	4279.74	6.51
Artocarpus heterophyullus	12	5	41.5	5763.65	6.20
Tamarindus indica	10	3	30	3714.87	4.48
Leucaena leucocephala	10	3.5	35	2274.32	4.10
Psidium guajava	12	4	33.25	514.87	4.07
Anthocephallus kadamba	8	2	25	2542.69	3.33
Aegle marmelos	10	2.5	25	779.15	3.31

TABLE 3.4 *(Continued)*

Trees under Agroforestry	Frequency (%)	Density (tree/ha)	Abundance (tree/ha)	Total Basal Area (m2/30 plots)	IVI
Cassia fistula	8	3	37.5	1497.31	3.21
Butea monosperma	8	2	25	1548.37	2.98
Tectona grandis	6	3.5	58.25	1605.33	2.90
Bauhinia variegata	8	2.5	31.25	264.12	2.64
Carica papaya	4	5.5	137.5	733.65	2.61
Musa paradisiaca	6	2.5	41.5	728.26	2.33
Punica granatum	2	3	150	612.41	1.45
Moringa oleifera	4	1	25	497.76	1.39
Ficus racemosa	2	0.5	25	1040.09	0.97
Morus alba	2	0.5	25	206.01	0.68

3.4.4   SOIL ORGANIC CARBON (SOC) UNDER EUCALYPTUS AND AONLA BASED AGROFORESTRY

SOC is the largest terrestrial carbon pool on earth. It plays a very significant role in the global carbon cycle and provides various ecosystem services to the society. Globally, the benefit of growing trees have been recognized, as it provides sustainable food production (AS), wood products and also improves soil carbon storage and soil quality (Benbi et al., 2012; Nair et al., 2009). In the present study, SOC content under *Eucalyptus* and *Emblica* based system was analyzed and results are presented in Figures 3.3–3.8. The results revealed that a significant difference in SOC was observed between the agroforestry system and soil depth under study. Several researchers have also observed similar results (Jobbagy and Jackson, 2000; Sundarapandian et al., 2016; Newaj et al., 2017).

3.4.5   SOC UNDER SOLE EUCALYPTUS SYSTEM

Under the sole eucalyptus forestry system, the SOC content varied from 0.62% to 0.82% with a mean value of 0.69% (Figure 3.3). Irrespective of the village, where samples were collected, the SOC content significantly decreased with increase in soil depth. This might be due to the fact that the contribution of fine roots biomass from the forest tree species decreased with increase in soil depth, which in turn resulted in lower organic matter input to the subsurface soil (Hendrick and Pregitzer, 1996). The mean SOC content was 0.77, 0.73, 0.65, 0.65, and 0.64% at 0–5, 5–15, 15–30, 30–60 and 60–90 cm soil depth, respectively. Overall significant differences were observed in the SOC content among the village's studied at their respective soil depth. Among the villages under the sole forestry system, soil carbon content in the surface layer (0–15 cm) were relatively greater in V2A (village), whereas the subsurface layer (15–90 cm) was greater in V1B (village name).

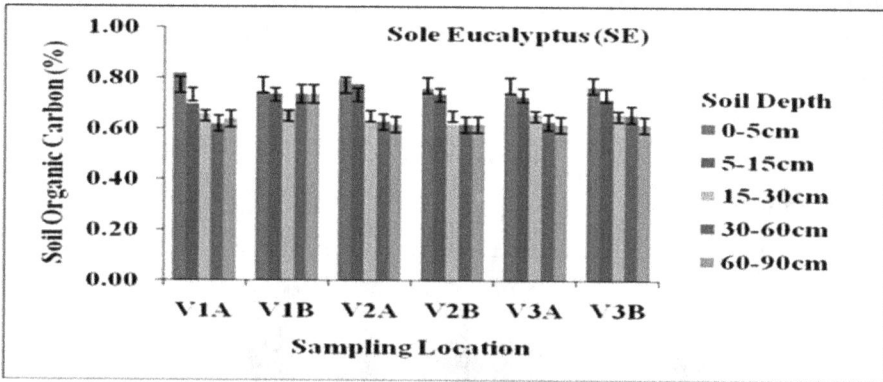

FIGURE 3.3 SOC status under sole eucalyptus system in Unchahar block of Raebareli district, Uttar Pradesh.

3.4.6 SOC UNDER SOLE AONLA SYSTEM

SOC content varied from 0.59 to 0.86% with a mean value of 0.68% under sole Aonla forestry system, (Figure 3.4). The mean SOC content was 0.83, 0.70, 0.65, 0.60, and 0.60% at 0–5, 5–15, 15–30, 30–60 and 60–90 cm soil depth, respectively. However, variation in SOC content under the sole Aonla system between the villages sampled was statistically non-significant at their respective soil depth, SOC content significantly decreased with increase in soil depth. Unlike under sole Eucalyptus system, soil carbon content in surface layer (0–15 cm) under sole Aonla was relatively greater in V1B.

3.4.7 SOC UNDER SOLE WHEAT SYSTEM

The SOC content under the sole wheat system varied from 0.43 to 0.66% with a mean value of 0.54%, and it significantly decreased with increase in soil depth from 0 to 30 cm (Figure 3.5). The mean SOC content was 0.64,

0.54, and 0.45% at 0–5, 5–15 and 15–30 cm, respectively. Irrespective of the soil depth, the SOC content under the sole wheat system was significantly lower than the sole Eucalyptus and Aonla system. Intensive cropping under sole agriculture system tends to have lower SOC content because of plant biomass removal and tillage operation. Repeated tillage operation year after year for raising subsequent annual crops will enhance organic matter decomposition process and thereby resulting in substantial carbon loss from the sole agriculture system (Sherrod et al., 2005). On the other hand, tillage operation and biomass removal is near minimal or absent in the agroforestry system; as a result, the SOC content is maintained higher and sustained under long term.

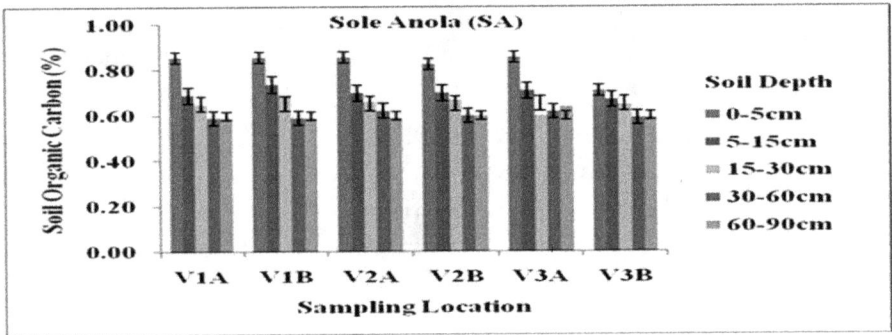

FIGURE 3.4   SOC status under sole aonla system in Unchahar block of Raebareli district, Uttar Pradesh.

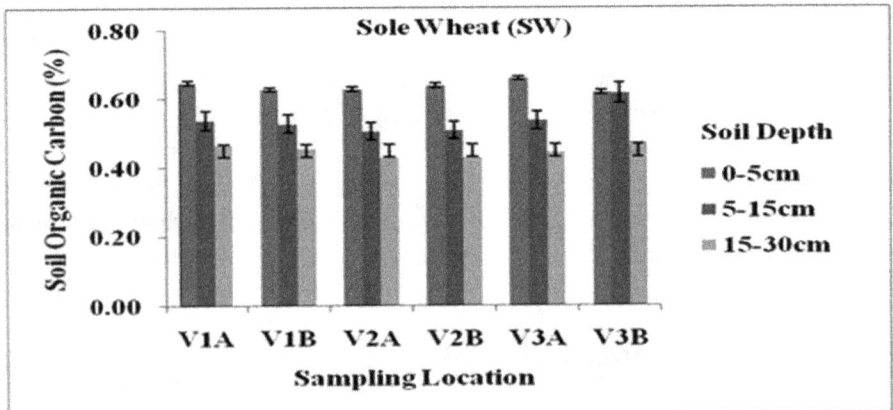

FIGURE 3.5   SOC status under sole wheat system in Unchahar block of Raebareli district, Uttar Pradesh.

3.4.8 SOC UNDER AGRI-SILVICULTURE (AS)

Based on a survey conducted at the study site, Raebareli district of UP, AS system is found to be the most predominant type of agroforestry system. Among the AS system followed at the study site, Aonla, and Eucalyptus is the most preferred woody perennials by the farmers along with wheat as annual rabi crop. The SOC content under Eucalyptus-wheat and Aonla-wheat AS system was not statistically significant, however, relatively higher content in SOC was observed in Aonla-wheat than Eucalyptus-wheat silviculture system (Figures 3.6 and 3.7). Similar to the sole forestry system (Eucalyptus/Aonla), SOC content under Agri-silviculture system (Eucalyptus-wheat and Aonla-wheat) is also decreased with increase in soil depth. The SOC content under Eucalyptus-wheat AS system varied from 0.60 to 0.92% with a mean value of 0.75%, whereas, it varied from 0.61 to 1.09% with a mean value of 0.76% under Aonla-wheat AS system.

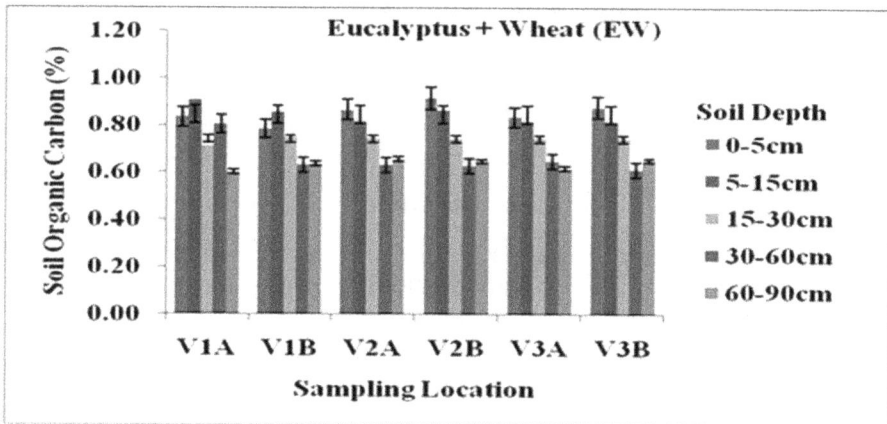

FIGURE 3.6 SOC status under eucalyptus-wheat agri-silviculture system in Unchahar block of Raebareli district, Uttar Pradesh.

Overall, the SOC content was significantly higher in AS system (Eucalyptus-wheat and Aonla-wheat) as compared to sole forestry (Eucalyptus/Aonla) and sole agriculture (wheat) system (Figure 3.8). In general soils under forest trees contains higher SOC than other land-use system (Berthold and Beese, 2002) and forest soils are considered the major carbon sinks on earth (Dey, 2005). Higher litter input and root biomass in the forest system might be the reason for higher SOC content under sole forestry system

(Eucalyptus/Aonla) than sole agriculture system (wheat). Further, the study reveals that irrespective of the system under study, SOC content decreased with increase in soil depth and the percent decrease was relatively higher in sole agriculture (wheat) system as compared to sole forestry (Eucalyptus/ Aonla) and AS (Eucalyptus-wheat/Aonla-wheat) system. This was likely due to root biomass contribution from agriculture system is only confined to surface soil rather than subsurface soil. Therefore, a greater decline in soil SOC with an increase in soil depth is more likely to occur in the sole agriculture system as compared to the sole forestry system or AS system (Cao, 2004; Jackson et al., 1996).

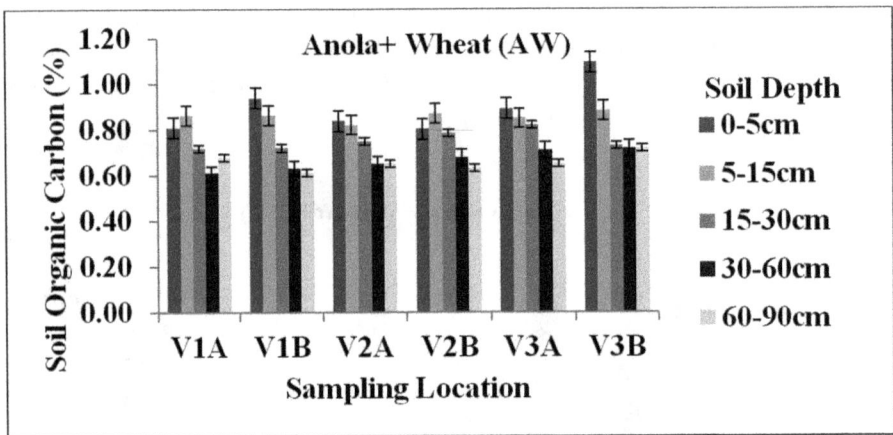

FIGURE 3.7   SOC status under aonla-wheat agri-silviculture system in Unchahar block of Raebareli district, Uttar Pradesh.

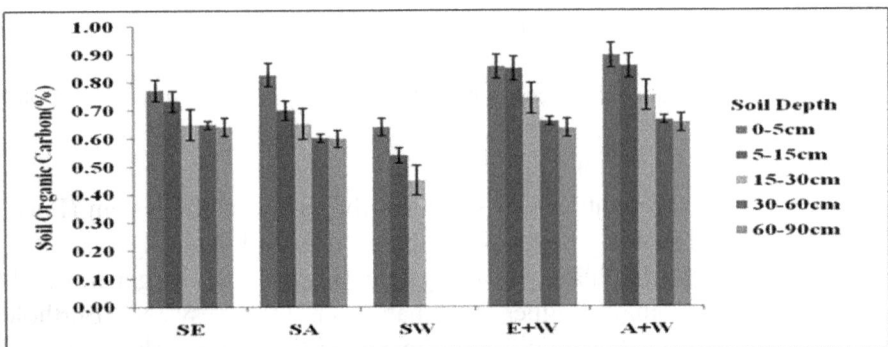

FIGURE 3.8   SOC status under different agroforestry system in Unchahar block of Raebareli district, Uttar Pradesh. *(Abbreviations: SE: sole eucalyptus; SA: sole aonla; SW: sole wheat; E+W: eucalyptus-wheat; A+W: aonla-wheat).*

The percent decrease in SOC from surface (0–15 cm) to subsurface soil (15–30 cm) was 23.7% in the sole wheat system as compared to 13.6%, 14.5%, 13.0%, and 14.3% in sole Eucalyptus, sole Aonla, Eucalyptus-wheat, and Aonla-wheat system, respectively. Moreover, SOC in the surface soil (0–5 cm) under AS (Eucalyptus-wheat) was 9.4% higher as compared to sole forestry system (Eucalyptus), while it was 24.7% higher as compared to sole agriculture system (wheat). Similarly, SOC in the surface soil (0–5 cm) under AS (Aonla-wheat) was 6.7% higher as compared to sole forestry system (Eucalyptus), while it was 28.1% higher as compared to sole agriculture system (wheat). Irrespective of the soil depth, mean SOC value (average of soil depth) was higher in Aonla-wheat system (0.76%) followed by Eucalyptus-wheat system (0.75%), sole Eucalyptus (0.69%), sole Aonla (0.68%) and sole wheat system (0.54%).

3.5 CONCLUSION

Traditional as well as planted agroforestry have precise roles to play in the livelihoods enhancement and industrial development in the study area. Many traditional trees are intentionally maintained by the farmers on the agricultural fields for preserving the tree species diversity, fulfilling domestic needs, and conserving plant genetic resources (PGRs). Planted/commercial agroforestry plays a consideration role in assuring alternate source of income as compared to the monocropping. Commercial cultivation of vegetables is also found to be common practice in the study area. *Eucalyptus spp., Dalbergia sissoo, A. nilotica, Emblica officinalis, Mangifera indica* are the most frequent species under agroforestry systems. Eucalyptus based bund/boundary system and Aonla based agroforestry is immersed as highly profitable and sustainable land-use system among the rural community. Additionally, SOC estimated under Aonla and Eucalyptus agroforestry revealed that the SOC content was significantly higher in AS system (*Eucalyptus*-wheat and Aonla-wheat) as compared to the sole tree (*Eucalyptus*/Aonla) and sole agriculture (wheat). However, agroforestry has specific roles to play in food security, livelihood security, and overall production and upsurges SOC.

ACKNOWLEDGMENT

The authors are thankful to the Director, IIFM, Bhopal, India, for encourage-ment and continuous support to research. We are also thankful to the farmer's assistance during the study, special to Mr. Ram Gopal Singh Chandel, of the village Barsawan. The authors are also thankful to the Ministry of Science and Technology-DST, India for providing the financial assistant via DST-INSPIRE fellowship (No. DST/INSPIRE Fellowship/2015/IF150338).

KEYWORDS

- **agroforestry**
- ***Emblica officinalis***
- **greenhouse gases**
- **importance value index**
- **soil organic carbon**
- **Uttar Pradesh**

REFERENCES

Benbi, D. K., Brar, K., Toor, A. S., Singh, P., & Singh, H., (2012). Soil carbon pools under poplar-based agroforestry, rice-wheat, and maize-wheat cropping systems in semi-arid India. *Nutr. Cycl. Agroecosyst., 92*, 107–118.

Berthold, D., & Beese, F., (2002). Carbon storage in soils after afforestation in relation to management practices. *Forst. Und Holz., 57*(13/14), 417–420.

Cao, F., (2004). *Ecological Basis for Ginkgo Agroforestry Systems.* PhD Thesis, Department of Forest Sciences, The University of British Columbia, Vancouver, BC, Canada.

Cardinael, R., Chevallier, T., Cambou, A., Béral, C., Barthès, B. G., Dupraz, C., & Chenu, C., (2017). Increased soil organic carbon stocks under agroforestry: A survey of six different sites in France. *Agriculture, Ecosystems and Environment, 236*, 243–255.

Curtis, J. T., & McIntosh, R. P., (1950). The interrelationship of certain analytic and synthetic phytosociological characters. *Ecology, 31*, 434–455.

Dey, S. K., (2005). A preliminary estimation of carbon stock sequestrated through rubber (*Hevea brasiliensis*) plantation in North Eastern regional of India. *Indian Forester, 131*(11), 1429–1435.

Hendrick, R. L., & Pregitzer, K. S., (1996). Temporal and depth-related patterns of fine root dynamics in northern hardwood forests. *Journal of Ecology*, 167–176.

Jackson, R. B., Canadell, J., Ehleringer, J. R., Mooney, H. A., Sala, O. E., & Schulze, E. D., (1996). A global analysis of root distributions for terrestrial biomes. *Oecologia, 108*, 389–411.

Jobbágy, E. G., & Jackson, R. B., (2000). The vertical distribution of soil organic carbon and its relation to climate and vegetation. *Ecological Applications, 10*(2), 423–436.

Kumar, A., Sinha, A. K., & Singh, D., (2003). Studies of eucalyptus plantations under the farm forestry and agroforestry systems of UP in Northern India. *Forests, Trees and Livelihoods, 13*(4), 313–330.

Lal, R., (2003). Global potential of soil carbon sequestration to mitigate the greenhouse effect. *Critical Reviews in Plant Sciences, 22*(2), 151–184.

Lal, R., (2004). Soil carbon sequestration impacts on global climate change and food security. *Science, 304*(5677), 1623–1627.

Nair, P. K. R., Fernandes, E. C. M., & Wambugu, P. N., (1985). Multipurpose leguminous trees and shrubs for agroforestry. *Agroforestry Systems, 2*(3), 145–163.

Nair, P. K. R., Mohan, K. B., & Nair, V. D., (2009). Agroforestry as a strategy for carbon sequestration. *Journal of Plant Nutrition and Soil Science, 172*(1), 10–23.

Newaj, R., Chaturvedi, O. P., Kumar, D., Prasad, R., Rizvi, R. H., Alam, B., Handa, A. K., et al., (2017). Soil organic carbon stock in agroforestry systems in western and southern plateau and hill regions of India. *Current Science, 112*(11), 2191–2193.

Phillips, E. A., (1959). *Methods of Vegetation Study*. A Holt Dryden Book Henry Hold and Co., Inc., New York.

Rai, P., Roy, R. D., & Rao, G. R., (1995). Evaluation of multipurpose tree species in rangeland under semi-arid condition of Uttar Pradesh. *Range Management and Agroforestry, 16*(2), 103–113.

Reddy, C. S., Rangaswamy, M., Pattanaik, C., & Jha, C. S., (2009). Invasion of alien species in wetlands of Samaspur Bird Sanctuary, Uttar Pradesh, India. *Asian Journal of Water, Environment and Pollution, 6*(3), 43–50.

Sherrod, L. A., Peterson, G. A., Westfall, D. G., & Ahuja, L. R., (2005). Soil organic carbon pools after 12 years in no-till dryland agroecosystems. *Soil Sci. Soc. Am. J., 67*, 1533–1543.

Singh, D., & Shukla, C. P. (2015). Phyto-diversity and traditional knowledge of Raebareli District, Uttar Pradesh. *Techno-Fame: A Journal of Multidisciplinary Advanced Research, 4*(1), 4–9.

Singh, P., (2014). Population and agroclimatic zones in India: An analytical analysis. *Procedia-Social and Behavioral Sciences, 120*, 268–278.

Srinivasarao, C., Lal, R., Kundu, S., Babu, M. P., Venkateswarlu, B., & Singh, A. K., (2014). Soil carbon sequestration in rainfed production systems in the semiarid tropics of India. *Science of the Total Environment, 487*, 587–603.

Sundarapandian, S. M., Amritha, S., Gowsalya, L., Kayathri, P., Thamizharasi, M., Dar, J. A., Srinivas, K., et al., (2016). Soil organic carbon stocks in different land uses in Pondicherry university campus, Puducherry, India. *Tropical Plant Research, 3*(1), 10–17.

Walkley, A., & Black, I. A., (1934). An examination of the Degtjareff method for determining soil organic matter and a proposed modification of the chromic acid titration method. *Soil Science, 37*(1), 29–38.

CHAPTER 4

Climber Community Structure in Relation to Environmental and Tree Attributes in Lowland Tropical Forests of Eastern Himalaya, India

SAMIRAN PANDAY,[1] DINESH SINGH RAWAT,[2] VIKAS KUMAR,[3] SUDHANSU SEKHAR DASH,[4] BIPIN KUMAR SINHA,[4] and PARAMJIT SINGH[5]

[1]*Department of Botany, Budge Budge College, 7-DBC Road, Budge Budge, South 24 Parganas–700137, West Bengal, India*

[2]*Department of Botany and Microbiology, HNB Garhwal University, Srinagar, Garhwal–246174, Uttarakhand, India*

[3]*High Altitude Biology, CSIR-Institute of Himalayan Bioresource Technology, Palampur–176061, Himachal Pradesh, India*

[4]*Botanical Survey of India, CGO Complex, Salt Lake City, Kolkata–700064, West Bengal, India, E-mail: ssdash2002@gmail.com (S. S. Dash)*

[5]*Department of Botany, Central University of Punjab, Bathinda–151001, Punjab, India*

ABSTRACT

The present study was conducted in the Namdapha National Park (NNP), Eastern Himalayas, which is India's largest *Dipterocarpus* dominant lowland tropical rainforest. The aim of this study was to examine the spatial pattern of climber species composition and its relationship with tree and environment attributes. Field data was collected by establishing sampling plots (each of a size 20 m × 20 m) in random stratified method. We divided testing variables into four classes to assess the changing pattern in climber attributes viz., altitude classes 300–400 m, 400–500 m, 500–600 m, and > 600 m; slope angle classes 0–15°, 15–30°, 30–45° and > 45°; slope aspects as east, west, north, and south while canopy cover classes were < 55%, 55–70%, 70–85% and > 85%. The differences in the climber attributes among the classes were tested applying non-parametric ANOVA (Kruskal-Wallis test). A total of 47 climber species belong to 44 genera and 32 families were recorded from 2.28 ha area (400 m^2 × 57 plots). Six communities were recognized through TWINSPAN classification using species matrix. *Ficus pumila, Thunbergia coccinea, Rhaphidophora hookeri* and *Erythropalum scandens* revealed as indicator taxa for different communities on the basis of indicator value (IV) scores. The present results revealed that the climber abundance, species richness (SR), and diversity dwindle with an increase in canopy cover and slope angles; with a decrease in altitude and tree SR.

4.1   INTRODUCTION

The climber is one of the important life forms, contribute significantly both in species diversity and structure especially in tropical and sub-tropical forest ecosystems (Schnitzer and Bongers, 2002; Bongers et al., 2005; Dalling et al., 2012). The richness and abundance of climbers vary among communities and depend on various factors such as climate, perturbation history, host specificity, and variables of community structure (Sfair and Martins, 2011). In many cases, distinct species compositions of climbers have been observed in relation to forest structure and composition (DeWalt et al., 2006). Assessments of climber community composition and structure are essential to understand the habitat specialization (Schnitzer et al., 2000), forest ecosystem dynamics and its functioning (Gentry, 1991; Schnitzer and Bongers, 2002), tree, and climber interactions and regeneration (Kadavul and Parthasarathy, 1999), maintenance of carbon budget and underneath microclimate (Dalling et al., 2012). In India, previous studies on liana diversity and community were confined mostly to the peninsular region (Kadavul and Parthasarathy, 1999; Muthuramkumar and Parthasarathy, 2000; Padaki and Parthasarathy, 2000; Chittibabu and Parthasarathy, 2001; Reddy and Parthasarathy, 2006; Mohandass and Davidar, 2014), while studies on the Himalayan region were rather scanty (Chetri et al., 2010; Barik et al., 2015).

The study site, NNP, is the largest protected area in North-east India, which was also declared as a Tiger Reserve in 1983, embedded in the Himalayan biodiversity hotspot. NNP has sustained India's largest *Dipterocarpus* dominant lowland tropical rainforest (Proctor et al., 1998). Literature survey particularly from NNP revealed that the comprehensive floristic account of the park was worked out by Chauhan et al. (1996), for vegetation and tree population structure by Nath et al. (2005) and Dash et al. (2021), for tree species gap phase performance, tree regeneration and seedling survival pattern by Deb and Sundriyal (2007, 2008), for leaf litter decomposition of dominant tree species by Barbhuiya et al. (2008) while Sarmah et al. (2006 a, b) documented the ethnobotanical knowledge and natural resource utilization pattern of the tribal dwells in and around the park. However, the climber community structure and composition of the region has never been characterized. In this study, we predicted that; (i) How many climber communities are exiting in Namdapha National Park (NNP); (ii) Do the diversity pattern vary among the communities; (iii) What are the dominant and indicator species of different communities; (iv) How the climber species richness (SR), abundance, and diversity related to the environmental factors.

4.2 MATERIALS AND METHODS

4.2.1 STUDY AREA

NNP, is located in Arunachal Pradesh, India, lying between 27°23' to 27°39' North latitude and 96°15' to 96°58' East longitude (Figure 4.1). It encompasses an area of 1985 km² with an altitudinal range from 200 to 4571 m asl. It is bounded by Myanmar in the east and south, Kamlang Wildlife Sanctuary (KWS) in the north while western fringe of the park is surrounded by fragmented human habitation with agricultural fields. NNP is harboring about 1119 plant taxa (Chauhan et al., 1996) and 1399 animal taxa (Ghosh, 1987) so far. The vegetation of the park is comprised of lowland, i.e., tropical forests, sub-tropical, temperate, and alpine meadows. The major rivers across the park are Noa-Dihing, Deban, Namdapha, and Burma Nala including rain-fed streams and streamlets (Nath et al., 2005). The park has a tropical climate and typical monsoon pattern with the prolonged rainy season (Deb and Sundriyal, 2007). At lower altitudes, temperature varies from 5° C to 35° C while it falls to 0° or below at higher elevations. The annual precipitation ranges from a minimum of 1400 mm to a maximum of 2500 mm, 75% of which falls between April and October (Arunachalam et al., 2004). The average relative humidity remains high (> 60%) throughout the year, and during the rainy season reaches up to (80–95%).

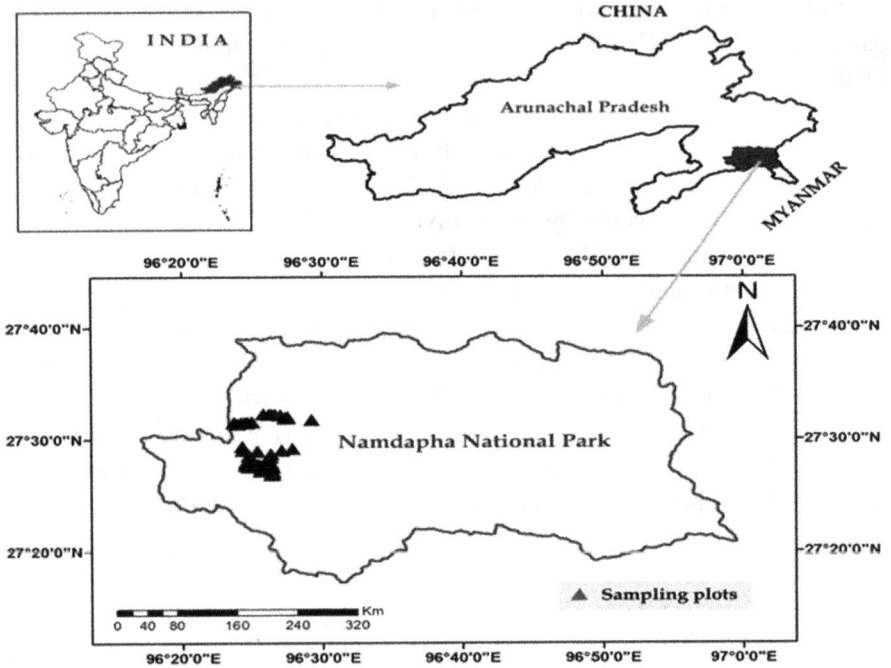

FIGURE 4.1 Location map of Arunachal Pradesh and NNP in India, and the position sampling plots (the black filled triangles).

4.2.2 FIELD SURVEY AND TRANSACTS SELECTION

In connection with a research project (NMHS/2015–16/LG-05), the authors have carried out floristic and ecological study of Namdhapa National Park since April 2016 with some noteworthy taxonomic explorations (Kumar et al., 2017). The present field data were collected during August-November, 2017 from the western part of the NNP that serve as an entrance to visitors, forest personnel or residents of village Gandhigram and Vijoynagar (both the villages are located at the eastern end of the park). In addition, this part of the park received comparatively higher anthropogenic disturbances (Nath et al., 2005). The high density of the understory vegetation like naturalized bamboos, bananas, zingibers, and ferns posed difficulties during systematic sampling. Thus, we adopted a stratified random sampling method to study climbers and trees along the three crisscross forest trails (as line transacts) of 6–7 km length viz., T1, T2 and T3. Of these, two trails (i.e., T1 and T2) located in the core zone area of the park and one (T3) in the buffer zone. The midpoint of T1 and T2 is at '25 mile base camp area,' as T1 moves from east to west direction while T2 downhill (riverbank) to uphill (goodbye point) direction. The midpoint of T3 is 'Hornbill base camp area' that starts from Haldibari to towards Bulbuliya area.

4.2.3 DATA COLLECTION, COMMUNITY IDENTIFICATION, AND SPECIES ESTIMATION

Twenty sampling plots of 20 m × 20 m (0.04 ha) size were demarcated along each trail (T1, T2 and T3), minimum distance between plots was 250 m. The field data like latitude, longitude, altitude, slope angle, aspect, canopy cover, and vegetation (trees and climbers) of each plot were recorded during the field survey. All the climbers (≥ 1.0 cm circumference at breast height (CBH) (Gerwing et al., 2006); at 1.3 m from rooting point) and trees (≥ 30 cm CBH; at 1.37 m above ground level) with a number of individuals of each species were counted within respective quadrats and noted on field data sheets (separate sheet for trees and climbers). Circumference was measured with the help of graduated tape or diameter with calipers. The habit of climber classified as liana (stem woody) and vines (stem herbaceous) while mode of climbing categorized into five main classes viz., hooked, rooted, scrambler, tendrilar, and twining climbers which were confirmed through relevant literature and herbarium. Species that occurred within each plot were collected, processed, and preserved as herbarium specimens. Botanical identity of each species was confirmed with the help of relevant literature (Chauhan et al., 1996; Chowdhery et al., 1996, 2008, 2009; Dash and Singh, 2017) and reference herbaria (ARUN, ASSAM, CAL).

Two-way indicator species analysis (TWINSPAN) was performed to differentiate 47 species and 57 plots into different climber communities (with similar species assemblages) using species matrix data. Software WinTWINS, version 2.3 (Hill and Šmilauer, 2005) was used for TWIN-SPAN with cut levels 0.0, 2.0, 5.0, 10.0 and 20.0. The species-area curve was obtained to determine the pattern of SR with increasing number of samples and to assess sampling efficiency using PC-ORD, version 4.34 (McCune and Mefford, 1999) with Sorensen (Bray-Curtis) distance. Non-parametric species estimators viz., Chao 2 (Chao, 1987), Jackknife 1 (Palmer, 1990) and Jackknife 2 (Palmer, 1991) were calculated to extrapolate the expected SR of the study area.

76 *Diversity and Dynamics in Forest Ecosystems*

4.2.4 DATA ANALYSIS

The ecological data of climbers and trees from each plot and community was computed for abundance, density, and basal area following standard methods (Misra, 1968; Muller-Dombois and Ellenberge, 1974). The importance value index (IVI) was calculated for each climber species within the community (Eqn. (1)). Species count in the plot or community was taken as SR (Phillips, 1959) of that particular plot or community. Individuals of climbers were divided into five size classes, i.e., CBH 1–2.9 cm, 3–4.9 cm, 5–6.9 cm, 7–8.9 cm, and > 9 cm. The dominance-diversity curves (d-d curves) generated to study the dominance pattern of species in a particular community. Various ecological indices viz., Shannon-Wiener diversity (Shannon and Wiener, 1963), dominance (Simpson, 1949), evenness (Buzas and Gibson, 1969), Margalef's (Margalef, 1958), equitability (Pielous, 1975) and Fisher alpha (Fisher et al., 1943) were calculated. Similarity index (Eqn. (2)) and beta diversity (Eqn. (3)) was calculated following Sorenson (1948) and Whittaker (1960) respectively. The indicator value (IV) (Dufrêne and Legendre, 1997) and total importance value (TIV) of each species in different community were calculated using Eqns. (4) and (5), respectively:

Importance value index (%) = RF + RD + RBA/3   (1)

Similarity index (%) = 2a/(2a + b + c) × 100   (2)

Beta diversity (%) = (a + b + c)/(2a + b + c)/2) – 1 × 100   (3)

Indicator value (IV_{ij} %) = RA_{ij} × RF_{ij} × 100   (4)

Total importance value (%) = (IVI + IV)/2   (5)

where; RF = relative frequency, RD = relative density, RBA = relative basal area; a = No. of common species between two plots, b = No. of species unique to first plot, c = No. of species unique to second plot; RA_{ij} = relative abundance of species i in each site group j [RA_{ij} was calculated dividing mean abundance of species i across sites of the group j (A_{ij}) by sum of the mean abundance of species i overall groups (A_i)], RA_{ij} = relative frequency of species i in each site group j [RA_{ij} was calculated dividing number of sites in group j where species i is present (S_{ij}) with the total number of sites in that group (S_i), IV = indicator value].

4.2.5 CLIMBER AND ENVIRONMENT RELATION ANALYSIS

Altitude, slope, aspects, and canopy cover were divided into four classes to investigate their relation with phytosociological attributes of climbers.

Altitude classes were 300–400 m, 400–500 m, 500–600 m, and > 600 m; slope angle 0–15°, 15–30°, 30–45° and > 45°; aspects as east, west, north, and south while canopy cover as < 55%, 55–70%, 70–85% and > 85%. The number of plots within the respective class was considered as groups and investigated for average species per plot, average abundance per plot, Simpson index, Shannon-Wiener diversity, Evenness index, and Margalef's index of SR. We have applied non-parametric ANOVA (Kruskal-Wallis test) to test the level of difference in climber attributes (SR and abundance) among the classes. The relation of climber attributes with environmental variables and tree parameters were based on Pearson correlation. Both ANOVA and Pearson correlation were calculated using IBM SPSS, version 23. Results of direct ordination method, i.e., canonical correspondence analysis (CCA) were generated through Software 'PAST, version 3.11' (Hammer, 2001) to study the overall relationship between species and environment.

4.3   RESULTS

4.3.1   FLORISTIC COMPOSITION

In the present study of NNP, a total of 57 sampling plots (each 400 m²), were established where 47 climber species reported belonging to 44 genera and 32 families (Annexure 4.1). Of which, 33 species (70.21%) were liana and 14 species (29.79%) were vines. Based on the climbing mechanisms, there were 15 twiner species (31.91%), followed by 9 species (19.15%) of scrambler, 7 species (14.89%) of tendrilar climber and 8 species (17.02%) in each of hooked and rooted climber. The infra-generic diversity was also high and of total genera, 93.18% (41 genera) were recorded with one species only, with two species each of the genera were *Ficus*, *Piper*, and *Rubus*. The dominant climber families of the region were Araceae, Menispermaceae, Mimosaceae, and Vitaceae, represented by three species each followed by Annonaceae, Asclepiadaceae, Moraceae, Olacaceae, Piperaceae Rosaceae, Rubiaceae, and Sabiaceae with two species each.

4.3.2   SPECIES RICHNESS (SR)

The climber species-accumulation curve (Figure 4.2) did not reach an asymptote in the present study. About 79% (37 species) of the total observed species richness (OSR) was recorded in the first 50.87% (29 plots) sampling efforts and rest 21.23% (10 species) of the OSR was covered by laying 28 additional plots (49.13%). The estimated SR found higher than that of the OSR (47 species). First-order jackknife and Chao 2 estimated 61.7 species while second-order jackknife estimated 70.5 species in the area. The OSR varied from 03 to 10 species among the plots with an average of 5.14 ± 1.86 species per plot.

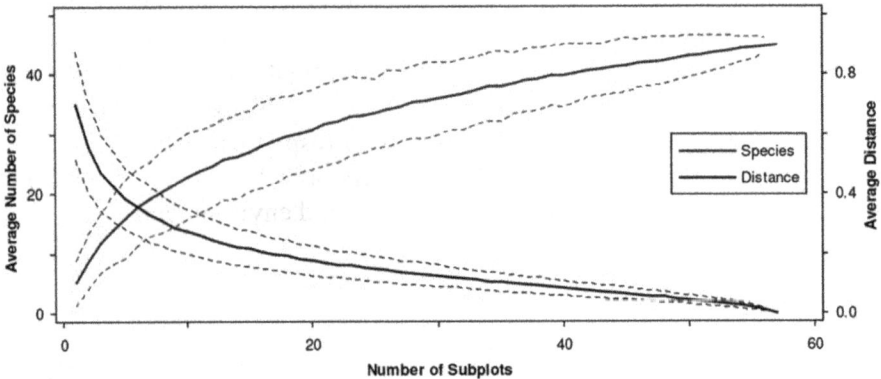

FIGURE 4.2 Species-accumulation curve of 47 climber species and 57 sampling plots in Namdapha National Park.

4.3.3 TWINSPAN CLASSIFICATIONS

TWINSPAN classified 47 species and 57 plots into 6 communities (groups) at fourth level of division (Figure 4.3) viz., G1 (*Ficus pumila, Hippocratea micrantha* and *Mucuna monosperma* community), G2 (*Schefflera venulosa, Byttneria aspera* and *Mikania scandens* community), G3 (*Sabia lanceolata, Combretum wallichii* var. *griffithii* and *Rhaphidophora hookeri* community), G4 (*Caesalpinia cucullata, Piper mullesua* and *Thunbergia coccinea* community), G5 (*Erythropalum scandens, Dioscorea alata* and *Sabia lanceolata* community) and G6 (*Chonemorpha fragrans* community). At the first level of division TWINSPAN isolated G6 (1 plot, *Chonemorpha fragrans* as indicator) from the rest of the groups (56 plots) that represented two rare species of the community, i.e., *C. fragrans* and *Derris cuneifolia*. At the second level of division, G1 (5 plots, *Ficus pumila* as indicator) is separated from the rest four groups. At the third and fourth level of division, the rest 51 plots were classified into G2 (19 plots), G3 (15 plots, *Sabia lanceolata* as indicator), G4 (12 plots) and G5 (5 plots, *Erythropalum scandens* as indicator). Group G6

was not included in further ecological analysis as the group was too small for
further ecological analysis as represented by only four taxa.

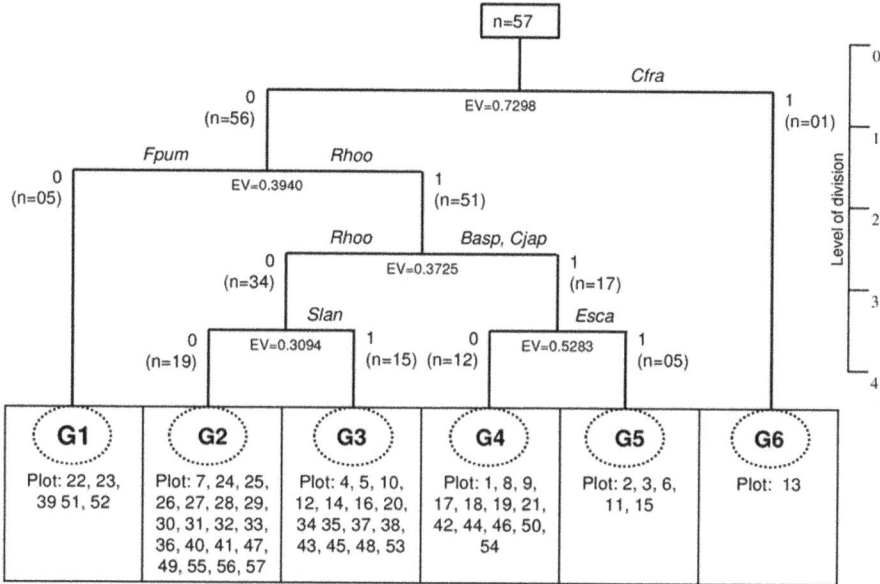

FIGURE 4.3 Climber communities (G) in NNP determined by the TWINSPAN using species
matrix. (Acronym according to Annexure 4.1).

4.3.4 COMMUNITY ANALYSIS

The consolidated phytosociological attributes and diversity indices for the five
climber communities (G1–G5) of the study area are shown in Table 4.1. The
SR among the communities varied from 12 to 28, Shannon-Wiener diversity
index 2.17 to 2.79, dominance index 0.08 to 0.17, and evenness index 0.54 to
0.83. The average value of Margalef's index, Equitability, and Fisher alpha
for different climber communities were recorded 4.04 ± 0.93, 0.85 ± 0.06
and 8.08 ± 2.44, respectively. The lowest Sørensen similarity index (29%)
and highest Whittaker beta diversity (71%) was observed between G1 and
G5. Total stem density of climbers ranged from 180 ha^{-1} to 225 ha^{-1} (mean
199 ± 21.06 ha^{-1}) and total basal area 450 cm^2 ha^{-1} to 1261 cm^2 ha^{-1} (mean
744 ± 317 cm^2 ha^{-1}) across the five communities. The distributional pattern
of climber stems into various size classes (CBH classes) are depicted in
Figure 4.4 in which the first two classes collectively represented 56% of the

total stem density of the area. The relative stem density slightly decreased towards higher CBH classes (from class 5.69 cm to ≥ 8.9 cm class).

TABLE 4.1   Phyto-Sociological Attributes of Climber Communities in NNP

Variables	Community Type					Statistics (n = 5)			
	G1	G2	G3	G4	G5	Min.	Max.	Mean	S.D.
Species count	16	28	16	22	12	12	28	18.8	6.26
No. of plots (size 400 m^2)	5	19	15	12	5	5	19	11.2	6.18
Total stem density (ha^{-1})	180	217	191.75	225	180	180	225	199	21.06
Total basal area (cm^2 ha^{-1})	450	612	1261	813	585	450	1261	744	317
Shannon-Wiener diversity	2.46	2.80	2.17	2.52	2.31	2.17	2.80	2.45	0.23
Dominance index	0.11	0.08	0.18	0.12	0.11	0.08	0.18	0.12	0.03
Evenness index	0.73	0.58	0.55	0.56	0.84	0.55	0.84	0.65	0.13
Margalef's index	4.19	5.29	3.16	4.49	3.07	3.07	5.29	4.04	0.93
Equitability index	0.89	0.84	0.78	0.81	0.93	0.78	0.93	0.85	0.06
Fisher alpha	11.04	9.68	5.05	8.35	6.30	5.05	11.04	8.08	2.44
Similarity index (%)									
G2	36	100	–	–	–	36	100	–	–
G3	61	40	100	–	–	40	100	–	–
G4	37	48	56	100	–	37	100	–	–
G5	29	35	48	35	100	29	100	–	–
Beta diversity (%)									
G2	64	0	–	–	–	0	64	–	–
G3	39	60	0	–	–	0	60	–	–
G4	63	52	44	0	–	0	63	–	–
G5	71	65	52	65	0	0	71	–	–

Ficus pumila (with IVI 16.81% at G1, IVI 44.12% at G2), *Byttneria aspera* (with IVI 20.09% at G4, IVI 16.01% at G5), and *Rhaphidophora hookeri* (with IVI 21.65% at G2) found as the dominant species of various communities (Annexure 4.1). Dominance-diversity curves (d-d curves) for five climber communities in present study were fit for lognormal situation (Figure 4.5). IV and total importance value (TIV) for each species under different community are given in Annexure 4.2. Maximum IV was recorded for *Ficus pumila* at G1 (51.7%) and G4 (48.6%), *Thunbergia coccinea* at G2 (30.2%), *Rhaphidophora hookeri* at G3 (64.9%) and *Erythropalum scandens* at G5 (94.7%). The TIV was recorded highest for *Ficus pumila* at

G1 (42.66%) and G3 (45.88%), followed by *Rhaphidophora hookeri* (30%) at G2, *Piper mullesua* (41.19%) at G4 and *Erythropalum scandens* (63.25%) at G5.

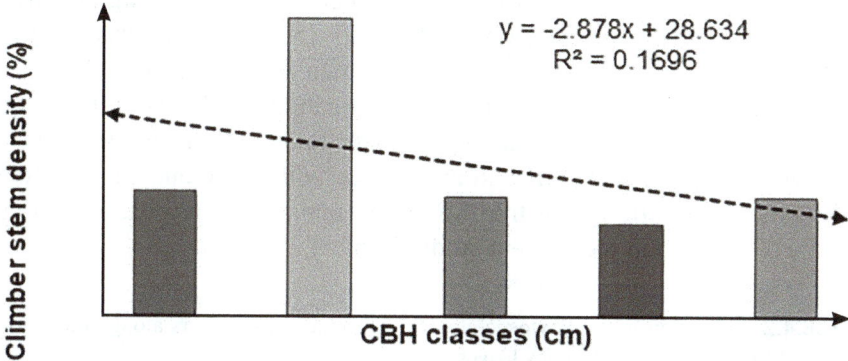

FIGURE 4.4   Distribution of climber individuals into different size classes in NNP.

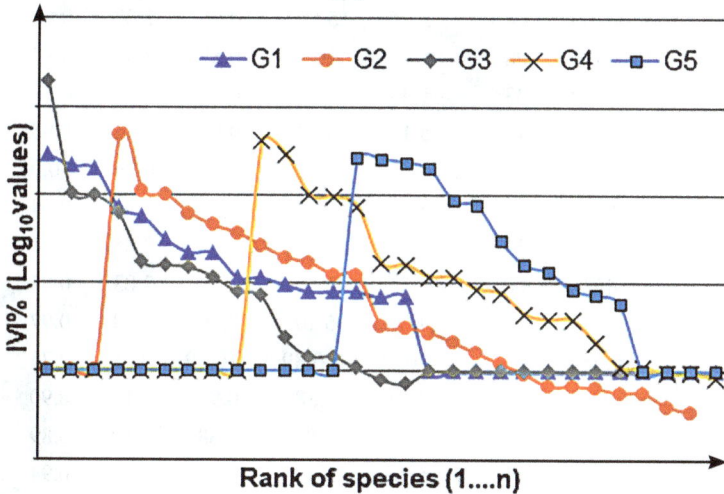

FIGURE 4.5   Dominance-diversity curves (d-d curves) for five climber communities in NNP.

4.3.5   GRADIENT ANALYSIS

The highest climber SR and abundance was observed in the altitudinal class of 400–500 m while 500–600 m class had represented the highest diversity (Table 4.2). Results showed that altitude had a significant impact on climber abundance (K-W test, $P = 0.016$) and SR ($P = 0.025$). The lowest slope class (0–15°) showed the highest diversity, SR and abundance however the slope had not influenced the climber SR ($P = 0.81$) and abundance significantly ($P = 0.103$). Aspect showed significant impact on the climber SR ($P = 0.011$) and abundance ($P = 0.039$) with the east aspect showing the highest diversity (SD, H,' MI). Highest climber diversity (SD, H,' MI) was found in the lowest canopy cover class (< 50%), which further decreased towards higher canopy class, though the influence of canopy on SR ($P = 0.722$) and abundance ($P = 0.861$) was not significant. Pearson correlation revealed negative relation for both SR and abundance with slope and canopy cover while positive relation with tree SR, tree abundance, tree basal area and altitude ($P > 0.05$) in the present study (Table 4.3).

TABLE 4.2 Variation in the Phytosociological Attributes of Climbers along Gradients of Altitude, Slope, Aspect, and Canopy Cover

Gradient	Gradient Classes	No. of Plots	Mean CSR	Mean Density	SD	H'	E	MI
Altitude	300–400 m	08	3.75	5.50	0.86	2.02	0.94	1.85
	400–500 m	08	6.12	10.12	0.87	2.04	0.97	1.59
	500–600 m	23	5.43	9.17	0.95	3.06	0.93	4.11
	>600 m	18	5.11	8.22	0.94	2.84	0.95	3.40
	Corr. (r)	–	0.33	0.35	0.82	0.76	–0.03	0.68
Slope angle	0–15°	20	5.80	10.70	0.94	2.94	0.94	3.60
	15–30°	19	4.27	7.90	0.94	2.85	0.91	3.54
	30–45°	08	5.73	7.73	0.86	2.03	0.95	1.72
	> 45°	10	4.20	6.20	0.89	2.28	0.97	2.13
	Corr. (r)	–	–0.40	–0.89	–0.79	0.85	0.75	–0.87
Aspects	E	26	4.63	7.38	0.95	3.15	0.90	4.71
	W	10	5.35	9.80	0.88	2.19	0.89	1.93
	N	11	5.36	8.47	0.90	2.33	0.94	2.24
	S	10	5.00	6.80	0.89	2.23	0.93	2.18
Canopy cover	< 55%	17	5.35	8.64	0.94	2.78	0.95	3.21
	55–70%	14	5.21	8.78	0.92	2.58	0.94	2.70
	70–85%	13	4.76	8.23	0.90	2.43	0.87	2.57
	> 85%	13	5.38	8.23	0.92	2.52	0.95	2.57
	Corr. (r)	–	–0.16	–0.18	–0.63	–0.81	–0.23	0.87

Abbreviations: SD = Simpson index; H' = Shannon-Wiener diversity; E = Evenness index; MI = Margalef's index.

TABLE 4.3 Pearson Correlation for Environmental Gradients and Phytosociological Attributes*

	CSR	CAB	TSR	TAB	TBA	CAN	ALT	SLOP
CSR	1	0.000	0.793	0.672	0.441	0.949	0.315	0.406
CAB	0.652	1	0.170	0.354	0.340	0.759	0.417	0.007
TSR	0.036	0.184	1	0.000	0.000	0.315	0.122	0.145
TAB	0.057	0.125	0.539	1	0.001	0.170	0.020	0.957
TBA	−0.104	0.129	0.830	0.435	1	0.145	0.780	0.002
CAN	−0.009	−0.041	−0.136	−0.184	−0.196	1	0.000	0.827
ALT	0.136	0.11	0.207	0.308	−0.038	−0.500	1	0.341
SLOP	−0.112	−0.352	−0.195	−0.007	−0.410	−0.03	0.129	1

Upper panel representing P value (significance) and lower panel representing r-value.
Abbreviations: CSR = climber species richness; CAB = climber abundance; TSR = tree species richness; TAB = tree abundance; TBA = tree total basal area; CAN = Tree canopy cover; ALT = altitude; SLOP = slope angle.

4.3.6 CCA ORDINATION

Eigenvalue of first 3 CCA axes were 0.346, 0.206 and 0.137, respectively. The first axis had a positive correlation with altitude (0.021), slope (0.119), aspects (0.502) and canopy cover (0.295) and negative correlation with TSR (−0.239), TAB (−0.181) and TBA (−0.548). The second axis had a negative correlation with slope (−0.411), aspects (−0.486), TAB (−0.067) while positive with altitude (0.243), canopy (0.050), TSR (0.411) and TAB (0.217). Plots under G5 and G6 were placed in the negative and positive part of axes 1 and 2, respectively. The majority of the plots of G1 and G2 belonged to the positive side of axis 1, while the plots of G4 to the negative side (Figure 4.6). Plots under the community G3 occupied central position. Species like *Cryptolepis dubia, Entada gigas, Hippocratea micrantha,* and *Mikania scandens* were associated positively with canopy and negatively with TAB and TBA. In contrast, *Argyreia argentea, Cissampelos pareira, Erythropalum scandens,* and *Myxopyrum smilacifolium* were negatively correlated with canopy. *A. argentea, Hoya globulosa, E. gigas, Epipremnum pinnatum,* and *Uncaria macrophylla* were positively associated with altitude and negatively with slope, while species like *Bauhinia scandens, Embelia floribunda, Ficus pumila,* and *Parabaena sagittata* were negatively correlated with altitude (Figure 4.7). The most frequent species of the community, *Byttneria aspera, Combretum wallichii* var. *griffithii, Pothos scandens, Thunbergia coccinea,* and *Tetrastigma leucostaphylum* were placed near the center, which showed the non-influence of any particular variable on their distribution.

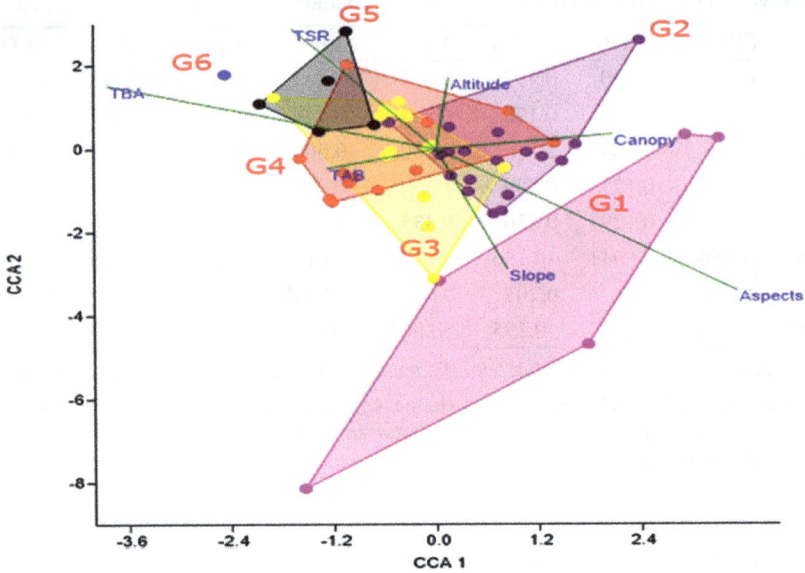

FIGURE 4.6　CCA ordination of climber communities (TWINSPAN groups) in NNP. Colored dot showing plots of particular community.

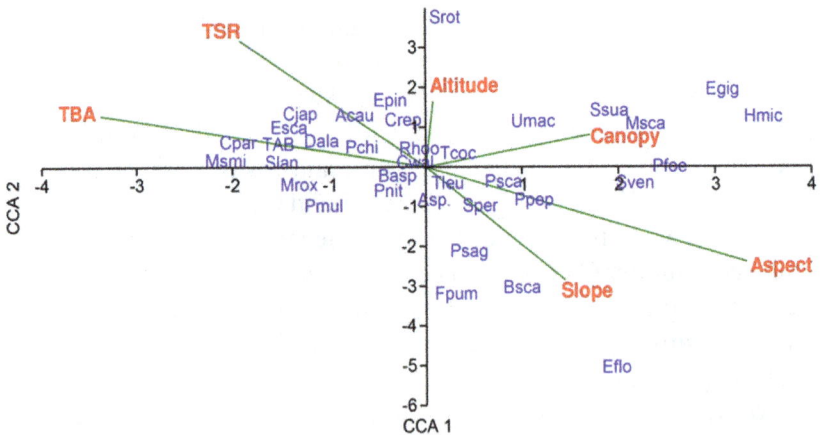

FIGURE 4.7　CCA ordination of climber species in NNP.

(Acronym according to Annexure 4.1 and Table 4.3).

4.4　DISCUSSION

The present study reported a total of 47 climber species (33 lianas and 14 vines) from 2.28 ha area (400 m² × 57 plots) of the tropical rainforest stands in western part of NNP. The reported SR for woody climbers in this study (33 species) is well within the range recorded from some other tropical forests of Andhra Pradesh (Mastan et al., 2015; Srinivas and Sundarapandian, 2017), Coromandel Coast (Reddy and Parthasarathy, 2003; Vivek and Parthasarathy, 2014), Eastern Ghats (Kadavul and Parthasarathy, 1999; Chittibabu and Parthasarathy, 2001; Naveenkumar et al., 2017), Peninsular India (Parthasarathy et al., 2004) and Western Ghats (Padaki and Parthasarathy, 2000) in Southern part of the India and from some selected forest types of Arunachal Pradesh, Sikkim, and Meghalaya (Chetri et al., 2010; Barik et al., 2015) in the northeastern India.

The species-accumulation curve did not reach an asymptote and expected SR by non-parametric species estimators was much high to those of the OSR. The sampling procedure could be the possible reason for this pattern of SR in the present study, because each next plot was established towards forward directions along three transects moving river bank to ridge top (T1), west to east direction (T2) and along mid-hill with gentle slopes (T3) that lead variations in the habit characteristics (like altitude, slope, aspects, canopy cover, vegetation composition, soil conditions). Thus, sampling in newer habitats added new species in the cumulative number of species, but the rate of new addition come down one species per 5 plots (after 46th plot and 45 species), in this standpoint, the sample size may be considered as adequate.

In this study, twiners were common in comparison to scrambler, tendrilar, hooked, and rooted climbers in term of SR, as similar to the findings from tropical forests of peninsular India (Chittibabu and Parthasarathy, 2001; Muthumperumal and Parthasarathy, 2010; Srinivas and Sundarapandian, 2017). Twiners, due to their higher climbing efficiency than other climbers, dominate in mature forests (Putz, 1984), while forests with earlier succession stages exhibit larger proportion of tendril climbers than twiners (Laurance et al., 2001; DeWalt et al., 2006; Santos et al., 2009). This study supports these statements, as Namdapha is one of the oldest and largest existing *Dipterocarpus* forests in India with diversity climax (Proctor et al., 1998).

The present work is the first of its kind in studying indicator species among climbers through TWINSPAN based classification, IV analysis, and TIV analysis in any tropical forest of India. The TWINSPAN recognized 6 communities (groups) using species matrix (Figure 4.2). Rarity (between groups) and commonness (within-group) are the main underlying gradients in classifying the communities, therefore the lowest similarity (29%) and highest beta diversity (71%) was observed in the most distant groups (i.e., G1–G5). TWINSPAN based indicator species for the community G1, G3, G5 and G6 are *Ficus pumila, Sabia lanceolata, Erythropalum scandens,* and *Chonemorpha fragrans,* respectively. No indicator was assigned to the community G2 and G4 communities at the 4th level of division, however, they fall in *Rhaphidophora hookeri* and *Byttneria aspera* group, respectively at the 3rd level of division.

The stem density in the study site varied between 180–225 ha[-1] with an average of 199 ± 21.06 ha[-1] which is similar to that of another tropical forest of Arunachal Pradesh (Barik et al., 2015) and the value was lower to comparison to the forests of peninsular India (Muthuramkumar and Parthasarathy, 2000; Pandian and Parthasarathy, 2016; Naveenkumar et al., 2017; Srinivas and Sundarapandian, 2017) and higher to that of in shola forests of Nilgiri Mountains (Mohandass and Davidar, 2014). The reverse J-shaped distribution of density into size classes is considered an indication of steady population having good regeneration (Saxena and Singh, 1984; Khan et al., 1987; Kumar et al., 2004; Rawat et al., 2018), which was also observed for climber in present study with more than half climber population were distributed to lower CBH classes (1st and 2nd class) while the rest in middle (3rd) and upper classes (4th and 5th) with rare variations. The density and basal area of climber in this study may be attributed species composition, age structure and succession stage of climber communities and availability of hosts (trees and larger shrubs).

Dominance-diversity curves (d-d curves) fit for lognormal situation with shallow (gentle-moderate) gradients indicated co-dominance among the species from high rank to lower rank. However, d-d curves for G2 and G3 had steep gradient due to the larger difference in IVI between first rank species and the second rank species, i.e., 33.88% between *F. pumila* and *P. mullesua* for G2, 11.19% between *R. hookeri* and *P. scandens* for G3. The steep gradient indicated low evenness in dominance (IVI) and assumes higher competition for niche space among the high-ranking species (Magurran, 2004; Odum, 1971).

The SR and abundance of both climbers and trees were positively correlated with altitude and negatively with tree basal area and canopy cover (Table 4.2). Altitude represents a complex combination of related climatic variables closely correlated with numerous other environmental properties (Ramsay and Oxley, 1997), this is one of the primary influential factors on the phytosociological attributes of climbers (Laurance et al., 2001; Parthasarathy et al., 2004). The result in the present work, supported the liana diversity pattern with respect to altitudes, reported from other sites of eastern Himalayas (Barik et al., 2015) and differs from the pattern observed in montane zones of Sikkim Himalaya (Chetri et al., 2010). By taking into account these observations along with ours, it can be concluded that the climber diversity and abundance in the forests of eastern Himalayas rises with the altitudinal increase up to subtropical level and decrease towards higher elevation.

The negative effect of canopy cover with climber diversity in NNP was found, which can be attributed to insolation and the exponential decrease in radiation intensity along downward forest strata. Younger trees with smaller diameter supports diverse group of climber leading to higher SR in comparison to old forest with larger diameter trees, which are mostly dominated by a few groups of climber (like twiners). In our study, the basal areas of the trees have a positive influence on the climber abundance, though it affects the SR negatively, which can be further correlated to the availability of space on host tree trunks for some abundant rooted climbers (viz., *Epipremnum pinnatum*, *Ficus* spp., *Piper* spp., *Pothos scandens* and *Rhaphidophora hookeri*). The influence of aspects in NNP on the climber SR and abundance was found significant with the highest diversity in east aspects.

CCA for different communities (Figure 4.6) showed relatedness to all the variables viz., TSR, TAB, TBA, canopy covers, altitude, slope angle and slope aspect to the climber community in the NNP. The TWINSPAN communities G2, G3, and G4 included nearly 90% of the study area where the climber diversity was influenced by altitude, slope, canopy cover, and tree abundance. Therefore, it can be understood that these factors are the primary forces in deriving a diversity pattern for climbers in NNP. Further, the communities G1 and G5 were placed in two extremities along both axis-1 and axis-2. The similarity between these two communities is lowest. The influencing factors for G1 communities were geographical, i.e., slope, and aspect, but tree SR attributed towards the diversity in G5. However, the maximum strength was recorded for slope aspect and TAB followed by canopy covers, TSR, slope angle, and TAB. CCA for the climber species (Figure 4.7) showed the regulating factors in the distribution of climber species in NNP. Altitude, though showed significant correlation, yet influenced the least towards the species distribution in comparison to other parameters. Distribution of a majority of climber species was attributed by tree basal area and slope of the forest. In natural forests, the diversity tends to increase as the environment becomes more favorable and predictable (Putman, 1994). A favorable environmental condition with a climber population at its climax was observed in NNP and still predictability of further taxonomic additions cannot be ignored completely, as a larger portion of this protected forest area is yet to be explored.

4.5 CONCLUSION

Studies on the community structure and composition of climbers in the Indian Himalayan Region (IHR) are scanty with some handpicked notes from eastern Himalayas. Realizing this gap, the present study was intended to observe not only the community structure but also the relation of climber attributes with environment and tree variables. TWINSPAN recognized 6 climber communities in the area viz., *Ficus pumila, Hippocratea micrantha* and *Mucuna monosperma* community (G1), *Schefflera venulosa, Byttneria aspera* and *Mikania scandens* community (G2), *Sabia lanceolata, Combretum wallichii* var. *griffithii* and *Rhaphidophora hookeri* community (G3), *Acacia* sp., *Piper mullesua* and *Thunbergia coccinea* community (G4), *Erythropalum scandens, Dioscorea alata* and *Sabia lanceolata* community (G5) and *Chonemorpha fragrans* community (G6) using species matrix (47 species and 57 plots). The expected numbers of species are high to those of the observed in this study. The present work is the first of its kind in studying indicator species among climbers through TWINSPAN based classification, IV analysis, and TIV analysis in any tropical forest of India. TWINSPAN based indicator species for the community G1, G3, G5, and G6 are *Ficus pumila, Sabia lanceolata, Erythropalum scandens,* and *Chonemorpha fragrans,* respectively. CCA for different communities showed relatedness to all the variables viz., TSR, TAB, TBA, canopy covers, altitude, slope angle, and slope aspect to the climber community in the NNP. The SR of twiners increases with an increase in tree community stability and maturity. On the basis of maximum IV, *Ficus pumila* (G1 and G4), *Thunbergia coccinea* (G2), *Rhaphidophora hookeri* (G3), and *Erythropalum scandens* at (G5) reveled as indicator species for long term monitoring of climber communities in the study area as the present result will serve as baseline data in this reference.

KEYWORDS

- canonical correspondence analysis
- climber community
- eastern Himalaya
- indicator value
- lowland tropical forest
- species richness

ANNEXURE 4.1   List of Climbers of NNP with Name of Family, Type of Habit, Mode of Climbing and Importance Value Index (%) in Different Communities

Name of Taxa	Taxa Code	Family	Habit	Climbing Mode	Importance Value Index (%)				
					G1	G2	G3	G4	G5
Caesalpinia cucullata Roxb.	Asp.	Caesalpiniaceae	Lina	Hook	3.40	4.09	2.74	3.42	—
Argyreia argentea (Roxb.) Sweet	Aarg	Convolvulaceae	Vine	Twiner	—	—	—	0.92	—
Artabotrys caudatus Wall. ex Hook.f. and Thomson	Acau	Annonaceae	Lina	Hook	—	—	—	1.46	—
Bauhinia scandens L.	Bsca	Caesalpiniaceae	Lina	Tendril	8.40	0.82	7.60	—	—
Byttneria aspera Collebr. ex Wall.	Basp	Sterculiaceae	Lina	Twiner	4.69	9.95	3.87	20.09	16.01
Cayratia japonica (Thunb.) Gagnep.	Cjap	Vitaceae	Vine	Tendril	—	—	—	4.08	8.65
Chonemorpha fragrans (Moon) Alston	Cfra	Apocynaceae	Lina	Twiner	—	—	—	—	—
Cissampelos pareira L.	Cpar	Menispermaceae	Lina	Twiner	—	—	2.61	—	—
Cissus repens Lam.	Crep	Vitaceae	Vine	Tendril	—	0.75	—	—	5.46
Combretum wallichii var. *griffithii* (Van Heurck and Müll. Arg.) M.G. Gangop. and Chakrab.	Cwal	Combretaceae	Lina	Scrambler	—	5.15	1.01	9.93	15.22
Cryptolepis dubia (Burm.f.) M.R. Almeida	Cdub	Asclepiadaceae	Lina	Twiner	3.40	—	—	—	—
Derris cuneifolia Benth.	Mrox	Fabaceae	Lina	Twiner	—	—	—	—	—
Dioscorea alata L.	Dala	Dioscoreaceae	Vine	Twiner	—	0.95	—	—	3.65
Embelia floribunda Wall.	Eflo	Myrsinaceae	Lina	Scrambler	—	0.80	—	—	—
Entada gigas (L.) Fawc. and Rendle	Egig	Mimosaceae	Lina	Tendril	5.60	1.65	—	—	—
Epipremnum pinnatum (L.) Engl.	Epin	Araceae	Lina	Root	—	—	0.87	2.76	13.96
Erythropalum scandens Blume	Esca	Olacaceae	Vine	Tendril	—	—	9.62	—	15.90

ANNEXURE 4.1 *(Continued)*

Name of Taxa	Taxa Code	Family	Habit	Climbing Mode	Importance Value Index (%)				
					G1	G2	G3	G4	G5
Ficus pumila L.	Fpum	Moraceae	Lina	Root	16.81	1.81	43.19	–	–
Ficus sarmentosa Buch.-Ham. ex Sm.	Fsar	Moraceae	Lina	Root	3.09	–	–	–	–
Fissistigma polyanthum (Hook.f. and Thomson)	Aaug	Annonaceae	Lina	Scrambler	–	–	–	0.92	–
Hippocratea micrantha Cambess.	Hmic	Celastraceae	Lina	Scrambler	7.55	–	–	–	–
Holboellia latifolia Wall.	Hlat	Lardizabalaceae	Lina	Scrambler	–	–	–	1.05	–
Hoya globulosa Hook.f.	Hglo	Asclepiadaceae	Lina	Twiner	–	–	–	1.96	–
Mikania scandens (L.) Willd.	Msca	Asteraceae	Vine	Twiner	13.96	6.06	–	2.91	–
Mucuna monosperma Wight	Mmon	Mimosaceae	Lina	Twiner	2.83	–	–	–	–
Myxopyrum smilacifolium (Wall.) Blume	Msmi	Oleaceae	Lina	Twiner	–	1.45	–	1.95	2.41
Paederia foetida L.	Pfoe	Rubiaceae	Vine	Hook	14.62	–	0.82	1.05	–
Parabaena sagittata Miers	Psag	Menispermaceae	Lina	Twiner	–	0.63	–	0.98	–
Pegia nitida Colebr.	Pnit	Anacardiaceae	Lina	Hook	–	1.26	–	0.98	–
Persicaria chinensis (L.) H. Gross	Pchi	Polygonaceae	Vine	Scrambler	2.83	–	1.18	–	3.97
Piper mullesua Buch.-Ham. ex D. Don	Pmul	Piperaceae	Vine	Root	4.73	–	9.92	16.89	–
Piper peploides (Kunth) Poir. ()	Ppep	Piperaceae	Vine	Root	–	3.46	–	–	–
Poikilospermum naucleiflorum (Roxburgh ex Lindl.) Chew	Pnau	Urticaceae	Lina	Root	–	0.55	–	–	–
Pothos scandens L.	Psca	Araceae	Lina	Root	2.83	10.46	1.51	4.00	–
Rhaphidophora hookeri Schott	Rhoo	Araceae	Lina	Root	–	21.65	3.32	9.72	–
Rubus lucens Focke	Rluc	Rosaceae	Lina	Hook	–	0.58	–	–	–

ANNEXURE 4.1 *(Continued)*

Name of Taxa	Taxa Code	Family	Habit	Climbing Mode	Importance Value Index (%)				
					G1	G2	G3	G4	G5
Rubus paniculatus Sm.	Rpan	Rosaceae	Lina	Hook	–	0.82	–	–	–
Sabia lanceolata Colebr.	Slan	Sabiaceae	Lina	Scrambler	2.63	–	4.05	–	9.22
Schefflera venulosa (Wight and Arn.) Harms	Sven	Araliaceae	Lina	Scrambler	–	4.40	–	–	–
Smilax perfoliata Lour.	Sper	Smilacaceae	Lina	Hook	–	3.50	2.74	0.87	–
Stemona tuberosa Lour.	Stub	Stemonaceae	Vine	Twiner	–	0.58	–	–	–
Stephania rotunda Lour.	Srot	Menispermaceae	Vine	Twiner	–	1.10	–	–	2.88
Stixis suaveolens (Roxburgh) Pierre	Ssua	Capparaceae	Lina	Scrambler	–	1.75	–	–	–
Tetrastigma leucostaphylum (Dennst.) Alston	Tleu	Vitaceae	Lina	Tendril	2.63	6.68	3.78	8.57	2.68
Thunbergia coccinea Wall.	Tcoc	Acanthaceae	Vine	Twiner	–	7.80	1.18	3.41	–
Trichosanthes tricuspidata Lour.	Ttri	Cucurbitaceae	Vine	Tendril	–	0.58	–	–	–
Uncaria macrophylla Wall.	Umac	Rubiaceae	Lina	Hook	–	0.75	–	2.10	–

ANNEXURE 4.2 Indicator Value (%) and Total Importance Value (%) of Climbers in Different Communities within NNP

Name of Taxa	Indicator Value (%)					Total Importance Value (%)				
	G1	G2	G3	G4	G5	G1	G2	G3	G4	G5
Caesalpinia cucullata Roxb.	5.57	10.30	5.57	6.47	—	6.19	9.24	5.53	6.66	—
Argyreia argentea (Roxb.) Sweet	—	—	—	8.33	—	—	—	—	5.09	—
Artabotrys caudatus Wall. ex Hook.f. and Thomson	—	—	—	8.33	—	—	—	—	5.63	—
Bauhinia scandens L.	27.30	4.73	1.52	—	—	22.05	3.19	8.36	—	—
Byttneria aspera Collebr. ex Wall.	10.50	9.14	14.00	34.70	25.10	9.94	14.52	10.87	37.44	28.56
Cayratia japonica (Thunb.) Gagnep.	—	—	—	16.60	35.50	—	—	—	12.38	26.40
Chonemorpha fragrans (Moon) Alston	—	—	—	—	—	—	—	—	—	—
Cissampelos pareira L.	—	—	6.67	—	—	—	—	5.95	—	—
Cissus repens Lam.	—	0.61	—	—	17.70	—	1.06	—	—	14.31
Combretum wallichii var. *griffithii* (Van Heurck and Müll. Arg.) M.G. Gangop. and Chakrab.	—	5.63	16.80	16.20	17.70	—	7.97	9.41	18.03	24.07
Cryptolepis dubia (Burm.f.) M.R. Almeida	20.00	—	—	—	—	13.40	—	—	—	—
Derris cuneifolia Benth.	—	—	—	—	—	—	—	—	—	—
Dioscorea alata L.	—	1.49	—	—	15.80	—	1.70	—	—	11.55
Embelia floribunda Wall.	—	5.26	—	—	—	—	3.43	—	—	—
Entada gigas (L.) Fawc. and Rendle	15.80	4.47	—	—	—	13.50	3.89	—	—	—
Epipremnum pinnatum (L.) Engl.	—	—	1.79	1.44	12.50	—	—	1.77	3.48	20.21
Erythropalum scandens Blume	—	—	3.51	—	94.70	—	—	11.38	—	63.25
Ficus pumila L.	51.70	8.95	5.38	—	—	42.66	6.29	45.88	—	—

ANNEXURE 4.2 *(Continued)*

Name of Taxa	Indicator Value (%)					Total Importance Value (%)				
	G1	G2	G3	G4	G5	G1	G2	G3	G4	G5
Ficus sarmentosa Buch.-Ham. ex Sm.	20.00	–	–	–	–	13.09	–	–	–	–
Fissistigma polyanthum (Hook.f. and Thomson)	–	–	–	8.33	–	–	–	–	5.09	–
Hippocratea micrantha Cambess.	40.00	–	–	–	–	27.55	–	–	–	–
Holboellia latifolia Wall.	–	–	–	8.33	–	–	–	–	5.22	–
Hoya globulosa Hook.f.	–	–	–	8.33	–	–	–	–	6.13	–
Mikania scandens (L.) Willd.	24.00	7.26	–	2.08	–	25.96	9.69	–	3.95	–
Mucuna monosperma Wight	20.00	–	–	–	–	12.83	–	–	–	–
Myxopyrum smilacifolium (Wall.) Blume	–	3.30	–	5.30	7.63	–	3.10	–	4.60	6.23
Paederia foetida L.	35.60	–	3.29	5.14	–	32.42	–	2.47	3.62	–
Parabaena sagittata Miers	–	2.04	–	5.11	–	–	1.65	–	3.54	–
Pegia nitida Colebr.	–	5.88	–	3.68	–	–	4.20	–	2.82	–
Persicaria chinensis (L.) H. Gross	5.45	–	1.21	–	10.90	5.56	–	1.79	–	9.42
Piper mullesua Buch.-Ham. ex D. Don	2.86	–	1.59	48.60	–	6.16	–	10.72	41.19	–
Piper peploides (Kunth) Poir. ()	–	21.10	–	–	–	–	14.01	–	–	–
Poikilospermum naucleiflorum (Roxburgh ex Lindl.) Chew	–	5.26	–	–	–	–	3.18	–	–	–
Pothos scandens L.	18.00	28.20	8.99	5.57	–	11.83	24.56	6.01	6.79	–
Rhaphidophora hookeri Schott	–	16.70	64.90	32.00	–	–	30.00	35.77	25.72	–
Rubus lucens Focke	–	5.26	–	–	–	–	3.21	–	–	–

ANNEXURE 4.2 *(Continued)*

Name of Taxa	Indicator Value (%)					Total Importance Value (%)				
	G1	G2	G3	G4	G5	G1	G2	G3	G4	G5
Rubus paniculatus Sm.	–	5.26	–	–	–	–	3.45	–	–	–
Sabia lanceolata Colebr.	20.00	–	27.30	–	18.20	12.63	–	17.70	–	18.32
Schefflera venulosa (Wight and Arn.) Harms	–	21.10	–	–	–	–	14.95	–	–	–
Smilax perfoliata Lour.	–	12.50	14.80	1.19	–	–	9.75	10.14	1.47	–
Stemona tuberosa Lour.	–	5.26	–	–	–	–	3.21	–	–	–
Stephania rotunda Lour.	–	3.63	–	–	13.10	–	2.92	–	–	9.43
Stixis suaveolens (Roxburgh) Pierre	–	15.80	–	–	–	–	9.65	–	–	–
Tetrastigma leucostaphylum (Dennst.) Alston	14.60	13.30	12.50	22.40	14.60	9.93	13.33	10.03	19.77	9.98
Thunbergia coccinea Wall.	–	30.20	5.69	5.34	–	–	22.90	4.03	6.08	–
Trichosanthes tricuspidata Lour.	–	5.26	–	–	–	–	3.21	–	–	–
Uncaria macrophylla Wall.	–	1.26	–	12.70	–	–	1.38	–	8.45	–

REFERENCES

Arunachalam, A., Sarmah, R., Adhikari, D., Majumder, M., & Khan, M. L., (2004). Anthropogenic threats and biodiversity conservation in Namdapha nature reserve in the Indian Eastern Himalayas. *Curr. Sci., 87,* 447–454.

Barbhuiya, A. R., Arunachalam, A., Nath, P. C., Khan, M. L., & Arunachalam, K., (2008). Leaf litter decomposition of dominant tree species of Namdapha. *J. For. Res., 13,* 25–34.

Barik, S. K., Adhikari, D., Chettri, A., & Singh, P. P., (2015). Diversity of lianas in eastern Himalayas and North-Eastern India. In: Parthasarathy, N., (ed.), *Biodiversity of Lianas* (pp. 99–121). Springer International Publishing, Switzerland.

Bongers, F., Parren, M. P. E., & Traoré, D., (2005). *Forest Climbing Plants of West Africa: Diversity, Ecology and Management* (p. 273). CAB International, Oxfordshire, Wallingford.

Buzas, M. A., & Gibson, T. G., (1969). Species diversity: Benthonic foraminifera in the Western North Atlantic. *Science, 163,* 72–75.

Chao, A., (1987). Estimating the population size for capture-recapture data with unequal catchability. *Biometrics, 43,* 783–791.

Chauhan, A. S., Singh, P. K., & Singh, D. K., (1996). *Contribution to the Flora of Namdapha, Arunachal Pradesh* (p. 422). Botanical Survey of India, Calcutta, India.

Chettri, A., Barik, S. K., Pandey, H. N., & Lyngdoh, M. K., (2010). Liana diversity and abundance as related to microenvironment in three forest types located in different elevational ranges of the Eastern Himalayas. *Plant Ecol. Divers., 3,* 175–185.

Chittibabu, C. V., & Parthasarathy, N., (2001). Liana diversity and host relationships in a tropical evergreen forest in the Indian Eastern Ghats. *Ecol. Res., 16*(3), 519–529.

Chowdhery, H. J., Giri, G. S., Pal, G. D., Pramanik, A., & Das, S. K., (1996). Materials for the flora of Arunachal Pradesh. *Ranunculaceae-Dipsacaceae: Flora of India, Series-2* (Vol. 1, p. 693). Botanical Survey of India, Kolkata, India.

Chowdhery, H. J., Giri, G. S., Pal, G. D., Pramanik, A., & Das, S. K., (2008). Materials for the flora of Arunachal Pradesh. *Asteraceae-Ceratophyllaceae: Flora of India, Series-2* (Vol. 2, p. 492). Botanical Survey of India, Kolkata, India.

Chowdhery, H. J., Giri, G. S., Pal, G. D., Pramanik, A., & Das, S. K., (2009). Materials for the flora of Arunachal Pradesh. *Hydrocharitaceae-Poaceae: Flora of India, Series-2* (Vol. 3, p. 491). Botanical Survey of India, Kolkata, India.

Dalling, J. W., Schnitzer, S. A., Baldeck, C., Harms, K. E., John, R., Mangan, S. A., Lobo, E., et al., (2012). Resource-based habitat associations in a neotropical liana community. *J. Ecol., 100*(5), 1174–1182.

Dash, S. S., & Singh, P., (2017). Flora of Kurung Kumey District, Arunachal Pradesh. *Botanical Survey of India* (p. 778). Kolkata.

Dash, S. S., Panday, S., Rawat, D. S., Kumar, V., Lahiri, S., Sinha, B. K., & Singh, P. (2021). Quantitative assessment of vegetation layers in tropical evergreen forests of Arunachal Pradesh, eastern Himalaya, India. *Curr. Sci., 120*(5), 850–858.

Deb, P., & Sundriyal, R. C., (2007). Tree species gap phase performance in the buffer zone area of Namdapha National Park, Eastern Himalaya, India. *Trop. Ecol., 48*(2), 209–225.

Deb, P., & Sundriyal, R. C., (2008). Tree regeneration and seedling survival patterns in old-growth lowland tropical rainforest in Namdapha National Park, north-east India. *Forest Ecol. Manag., 255,* 3995–4006.

DeWalt, S. J., Ickes, K., Nilus, R., Harms, K. E., & Burslem, D. F. R. P., (2006). Liana habitat associations and community structure in a Bornean lowland tropical forest. *Plant Ecol., 186,* 203–216.

DeWalt, S. J., Schnitzer, S. A., & Denslow, J. S., (2000). Density and diversity of lianas along a chronosequence in a central Panamanian lowland forest. *J. Trop. Ecol., 16*(1), 1–9.

Dufrêne, M., & Legendre, P., (1997). Species assemblages and indicator species: The need for a flexible asymmetrical approach. *Ecol. Monogr., 67,* 345–366.

Fisher, R. A., Corbet, A. S., & Williams, C. B., (1943). The relation between the number of species and the number of individuals in a random sample of an animal population. *J. Anim. Ecol., 12,* 42–58.

Gentry, A. H., (1991). The distribution and evolution of climbing plants. In: Putz, F. E., & Mooney, H. A., (eds.), *The Biology of Vines* (pp. 3–49). Cambridge University Press, Cambridge.

Gerwing, J. J., Schnitzer, S. A., Burnham, R. J., Bongers, F., Chave, J., DeWalt, S. J., Ewango, C. E. N., et al., (2006). A standard protocol for liana censuses. *Biotropica, 38,* 56–261.

Ghosh, A. K., (1987). *Qualitative Analysis of Faunal Resources of Proposed Namdapha Biosphere Reserve.* Zoological Survey of India, Calcutta, India.

Hammer, O., Harper, D. A. T., & Ryan, P. D., (2001). Past: Paleontological statistics software package for education and data analysis. *Palaeontol. Electron., 4,* 1–9.

Hill, M. O., & Šmilauer, P., (2005). *TWINSPAN for Windows Version 2.3.* Centre for Ecology and Hydrology and University of South Bohemia, Huntingdon and Ceske Budejovice.

Kadavul, K., & Parthasarathy, N., (1999). Plant biodiversity and conservation of tropical semi-evergreen forest in the Shervarayan hills of Eastern Ghats, India. *Biodivers. Conserv., 8*(3), 419–437.

Khan, M. L., Rai, J. P. N., & Tripathi, R. S., (1987). Population structure of some tree species in disturbed and protected subtropical forests of northeast India. *Acta Oecol., 8*(3), 247–255.

Kumar, M., Sharma, C. M., & Rajwar, G. S., (2004). A study on the community structure and diversity of a sub-tropical forest of Garhwal-Himalaya. *Indian For., 130*(2), 207–214.

Kumar, V., Dash, S. S., Panday, S., Lahiri, S., Sinha, B. K., & Singh, P., (2017). Akaniaceae: A new family record for flora of India and lectotypification of the name *Bretschneidera sinensis. Nelumbo, 59*(1), 1–9.

Laurance, W. F., Pérez-Salicrup, D., Delamônica, P., Fearnside, P. M., D'angelo, S., Jerolinski, A., Pohl, L., & Lovejoy, T. E., (2001). Rain forest fragmentation and structure of Amazonian liana communities. *Ecology, 82*(1), 105–116.

Magurran, A. E., (2004). *Measuring Biological Diversity* (p. 256). Blackwell Science, Oxford.

Margalef, R., (1958). Information theory in ecology. *Gen. Sys., 3,* 36–71.

Mastan, T., Parveen, S. N., & Reddy, M. S., (2015). Liana species inventory in a tropical dry forest of Sri Lankamalla Wildlife Sanctuary, Andhra Pradesh, India. *J. Environ. Res. Devel., 9*(3A), 1024–1030.

McCune, B., & Mefford, M. J., (1999). *PC-ORD. Multivariate Analysis of Ecological Data.* Version 4.34. MjM Software, Gleneden Beach, Oregon.

Misra, R., (1968). *Ecology Workbook* (p. 242). Oxford and IBH Publication Company, Calcutta, India.

Mohandass, D., & Davidar, P., (2014). Floristic composition and reproductive traits of lianas in Shola forests of the Nilgiri Mountains, Western Ghats, India. *Int. J. Sci. Nat., 5*(1), 131–140.

Mueller-Dombois, D., & Ellenberg, H., (1974). *Aims and Methods of Vegetation Ecology* (p. 547). John Wiley and Sons, New York.

Muthumperumal, C., & Parthasarathy, N., (2010). A large-scale inventory of liana diversity in tropical forests of South-Eastern Ghats, India. *Syst. Biodivers., 8*(2), 289–300.

Muthuramkumar, S., & Parthasarathy, N., (2000). Alpha diversity of lianas in a tropical evergreen forest in the Anamalais, Western Ghats, India. *Divers. Distrib., 6*(1), 1–14.

Nath, P. C., Arunachalam, A., Khan, M. L., Arunachalam, K., & Barbhuiya, A. R., (2005). Vegetation analysis and tree population structure of tropical wet evergreen forests in and around Namdapha national park, north-east India. *Biodivers. Conserv., 14,* 2109–2136.

Naveenkumar, J., Arunkumar, K. S., & Sundarapandian, S., (2017). Biomass and carbon stocks of a tropical dry forest of the Javadi Hills, Eastern Ghats, India. *Carbon Manag., 8*(5/6), 351–361.

Odum, E. P., (1971). *Fundamentals of Ecology* (p. 574). Saunders Company, Philadelphia, USA.

Padaki, A., & Parthasarathy, N., (2000). Abundance and distribution of lianas in tropical lowland evergreen forest of Agumbe central Western Ghats India. *Trop. Ecol., 41*(2), 143–154.

Palmer, M. W., (1990). The estimation of species richness by extrapolation. *Ecology, 71,* 1195–1198.

Palmer, M. W., (1991). Estimating species richness: The second-order Jackknife reconsidered. *Ecology, 72,* 1512–1513.

Parthasarathy, N., Muthuramkumar, S., & Reddy, M. S., (2004). Patterns of liana diversity in tropical evergreen forests of peninsular India. *Forest Ecol. Manag., 190*(1), 15–31.

Philips, E. A., (1959). *Method of Vegetation Study* (p. 55–77). Holt. Reinhart Winston Len., New York.

Pielou, E. C., (1975). *Ecological Diversity* (p. 165). Wiley Interscience, New York.

Proctor, J., Haridasan, K., & Smith, G. W., (1998). How far does lowland evergreen tropical rainforest go? *Glob. Ecol. Biogeogr. Letters, 7,* 141–146.

Putman, R. J., (1994). *Community Ecology* (p. 178). Chapman and Hall, London.

Putz, F. E., (1984). The natural history of lianas on Barro-Colorado Island, Panama. *Ecology, 65*(6), 1713–1724.

Ramsay, P. M., & Oxley, E. R. B., (1997). The growth form composition of plant communities in the Ecuadorian Páramos. *Plant Ecol., 131*(2), 173–192.

Rawat, D. S., Dash, S. S., Sinha, B. K., Kumar, V., Banerjee, A., & Singh, P., (2018). Community structure and regeneration status of tree species in eastern Himalaya: A case study from Neora Valley National Park, West Bengal, India. *Taiwania, 63*(1), 16–24.

Reddy, M. S., & Parthasarathy, N., (2003). Liana diversity and distribution in four tropical dry evergreen forests on the Coromandel coast of south India. *Biodivers. Conserv., 12,* 1609–1627.

Reddy, M. S., & Parthasarathy, N., (2006). Liana diversity and distribution on host trees in four inland tropical dry evergreen forests of peninsular India. *Trop. Ecol., 47*(1), 109–123.

Santos, K., Kinoshita, L. S., & Rezende, A. A., (2009). Species composition of climbers in seasonal semideciduous forest fragments of Southeastern Brazil. *Biota Neotrop., 9*(4), 175–188.

Sarmah, R., Adhikari, D., Majumder, M., & Arunachalam, A., (2006a). Indigenous technical knowledge of Lisus with reference to natural resource utilization in the far-eastern villages of Arunachal Pradesh, India. *Indian J. Tradit. Know., 5*(1), 51–56.

Sarmah, R., Arunachalam, A., Majumder, M., Melkania, U., & Adhikari, D., (2006b). Ethno-medicobotany of Chakmas in Arunachal Pradesh, India. *Indian For., 132*(4), 474–484.

Saxena, A. K., & Singh, J. S., (1984). Tree population structure of certain Himalayan forest associations and implications concerning their future composition. *Vegetatio, 58*(2), 61–69.

Schnitzer, S. A., Dalling, J. W., & Carson, W. P., (2000). The impact of lianas on tree regeneration in tropical forest canopy gaps: Evidence for an alternative pathway of gap-phase regeneration. *J. Ecol., 88,* 655–666.

Schnitzer, S., & Bongers, F., (2002). The ecology of lianas and their role in forests. *Trends Ecol. Evol., 17*(5), 223–230.

Sfair, J. C., & Martins, F. R., (2011). The role of heterogeneity on climber diversity: Is liana diversity related to tree diversity? *Global Journal of Biodiversity Sci. Manage., 1*(1), 1–10.

Shannon, C. E., & Wiener, W. E., (1963). *The Mathematical Theory of Communities* (p. 117). University of Illinois Press, Urbana, USA.

Simpson, E. H., (1949). Measurement of diversity. *Nature, 163,* 688.

Sørensen, T. A., (1948). A method of establishing groups of equal amplitude in plant sociology based on similarity of species content, and its application to analyses of the vegetation on Danish commons. Kong. Dan. *Vid. Seisk. Biol. Skr., 5,* 1–34.

Srinivas, K., & Sundarapandian, S., (2017). Diversity and composition of climbing plants in tropical dry forests of Northern Andhra Pradesh, India. *Int. Res. J. Nat. Appl. Sci., 4*(10), 31–47.

Vivek, P., & Parthasarathy, N., (2017). Patterns of tree-liana interactions: Distribution and host preference of lianas in a tropical dry evergreen forest in India. *Trop. Ecol., 58*(3), 591–603.

Whittaker, R. H., (1960). Vegetation of the Siskiyou Mountains, Oregon and California. *Ecol. Monogr., 30,* 279–338.

CHAPTER 5

Vegetation Analysis and Regeneration Pattern of Dominant Tree Species in Timberline Zone of Nanda Devi Biosphere Reserve, Western Himalaya, Uttarakhand, India

AJAY MALETHA,[1] R. K. MAIKHURI,[2,3] and S. S. BARGALI,[3,4]

[1]*Amity Institute of Forestry and Wildlife, Amity University, Noida, Uttar Pradesh, India, E-mail: maletha.jay@gmail.com*

[2]*G.B. Pant Institute of Himalayan Environment and Development, Garhwal Unit, Post Box No.–92, Srinagar, Garhwal, Uttarakhand, India*

[3]*Department of Environmental science, H.N.B. Garhwal University, Srinagar Garhwal, Uttarakhand, India*

[4]*Department of Botany, DSB Campus, Kumaun University, Nainital–263001, Uttarakhand, India*

ABSTRACT

The present study analyzes the structure and regeneration pattern of dominant tree species in the timberline zone of Nanda Devi Biosphere Reserve (NDBR), in Indian Central Himalaya. A total of 278 plant species belonging to 167 genera and 66 families of angiosperm, gymnosperms, and pteridophytes have been identified and documented. The tree density varies from 20 to 724 tree/ha at different sites. Diversity index (H) for tree layer was recorded between 1.65 to 1.93, while concentration of dominance (CD) ranged between 0.18 and 0.26 at Tolma and Ghangaria sites in the buffer zone of NDBR. Out of 12 tree species recorded, 56% exhibited fair regeneration, 22% good, 11% poor, and the remaining 11% species showed no regeneration at Ghangaria site. However, on the other hand, at Tolma site, 40% tree species represented fair regeneration, 40% good and rest 20% exhibited no regeneration. In the tree layer, the majority of the species (78%) were distributed randomly and only about 22% showed contagious distribution at Ghangaria site. In Tolma site, 89% trees were exhibited contagious distribution pattern and only 11% tree showed random distribution pattern. However, none of the tree species showed a regular distribution pattern in both the sites. The investigation was also aimed to generate baseline information on timberline ecotone's ecological characteristics in NDBR, especially the extent and changes in the composition of certain tree species and their response to augment in plant density at multiple directions under the changing climatic scenarios. The overall findings of the present study would be highly useful in designing strategies and action plan for sustainable management and conservation of timberline ecotone of the region.

5.1   INTRODUCTION

High altitude ecosystems of Himalaya are most vulnerable geographic regions of the world to climate change outside the polar region (Cavaliere, 2009; Xu et al., 2009). High altitude vegetation distribution strongly influenced by the climatic parameters (temperature, precipitation, wind, and solar, etc.). Including that the other factors, i.e., topography, soils, postglacial succession, as well as human disturbances also affect the pattern of the vegetation (Krauchi et al., 2000; Dolezal and Srutek, 2002). It is the most important process to maintain the stable age structure of the plant species in a community, affected directly/indirectly by various climatic and edaphic factors (Singh and Singh, 1992). Coincidentally, major human settlements are located around the biosphere reserve or forest fringes, inhabitants of which use the nearby forests for extraction of timber, leaf litter for cattle bedding, firewood, and non-timber forest produce for food. Forest sites are also often used for livestock grazing, and thereby, forests become the principal source of their livelihood. Many studies have been conducted for the assessment of both qualitative and quantitative parameters of the vegetation in Western Himalaya (Saxena and Singh, 1982; Rawal and Pangtey, 1994; Baduni and Sharma, 1996; Dhar et al., 1997; Rawat et al., 2001; Adhikari et al., 2004; Joshi and Samant, 2004; Maletha et al., 2020) and other parts of the world (Shaw, 1909; Goldsmith and Smith, 1926; Griggs, 1938; Allen and Walsh, 1996; Holtmeier, 2003; Schickhoff, 2005). The potential of regeneration status of trees can be predicted by the age structure of the population (Marks, 1974; Pritts and Hancock, 1983; Saxena et al., 1984; Khan et al.,

1987) including number of seedling, sapling, and young in given population (Saxena and Singh, 1984). The pattern of population dynamics of seedlings, saplings, and adults can exhibit the regeneration status (Bekele, 1994; Teketay, 1996). However, regeneration of tree species is greatly influenced by biotic and abiotic factors of the surrounding environment (Uma, 2001). All these factors may directly affect the recruitment, survival, and growth of tree seedlings and sprouts. The future composition of the forests largely depends on the potential regenerative status of tree species within a forest stand in space and time (Henle et al., 2004).

In the present study, we investigated the floristic composition, structural traits, and regeneration status of timberline forests of NDBR. Our specific goal was to assess the regeneration dynamics of dominant trees based on their population structure and seedling/sapling densities. Such ecological knowledge is fundamental for conservation, sustainable utilization, and linking complex altitudinal and climatic gradients with human activities. Such an analysis may provide important information for the policymakers for developing management plans for fragile and sensitive Himalayan mountain ecosystems. The study can be beneficial for the foresters and the villagers to make them aware of the present condition and status of timberline forests under the scenarios of global change.

5.2   MATERIALS AND METHODS

5.2.1   STUDY SITE

The study was conducted at two different sites of NDBR timberline ecotone viz., (A) Ghangaria and (B) Tolma region (buffer zone), covering an area between 30°16' N to 30° 41' N and 79° 33' E to 79° 44' E with elevation ranges from 3000 m to 3600 m asl in the Central Himalaya (Figure 5.1). This NDBRI recognized as a World Heritage Site in 1992 and included in the UNESCO's world network of Biosphere Reserves in 2004. The area of NBDRI covers about 5860 km^2 with two core zones: (i) the Nanda Devi National Park (625 km^2) and the Valley of Flowers National Park (88 km^2). A large area of NDBRI lies above the timberline area and remain under snow cover for a period more than 6 months in a year. Being an inner part of the Himalayan valley, the NDBRI has its different microclimate. The area has generally dry and low annual precipitation, but heavy rainfall occurs from end of the June to early of September month. In the monsoon period, prevailing mist and low cloud maintain soil moisture and also keep greenery instead of drier inner Himalayan valleys. The monthly temperature ranges between maximum (15.3°C to 27.2°C) and minimum (2.2°C to 16.0°C). Total annual rainfall reported about 936.6 mm (Figure 5.2), of which 43% of annual rainfall occurs only in July and August. Snow accumulates during winter and does not some time melt completely until the end of April or mid-May.

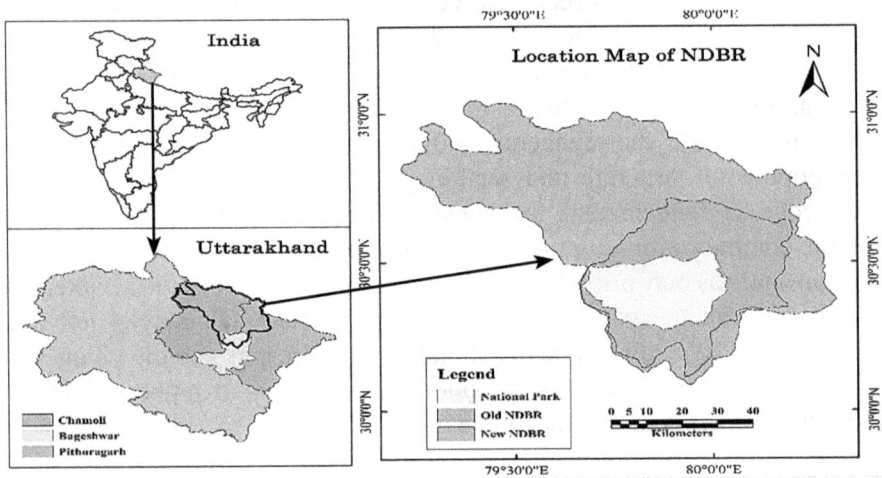

FIGURE 5.1   Location map of study area.

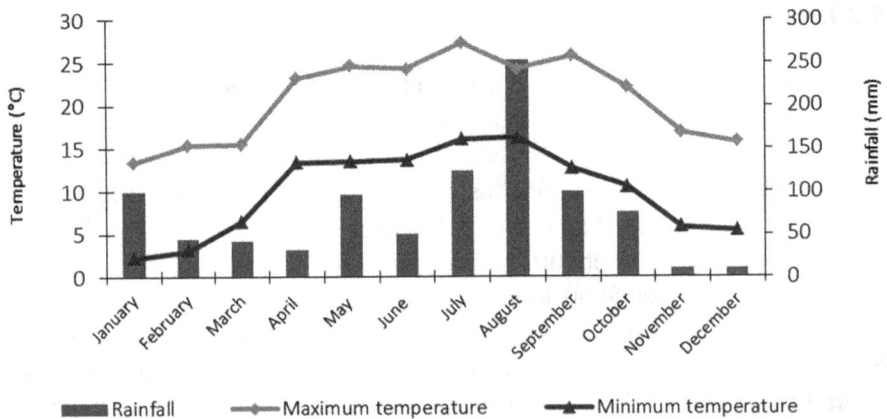

FIGURE 5.2   Climatic data of study area (Nanda Devi Biosphere Reserve).

5.2.1.1 DISTURBANCES AND THREATS

Timberline forests of the study area exhibit a high level of degradation caused due to anthropogenic pressure such as a collection of firewood, timber, and medicinal and aromatic plant species (MAPs), besides livestock grazing. Harsh climatic conditions and longer period of snow accumulation damage trees and their regeneration particularly contain of certain rare, endangered, and threatened (RET) species and thus authorities of Biosphere Reserve have to give some specific management strategies to overall protection.

5.2.2 VEGETATION SAMPLING

Vegetation analysis was carried out by using a stratified random sampling design. Sampling was carried out at two sites of timberline area in NDBR. A total of 50 plots (25 plots of 10 m × 10 m in each site) were established at an altitudinal range from 3,000 to 3,550 m asl. Trees in each quadrate were surveyed for height and the diameter at breast height of all individuals exceeding height of 2 m. Similarly, the number of all saplings and seedlings were recorded from each 5 m × 5 m and 1 m × 1 m plots nested at two opposite corners inside the 10 m × 10 m plots. Due to the very slow growth rate and low stature of trees along the timberline, a tree having CBH more than 20 cm was considered as a tree, and girth classes was prepared with the interval of 10 cm. All the classes were divided into seedling, sapling, and different girth classes viz., 21–30 cm, 31–40 cm, 41–50 cm, 51–60 cm, 61–70 cm, 71–80 cm, 81–90 cm, 91–100 cm, 101–110 cm, and above >111 cm. In each quadrate, all trees (≥ 20 cm cbh) and saplings (10–20 cm cbh) were individually measured at breast height, i.e., 1.37 m from the ground, individuals having ≤10 cm collar height circumference considered as seedlings. Field data were analyzed for abundance, density, and frequency. The sum of the relative values, viz., frequency, density, and dominance was used for calculating importance value index (IVI) (Curtis, 1959). The abundance to frequency ratio was done using formula laid by Curtis and Cottam (1956). Regeneration status (seedlings and saplings) of species was based on population size (Khan et al., 1987; Uma, 2001). The regeneration was considered "good" when seedling density > sapling density > adult tree density, "fair" when seedling density > sapling density ≤ adult density, "poor" when the species survived in only the sapling stage but not in the seedling stage, "none," for species with no sapling or seedling stages but present as adult trees, and "new" when adults of a species were absent but sapling and/or seedling stage(s) were present (Paul, 2008). When a species frequency distribution fits a "reverse J" pattern (when the gradual declining of individuals numbers from seedling to the sapling and further decline in small tree and mature tree), species was recognized as the dominant in the whole stand.

5.3   RESULTS

5.3.1   FLORAL DIVERSITY

A total of 278 species of vascular plants (angiosperm, gymnosperm, and pteridophytes) were identified within the study area. These species belong to 67 families and 167 genera. Figures 5.3 and 5.4 show the population structure of the various forest types along the timberline ecotone. Table 5.1 shows the composition of timberline forests in NDBR based on the species diversity. *Betula utilis* and *Abies pindrow* were the principle species found at Ghangaria and Tolma sites, respectively. Density, basal cover, and importance values index of these species have shown in Table 5.2.

5.3.2   SPECIES RICHNESS (SR) AND DIVERSITY INDEX

The total species richness (SR) [tree and shrub] was found higher in Ghangaria as compared to the Tolma site. Most of the tree and shrub species were common at both the sites, resulting in 90% similarity reported in the tree layer and 60% similarity in the shrub layer. In Ghangaria site, the diversity index for different layers reported 1.93 (trees), 1.99 (for all saplings, seedlings, and shrubs layers). The CD values for trees, saplings, seedlings, and shrubs were reported as 0.18, 0.16, 0.15, and 0.17, respectively. In Totlma site, the values of diversity for trees, saplings, seedling, and shrubs reported 1.65, 1,87, 1.77 and 2.55 respectively however; CD values for trees, saplings, seedlings, and shrubs reported 0.26, 0.20, 22, 0.08, respectively (Table 5.1).

5.3.3   POPULATION STRUCTURE

The total tree density (1632 trees/ha) and total basal cover (TBC) (67.39 m^2/ha) was found maximum at Tolma site as compared to the Ghangaria site (1016 trees/ha and 25.18 m^2/ha respectively). Among the trees, the highest density (724 trees ha^{-1}) was recorded for *Betula utilis* and lowest (20 trees/

ha) for *Acer caesium* at Tolma site (Table 5.2). *Betula utilis* exhibited higher dominance (IVI-114.35) at the same site. *Cedrus deodara* was found dominant tree species at Ghangaria site (IVI-63.43). However, *Pinus wallichiana* and *Abies pindrow* were found most important species with IVI values of 70.84 and 58.42 at Tolma site respectively (Table 5.2). The total sapling density recorded at Ghangaria and Tolma sites were 1364 and 1524 saplings/ha, respectively. The higher density was recorded for *R. campanulatum* (320 saplings/ha) followed by *Betula utilis* (232 saplings/ha) at Ghangaria site, while at Tolma site, higher density was found for *Betula utilis* (544 saplings/ha) and *Salix sikkimensis* (268 saplings/ha). *Abies pindrow* was another associate species exhibited higher sapling density at both sites. In the seedling stratum, the maximum total density was recorded at Tolma site (2328 seedlings/ha) as compared to the Ghangaria site (2032 seedling/ha). Among the species, higher density was recorded for *Betula utilis* at both the sites with maximum density (920 seedlings/ha) at Tolma site and minimum (456 seedlings/ha) at Ghangaria site followed by *Rhododendron campanulatum* (428 seedlings/ha) and *Abies pindrow* (364 seedlings/ha) in timberline of NDBR (Table 5.4).

TABLE 5.1 Diversity Index (H') and Concentration of Dominance (CD) for Trees, Saplings, Seedlings, and Shrubs in Ghangaria and Tolma Forest Area of NDBR

Site	Parameter	Forest Layer	Values	Dominant Tree Species
Ghangaria site	H'	Tree	1.93	*Betula utilis, Cedrus deodar, Abies pindow, Taxus baccata, Acer caesium, Salix sikkimensis, Rhododendron campanulatum*
		Sapling	1.99	
		Seedling	1.99	
		Shrub	1.99	
	CD	Tree	0.18	
		Sapling	0.16	
		Seedling	0.15	
		Shrub	0.17	
Tolma site	H'	Tree	1.65	*Betula utilis, Abies pindrow, Pinus wallichiana, Salix sikkimensis, Cedrus deodar, Populus ciliata, Picea smithiana, Acer caesium*
		Sapling	1.87	
		Seedling	1.77	
		Shrub	2.55	
	CD	Tree	0.26	
		Sapling	0.20	
		Seedling	0.22	
		Shrub	0.08	

TABLE 5.2   Phytosociological Parameters of Tree Species around Timberline Zone of NDBR

Tree Species	Ghangaria Site			Tolma Site		
	Density (trees ha⁻¹)	TBC (m² ha⁻¹)	IVI	Density (trees ha⁻¹)	TBC (m² ha⁻¹)	IVI
Betula utilis	368	6.7	85.68	724	12.8	114.35
Abies pindrow	128	4.8	48.70	248	41.3	58.42
Cedrus deodara	156	8.5	63.43	44	0.45	7.20
Pinus wallichiana	–	–	–	384	11.9	70.83
Taxus baccata	132	3.2	38.93	60	0.19	11.51
Acer caesium	80	0.1	22.64	20	0.04	5.84
Sorbus foliolosa	36	0.2	10.23	–	–	–
Rhododendron campanulatum	44	0.3	12.33	–	–	–
Salix sikkimensis	20	0.3	5.690	68	0.2	12.63
Populus ciliata	–	–	–	60	0.36	13.73
Picea smithiana	–	–	–	24	0.11	5.44
Syzygium cummini	52	1.0	12.33	–	–	–
Total	**10.16**	**25.18**	**300**	**16.32**	**67.39**	**300**

In the shrub layer, *Ribes alpestre* (396 individuals/ha), *Origanum* spp. (332 individuals/ha), *Spiraea bella* (224 individuals/ha), *Polygonum polystachyum* (176 individuals/ha), and *Rosa webbiana* (132 individuals/ha) showed higher density at the Ghangaria site, while at Tolma, highest density was represented by *Berberis jaschiana* (228 individuals/ha) followed by *Salix sikkimensis* (176 individuals/ha), *Indigofera heterantha* (152 individuals/ha) and *Sorbus microphylla* (120 individuals/ha) (Table 5.3). In the herbaceous layer at Ghangaria site, the dominant species were found in following sequence-*Fragaria nubicola* (844 individuals/ha)> *Oxalis corniculata* (812 individuals/ha)> *Geranium wallichianum* (780 individuals/ha)> *Anaphalis triplinervis* (736 individuals/ha)> *Impatiens sulcata* (364 individuals/ha). The other co-dominant species were recorded includes; *Potentilla atrisanguinea*, *Fragaria nubicola*, *Anemone obtusifolia*, *Ligularia amplexicaulis*, *Origanum vulgare*, *Senecio graciliflorus*, etc. At Tolma site, maximum density was recorded for *Geranium himalayense* (1384 individuals/ha), followed by *Oxalis corcunata* (792 individuals/ha) and *Fragaria nubicola* (756 individuals/ha) and minimum for *Angelica glauca* (48 individuals/ha), *Bergenia ciliata* (112 individuals/ha), and *Malaxis muscifera* (140 individuals/ha). Some other important associated prominent species found in the area were *Berginia starchy, Bistorta affinis, Geum elatum, Impatiens devendrae, Taraxacum officinale, Origanum vulgare*, etc. The distribution pattern of various woody species at different layers at two different sites indicated the species were distributed randomly and contagiously; however, the regular distribution was not appeared at both the sites. Contagious distribution was observed more common as compared to the random distribution. Almost all woody layer species (sapling, seedling, and shrub) were distributed in the contagious pattern at both sites, while 78% of the tree species were distributed randomly at Ghangaria and 11% species at Tolma site.

TABLE 5.3 Phytosociological Parameters of Shrub Species around Timberline Zone of NDBR

Species	Ghangaria Site			Tolma Site		
	Density (ind/ha)	A/F Ratio	IVI	Density (ind/ha)	A/F Ratio	IVI
Astragalus chlorostachys	–	–	–	108	0.421	11.96
Berberis aristata	–	–	–	60	0.234	8.81
Berberis jaeschkeana	72	0.18	14.36	228	0.291	23.49
Colquhounia coccinea	–	–	–	88	0.22	11.87
Elsholtzia fruticosa	–	–	–	88	0.22	11.87
Fren spp.	80	1.25	9.5	–	–	–
Indigofera heterantha	–	–	–	152	0.38	16.07
Lonicera microphylla	64	0.44	10.15	–	–	–
Oraganum vulgure	332	0.32	38.88	–	–	–
Prinspia utilis	64	0.25	11.96	88	0.22	11.87
Polygonum polystachyum	176	0.13	22.64	–	–	–
Ribies alpester	396	0.31	35.39	92	0.159	13.35
Rosa microphylla	80	0.1	14.59	112	0.143	15.886
Rosa webbiana	132	0.23	14.58	–	–	–
Rosa serica	–	–	–	80	0.139	12.56
Rubus niveus	–	–	–	56	0.072	12.21
Sorbus microphylla	112	0.14	11.94	120	0.092	18.84
Sorbaria tomentosa	32	0.13	9.62	76	0.132	12.30
Spiraea bella	224	0.22	7.46	–	–	–
Salix lindleyana	–	–	–	176	0.306	18.86
Total	**1364**	**3.35**	**200**	**1524**	**3.029**	**200**

5.3.4 REGENERATION POTENTIAL

In all 12 tree species were found along an elevational gradient between 3000–3600 m asl at both sites of NDBR. Among all these tree species, 56% showed fair regeneration, 22% showed good regeneration, 11% exhibited poor and remaining (11%) indicated no regeneration at Ghangaria site; while at Tolma site, 40% species showed fair regeneration, 40% had good regeneration and remaining 20% exhibited no regeneration. However, at Ghangaria site two species viz., *Salix sikkimensis* and *Rhododendron campanulatum* showed good regeneration as both the species represented by a good number of seedlings and saplings. Besides, about 5 plant species showed fair regeneration those includes; *Betula utilis, Abies pindrow, Taxus baccata, Cedrus deodara* and *Sorbus tomentosa*, whereas species such as *Acer caesium* showed poor regeneration and *Sygiyum cummini* exhibited no regenerating. At Tolma forest site, about 4 species (viz., *Salix sikkimensis, Populus ciliate, Taxus baccata,* and *Cedus deodara*) showed good regeneration (Table 5.4). The density diameter class distribution of tree species at both the sites showed a decline in density from small diameter class to higher diameter class. It indicated that individual plant species with smaller diameter class though were higher in number, but only a less number of individual in seedlings and saplings classes usually survive and form the larger tree classes (Figures 5.3 and 5.4). In general, the forests of both the sites showed regeneration, but it was found better in Tolma site because of the high density of seedlings and saplings. Human presence for pilgrimage and nature tourism and livestock grazing, along with cutting, lopping, and debarking of trees and extraction of other bio-resources such as medicinal plants which might have affected the regeneration of species at Ghangaria site.

5.4 DISCUSSION

In the Western Himalaya, timberline ecotone zone considered highly sensitive. Due to the harsh climatic condition of the zone, the species struggle for their survival, growth, and regeneration due to harsh climatic conditions. In the present study area, the regeneration of the dominant species was quite moderate except for the *Betula utilis* and *Rhododendron campanulatum*. The higher seedling and sapling density of *Betula utilis* along the timberline is predicted better growth, survival, and the advancement, but negligible in the near settlement and highly disturbed area (Rai et al., 2012). The population structure of the forest in the present study reveals that dominant species of the timberline are distributed in all the life stage classes. However, the percentage density declines progressively from seedling to adult tree stages along the timberline ecotone have been clearly indicated in Figures 5.3 and 5.4. Among the plant species, *Betula utilis* had the highest tree density (724 and 324 trees/ha) and contributed 44% and 36% of the total population at Tolma and Ghangaria sites followed by *Pinus wallichiana* (24%) and *Cedrus deodara* (15%), respectively. The average total basal area was recorded 2.77 m²/ha at Ghangaria and 7.48 m²/ha at Tolma site. The study also reveals that shrubs and herbs density was found higher in the Ghangaria site where anthropogenic disturbance was lower than Tolma site of the timberline. Besides, both the study areas showed a good number of seedlings and saplings population. Among the tree species, higher density was represented by *Betula utilis* at both the sites with maximum density (2476 trees/ha) were recorded at Tolma site and lower for *Vibernum* spp. (100 trees/ha) followed by *Syzium cummini* (50 trees/ha). Some species viz., *Populus ciliata*, *Picea smithiana* and *Pinus wallichiana* were not found at Ghangaria site but present at Tolma site (Table 5.1). The density diameter curve of the tree population of both sites resembled a 'reverse J' shape (Figures 5.3 and 5.4). Thus, in general, it was observed that both the forests are regenerating, however, regeneration was found higher at Tolma site.

TABLE 5.4   Regeneration Status (Density/ha) of Trees Species around Ghangaria and Tolma Sites, NDBR

Dominant Trees	Ghangaria site				Tolma site			
	Tree	Sapling	Seedling	Status	Tree	Sapling	Seedling	Status
Betula utilis	484	232	456	F	852	544	920	F
Abies pindrow	192	168	340	F	392	244	364	F
Salix sikkimensis	76	148	196	G	68	268	344	G
Pinus wallichiana	–	–	–	SA	464	100	200	F
Acer caesium	80	156	104	P	20	68	–	NR
Populus ciliata	–	–	–	SA	60	80	100	G
Taxus baccata	196	164	300	F	60	68	176	G
Picea smithiana	–	–	–	SA	24	–	–	NR
Cedus deodara	180	60	100	F	44	96	104	G
Rhododendron campanulatum	172	320	428	G	108	56	120	F
Sorbus tomentosa	36	140	108	F	–	–	–	SA
Syzygium cummini	52	–	–	NR	–	–	–	SA
Total	**1468**	**1388**	**2032**		**2092**	**1524**	**2328**	–

Abbreviations: F = Fair regeneration; G = Good regeneration; P = Poor regeneration; NR = No regeneration; SA = Species absent.

Diversity and Dynamics in Forest Ecosystems

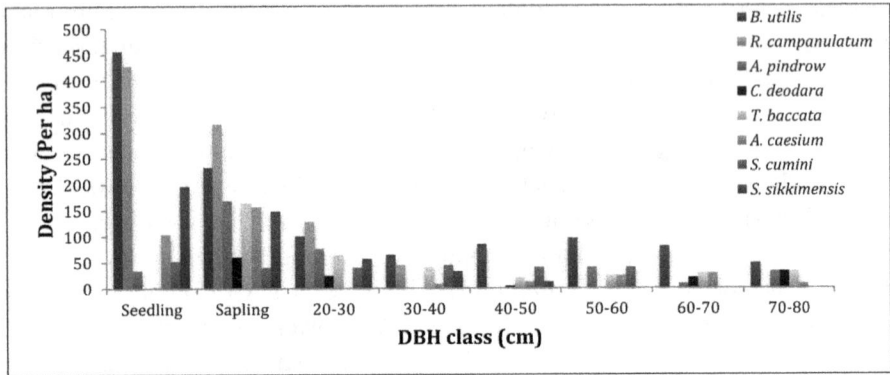

FIGURE 5.3   Density-diameter curve of *Betula utilis* forests in Ghangaria site, NDBR.

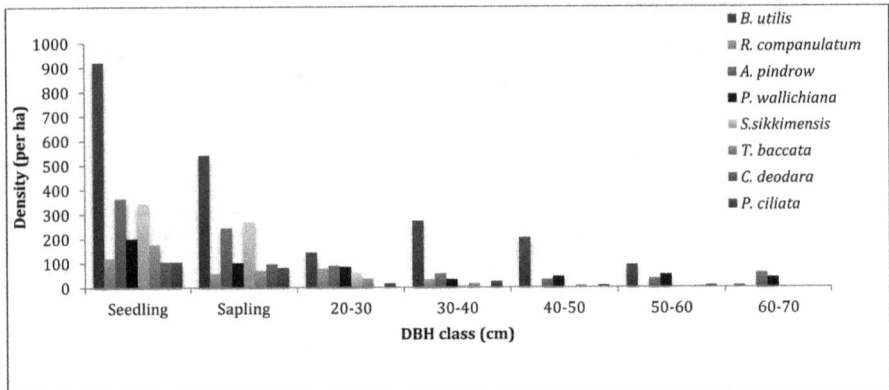

FIGURE 5.4   Density-diameter curve of *Betula utilis* forests in Tolma site, NDBR.

Similar observations were also reported by Shrestha et al. (2007) while working in trans-Himalayan forests of dry Manang valley (Central Nepal) on regeneration of *Betula utilis* in mixed and pure forests. In another study conducted in Bhutan on similar ecological zone and recorded three types of population structure, i.e., unimodal, sporadic, and reverse-J shaped among the dominant species along the altitudinal gradient (Wangda and Ohsawa, 2006). A study carried out in sub-alpine zone of old growth coniferous forest in Central exhibited 'reverse J' shaped distribution of major tree species (Mi-Yadokoro et al., 2003). Reverse "J" distribution is considered as an indication of stable population structure with fairly good regeneration status (Vetaas, 2000; Tesfaye, 1997, 2002, 2010; Maletha et al., 2021). The study revealed that intense recruitment of seedlings and saplings of *Betula utilis* and *Rhododendron campanulatum* near-natural treeline projected the great potential for future treeline advancement. Augmentation ability of the seedlings and saplings above the treeline within *krummholz* zone and scrub heath is also an indicator of the shifting of plant species in upward direction/higher elevation. This study also shows the continuous decrease in the CBH and increase in the number of seedlings of *Betula utilis* and *Rhododendron campanulatum* along an altitudinal gradient, which is a clear indication that these species may move upward to the higher altitude. The Himalayan ecosystem are very sensitive to the climate warming and have a worldwide distribution, it is vital that their biological and physical component to be monitored to provide indication of climate change and its effect. A study carried out by Dimri and Dash (2011) revealed that temperature has increased by 0.6 and 1.3°C between 1975 and 2006 in the Himalayan alpine zone. If we look the temperature rise in Western Himalaya (Singh et al., 2005; Bhutiyani et al., 2007; Arora et al., 2008; Shekhar et al., 2010) it may be assumed that variability in the climate may facilitate the expansion of species towards the higher elevations (Beerling and Woodward, 1996; Sykes et al., 2008; Vijayaprakash and Ansari, 2009). Remote sensing investigation by Singh et al. (2012) indicated that the treeline shifted 388 ± 80 m upwards in the Himalayan region between 1970 and 2006. Several studies have been carried out to assess the shift of timberline ecotone in the different parts of the Himalayan region due to macro and micro climatic condition and climate variability (Dubey et al., 2003; Baker and Moseley, 2007; Singh et al., 2011; Dutta et al., 2013; Maletha et al., 2021). The most treeline vegetation change is growth and regeneration enhancement rather than shift. The abrupt recruitment of *Betula utilis* and *Rhododendron campanulatum* was observed in many localities at both study areas may be due to the more favorable climatic conditions for growth during past decades, and/or to land-use changes in the high-altitude regions (Maletha et al., 2021). The timberline due to global warming would be expected upwards movement of vegetation, but yet the changes observed limited in both the study areas. Thus to confirm the shifting of timberline in NDBR, the establishment of long-term studies plots are required with weather stations from these areas since no long-term scientific data available on the one hand and at the same time as the region has been under high anthropogenic pressure during recent past.

ACKNOWLEDGMENT

The authors wish to acknowledge the Director, G.B. Pant National Institute of Himalayan Environment (NIHE), GBPIHED, Garhwal Regional Centre for providing necessary facilities and Uttarakhand State Forest Department for granting permission to work in the remote area. The authors also acknowledge the support of DST, SERB division, Govt. of India for funding this study. Sincere appreciation goes to the local inhabitants for their cooperation and help.

KEYWORDS

- **Western Himalaya**
- **population dynamics**
- **regeneration status**
- **timberline ecotone**

REFERENCES

Adhikari, B. S., (2004). Ecological attribute of vegetation in and around Nanda Devi National Park. In: *Biodiversity Monitoring Expedition Nanda Devi-2003: A Report to the Ministry of Environment and Forests* (pp. 30–35). Govt. of India, Uttaranchal State Forest Department, Dehradun.

Allen, T. R., & Walsh, S. J., (1996). Spatial and compositional pattern of alpine treeline, Glacier National Park, Montana. *Photogrammetric Engineering and Remote Sensing, 62*(11), 261–268.

Arora, M., Singh, P., & Goel, N. K., (2008). Climate variability influences on hydrological responses of a large Himalayan basin. *Water Resource Management, 22,* 1461–1475.

Baduni, N. P., & Sharma, C. M., (1996). Effect of aspect on the structure of some natural stands of *Cupressus torulosa* in Himalayan moist temperate forest. *Proc. Ind. Nat. Sci. Acad., B62,* 345–352.

Baker, B. B., & Moseley, R. K., (2007). Advancing treeline and retreating glaciers: Implications for conservation in Yunnan, P. R., China. *Artic, Antarctic, and Alpine Research, 39,* 200–209.

Beerling, D. J., & Woodward, F. I., (1996). Palaeo-ecophysiological perspectives on plant responses to global change. *Trends Ecol. Evol., 11,* 20–23.

Bekele, T., (1994). *Studies on Remnant Afromontane Forests on the Central Plateau of Shewa, Ethiopia.* PhD thesis, Uppsala University, Uppsala.

Bhutiyani, M. R., Kale, V. S., & Pawar, N. J., (2007). Long-term trends in maximum, minimum, and mean annual air temperatures across the Northwestern Himalaya during the twentieth century. *Climatic Change, 85,* 159–177.

Cavaliere, C., (2009). The effect of climate change on medicinal and aromatic plants. *Herbal Gram, 81,* 44–57.

Curtis, J. T., & Cottom, G., (1956). *Plant Ecology Work Book* (p.163). Laboratory field reference manual, Burgess Publication Co., Minnesota.

Curtis, J. T., (1959). *The Vegetation of Wisconsin: An Ordination of Plant Communities* (p.657). University Wisconsin Press, Madison Wisconsin.

Dhar, U., Rawal, R. S., & Samant, S. S., (1997). Structural diversity and representativeness of forest vegetation in a protected area of Kumaun Himalaya, India: Implications for conservation. *Biodiversity and Conservation, 6,* 1045–1062.

Dimri, A. P., & Dash, S. K., (2011). Wintertime climatic trends in the western Himalayas. *Climate Change, 111*(3, 4), 775–800.

Dolezal, J., & Srutek, M., (2002). Altitudinal changes in composition and structure of mountain-temperate vegetation: A case study from the Western Carpathians. *Plant Ecology, 158,* 201–221.

Dubey, B., Yadav, R. R., Singh, J., & Chaturvedi, R., (2003). *Current Science, 85*(8), 1135–1136.

Dutta, P. K., (2007). *Documentation of High-Altitude Wetlands of Western Arunachal Pradesh.* Project Report submitted to WWF-India, Saving Wetland Sky High Program (Unpublished).

Goldsmith, G. W., & Smith, J. H. C., (1926). Some Physico-chemical properties of spruce sap and their seasonal and altitudinal variation. *Colorado College Pub. Stud. Gen. Ser., 137,* 171.

Griggs, R. F., (1938). Timberline in the northern Rocky Mountains. *Ecology, 19,* 548–564.

Henle, K., Davies, K. F., Kleyer, M., Margules, C., & Settele, J., (2004). Predictors of species sensitivity to fragmentation. *Biodiversity and Conservation, 13,* 207–251.

Holtmeier, F. K., (2003). *Mountain Timberlines-Ecology, Patchiness, and Dynamics.* Advances in global change research. Kluwer Academic Publishers, Dordrecht, Boston, London.

Joshi, H. C., & Samant, S. S., (2004). Assessment of forest vegetation and conservation priorities of communities in a part of Nanda Devi Biosphere Reserve, West Himalaya: Part 1. *International Journal of Sustainable Development and World Ecology, 11,* 326–336.

Khan, M. L., Rai, J. P. N., & Tripathi, R. S., (1987). Population structure of some tree species in disturbed and protected subtropical forests of north-east India. *Acta Oecologica, 8*(3), 247–255.

Krauchi, N., Brang, P., & Schonenberger, W., (2000). Forests of mountainous regions: Gaps in knowledge and research needs. *Forest Ecology and Management, 132,* 73–82.

Maletha, A., Maikhuri, R.K. and Bargali, S.S. (2020). Criteria and Indicator for assessing threat on Himalayan birch (B. utilis) at timberline ecotone of Nanda Devi Biosphere Reserve: a world heritage site, Western Himalaya, India. *Environmental and Sustainability Indicators 8* (2020) 100086.

Maletha, A., Maikhuri, R.K., & Bargali, S.S., (2021). Population Structure and Regeneration Pattern of Himalayan Birch (Betula utilis D.Don) at Timberline zone of Nanda Devi Biosphere Reserve, Western Himalaya, India. *Geology, Ecology, and Landscapes. doi.org/ 10.1080/24749508.2021.195276*

Marks, P. L., (1974). The role of pine cherry (*Prunus Pennsylvania* L.) in the maintenance of stability in northern hardwood ecosystems. *Ecological Monographs, 44,* 73–88.

Miyadokoro, T., Nishimura, N., & Yamamoto, S., (2003). Population structure and spatial patterns of major trees in a subalpine old-growth coniferous forest, central Japan. *Forest Ecology and Management, 182*(1–3), 259–272.

Paul, A., (2008). Studies on diversity and regeneration ecology of rhododendrons in Arunachal Pradesh. PhD thesis, Assam University, Silchar, India phytosociological characters. *Ecology, 31,* 434–455.

Pritts, M. P., & Hancock, J. F., (1983). The effect of population structure on growth patterns of the weedy goldenrod Solidago pauciflos close. *Canadian Journal of Botany, 61,* 1955–1958.

Rai, I. D., Adhikari, B. S., Rwat, G. S., & Bargali, K., (2012). Community structure along timberline ecotone in relation to micro-topography and disturbances in western Himalaya. *Nat. Sci. Biol., 4*(2), 41–52.

Rawal, R. S., & Pangtey, Y. P. S., (1994). High altitude forest vegetation with special reference to timberline in Kumaun central Himalaya. In: Pangtey, Y. P. S., & Rawal, R. S., (eds.), *High Altitudes of the Himalaya* (pp. 353–399). Gyanodaya Prakashan, Nainital, India.

Rawat, G. S., Kala, C. P., & Uniyal, V. K., (2001). Plant species diversity and community composition in the valley of flowers, National Park, Western Himalaya. In: Pande, P. C., & Samant, S. S., (eds.), *Plant Diversity of the Himalaya* (pp. 277–290). Gyanodaya Prakashan, Nainital.

Saxena, A. K., & Singh, J. S., (1982). A phytosociological analysis of woody species in forest communities of a part of Kumaun Himalaya. *Vegetatio., 50*, 3–22.

Saxena, A. K., & Singh, J. S., (1984). Tree population structure of certain Himalayan forest associations and implications concerning their future composition. *Vegetatio., 58*, 61–69.

Schickhoff, U., (2005). The upper timberline in the Himalayas, Hindu Kush and Karakorum: A review of geographical and ecological aspects. In: Broll, G., & Keplin, B., (eds.), *Mountain Ecosystems: Studies in Treeline Ecology* (pp. 275–354). Springer, Berlin/Heidelberg/New York.

Shankar, U., (2001). A case of high tree diversity in a Sal (*Shorea robusta*) dominated lowland forest of Eastern Himalaya: Floristic composition, regeneration and conservation. *Current Science, 81*, 776–86.

Shaw, C. H., (1909). The causes of timberline on mountains. *Plant World, 12*, 169–181.

Shekhar, M. S., Chand, H., Kumar, S., Srinivasan, K., & Ganju, A., (2010). Climate change studies in the western Himalaya. *Annals of Glaciology, 51*(54), 105–112.

Shrestha, B. B., Ghimire, B., Lekhak, H. D., & Jha, P. K., (2007). Regeneration of treeline Birch (*Betula utilis* D. Don) forest in a trans-Himalayan dry valley in central Nepal. *Mountain Research and Development, 27*(3), 259–267.

Singh, C. P., Panigrahy, S., Thapliyal, A., Kimothi, M. M., Soni, P., & Parihar, J. S., (2012). Monitoring the alpine treeline shift in parts of the Indian Himalayas using remote sensing. *Current Science, 102*, 559–562.

Singh, J. S., & Singh, S. P., (1992). *Forests of Himalaya: Structure, Functioning and Impact of Man.* Gynodaya Prakashan, Nainital, India.

Singh, P., & Bengtsson, L., (2005). Impact of warmer climate on melt and evaporation for the rainfed, snow-fed and glacier-fed basins in the Himalayan region. *Journal of Hydrology, 300*(1–4), 140–154.

Singh, S. P., Bassignana-Khadka, I., Karky, B. S., & Sharma, E., (2011). *Climate Change in the Hindu Kush-Himalayas: The State of Current Knowledge.* Kathmandu, Nepal: ICIMOD.

Sykes, M. T., & Zimmerman, N. E., (2008). Predicting global change impacts on plant species distributions: Future challenges. *Perspectives in Plant Ecology, Evolution and Systematics, 9*, 137–152.

Sykes, M. T., (2001). Modeling the potential distribution and community dynamics of lodgepole pine (*Pinus contorta* Dougl. ex. Loud.) in Scandinavia. *Forest Ecology and Management, 141*, 69–84.

Teketay, D., (1996). *Seed Ecology and Regeneration in Dry Afromontane Forests of Ethiopia.* PhD thesis, Swedish University of Agricultural Sciences, Umea.

Teketay, D., (1997). Seedling populations and regeneration of woody species in dry Afromontane forests of Ethiopia. *Forest Ecology and Management, 98*(2), 149–165.

Tesfaye, G., Teketay, D., & Fetene, M., (2002). Regeneration of fourteen tree species in Harenna forest, southeastern Ethiopia, *Flora-Morphology, Distribution, Functional Ecology of Plants, 197*(6), 461–474.

Tesfaye, G., Teketay, D., Fetene, M., & Beck, E., (2010). Regeneration of seven indigenous tree species in a dry Afromontane forest, southern Ethiopia. *Flora-Morphology, Distribution, Functional Ecology of Plants, 205*(2), 135–143.

Veblen, T. T., & Stewart, G. H., (1980). Comparison of forest structure and regeneration on Bench and Stewart Islands. *New Zealand Journal of Ecology, 3,* 50–68.

Vetaas, O. R., (2000). The effect of environmental factors on regeneration of *Quercus semecarpifolia* Sm. In central Himalaya, Nepal. *Plant Ecology, 146,* 137–144.

Vijayaprakash, V., & Ansari, A. S., (2009). *Climate Change and of Abies Spectabilis D.Don in the Treeline Areas of Gwang Kharqa in Sankhuwasava District of Eastern Nepal.* MSc Dissertation, Forest and Landscape: Division for Forest Genetic Resources University of Copenhagen.

Wangda, P., & Ohsawa, M., (2006). Structure and regeneration dynamics of dominant tree species along altitudinal gradient in a dry valley slopes of the Bhutan Himalaya. *Forest Ecology and Management, 230*(1–3), 136–150.

Xu, G. J. R. E., Shrestha, A., Eriksson, M., Yang, X., Wang, Y., & Wilkes, A., (2009). The melting Himalayas: Cascading effects of climate change on water, biodiversity and livelihoods. *Conservation Biology, 23,* 520–530.

CHAPTER 6

Herbaceous Diversity Along the Altitudinal Gradient in a Protected Area

JAHANGEER A. BHAT,[1,2] ZUBAIR A. MALIK,[3] MUNESH KUMAR,[4]
MANZOOR A. DAR,[5] A. K. NEGI,[4] and N. P. TODARIA[4]

[1]*Department of Forestry, College of Agriculture, Fisheries and Forestry, Koronivia, PO Box–1544, Nausori, Fiji National University, Republic of Fiji Islands, Fax: +679 340 0275*

[2]*Presently at College of Horticulture and Forestry, Rani Lakshmi Bai Central Agricultural University, Jhansi–284003 (UP), India*

[3]*Department of Botany, Government HSS Harduturoo, Anantnag, Jammu and Kashmir–192201, India*

[4]*Department of Forestry and Natural Resources, HNB Garhwal University, (A Central University), Srinagar, Garhwal, Uttarakhand, India*

[5]*Doon (PG) College of Agriculture, Science and Technology, Camp Road, Central Hope Town–248197, Dehradun, Uttarakhand, India, E-mail: manzoorhmd5@gmail.com*

ABSTRACT

By establishing a huge network of protected areas, a number of initiatives have been taken in order to preserve the biological resources of the Indian Himalayan Region (IHR). The present study was conducted in the Madhmeshwar area of Kedarnath Wildlife Sanctuary (KWLS), which is an integral part of Western Himalaya and is one of the largest protected areas in Garhwal Himalayas. The present study analyzes the diversity of herbaceous flora along the altitudinal gradient in a protected area. The Asteraceae family was observed dominant, representing 58 herb species followed by the family Lamiaceae. The highest density was found in the altitudinal zone-I and lowest in the altitudinal zone-IV. The highest values for Shannon index (H) and Simpson index (Cd) were recorded in the altitudinal zone-V and lowest in the altitudinal zone-IV. The acquaintance of the floristic configuration of a plant community is a requirement to understand the function and structure of any ecosystem. Therefore, the diversity in the present study was observed decreasing with the increasing altitude.

6.1   INTRODUCTION

Mountains, the vulnerable and insecure regions, are the rich repositories of biodiversity. 'Mountain Biodiversity,' one of the initiatives of the Convention on Biological Diversity (CBD) had aimed to minimize the biodiversity loss in the mountains at various levels ranging from regional to global. However, a number of obstacles including those related to climate change come in this initiative of conservation (Nogs-Bravo et al., 2007). A number of conservative strategies have been enforced in the IHR to preserve the biodiversity and minimize its loss. For this purpose, a huge protected area network (PAN) comprising of 5 biosphere reserves, 28 national parks, and 98 wildlife sanctuaries has been established in this region (Mathur et al., 2000).

Kedarnath Wildlife Sanctuary (KWLS) is an integral part of Western Himalaya and is one of the largest protected areas in Garhwal Himalaya. It covers a variety of forest types varying with altitude, and each altitude is associated with several other factors such as; aspect, slope, temperature, and soil type and forest composition. It comprises a wide variety of climates ranging from sub-tropical to alpine and encompasses different altitudes and aspects that give rise to different forest types. Altitude is the main controlling factor for the distribution of different forest types. Other important factors include geology, soils, orientation of valleys, and other biotic and abiotic stresses (Champion and Seth, 1968). The biodiversity and climate change are interconnected, and the later has been declared as one of the main factors that have a negative impact on the former (MEA, 2005). A few studies from Himalaya have also confirmed the fact that climate change has unpleasant effects on biodiversity and its services in Himalayan (Shrestha et al., 1999; Liu and Chen, 2000). Protected areas (PAs) are the primary approach for conservation of biodiversity, and local communities often have to change their behavior after its established, because natural resources become off-limit as they used earlier (Stevens, 1997). During the establishment and management of PAs, communities often have been disregarded by sponsoring organizations or individuals (West and Brechin, 1991; Western and Wright, 1994; Stevens, 1997). Conflicts may arise as a result of restrictions on natural resource use as well as from forceful evictions or other negative relations with PA staff, lack of local participation in conservation, and the absence of open communication and full disclosure of PA related information (Hough, 1988). Conflicts are manifested by a range of behaviors; from local expressions of anti-PA sentiments to intentional burning in PAs and threats of or actual bodily harm to PA staff (Ite, 1996; Brandon et al., 1998; Tello et al., 1998; Peters, 1999). PA managers face the dilemma of managing both conservation of the biodiversity within their jurisdiction and local community interests and resource needs. People living in and around PAs often impact the ability of the PA to meet conservation objectives. The attitudes of local residents as well as the interactions, level of local participation and conflicts between people and PAs have become a concern of PA effectiveness (West and Brechin, 1991; Wells and Brandon, 1992; Western and Wright, 1994; Pimbert and Pretty, 1995; Kramer et al., 1997; Stevens, 1997; Brandon et al., 1998).

The structure of vegetation of an area is based on the quality of floristic elements. The dominating plant species determine the structure of a community (Hanson and Churchill, 1961). The species that exert major controlling influence by their number, size, distribution pattern, IVI (importance value index) or other activities are described as dominants (Odum, 1971). The versatility of the species contributes to the magnitude of the diversity in a particular habitat. Diversity of species along with population size and dimension constitute the remarkable biological elements of any ecosystem. Each constituent species has not only its own ecological amplitude but also its particular relationship to the environment and to the associate species (Bliss, 1962). There are several ecological studies on plant communities in Garhwal Himalayas but the studies pertaining to high altitudinal zones of protected areas are meager. Therefore, keeping in view the above said factors, the present study, diversity of herbaceous flora along the altitudinal gradient in a protected area (KWLS) was carried out.

6.2 METHODOLOGY

KWLS was established in 1972 and located in the Northeastern part of Garhwal Himalayas and geographically it is situated between 30°25'–30°41' N, 78°55'–79°22' E (Bhat et al., 2013a). The KWLS is among the largest PA (975 km²) 25293.70 ha situated in the district Chamoli of 72224.10 ha in Rudraprayag district in the Western Himalayas. The sanctuary is located in the upper catchment of Alaknanda and Mandakini Rivers are the tributaries of river Ganges (Malik et al., 2014). The area is surrounded by high peaks of mountains, i.e., Mandani (6193 m), Kedarnath (6940 m) and Chaukhamba (7068 m) with alpine meadows such as Trijuginarayan, Kham, Mandani, Pandavshera, Manpai, and Bansinarayan in the north, and several dense broad leave Oak mixed forests in the south (Bhat et al., 2012, 2013b). The study was carried out from the base area of mountain to top peak Madhmaheshwer area between the coordinates (30° 35' 42"–30° 38' 12" N, 79° 10' 00"–79° 13' 00" E).

Based on the reconnaissance and preliminary survey in KWLS, Madhmeshwar area was selected for the study. During field surveys, topo-sheets, and maps were used to verify the altitudinal gradient of the area. On the basis of altitudinal gradient, five altitudinal zones above mean sea level were selected for herbaceous diversity. (i) Altitudinal Zone-I, i.e., AZ-I (1550–1750 m), (ii) Altitudinal Zone-II, i.e., AZ-II (2000–2200 m), (iii) Altitudinal zone-III, i.e., AZ-III (2450–2650 m), (iv) Altitudinal zone-IV, i.e., AZ-IV (2900–3100 m) and (v) Altitudinal Zone-V, i.e., AZ-V (3350–3550 m). Stratified random sampling technique was used, placing random quadrats for analysis. The minimum sample size was used based on the species-area curve method. An altitudinal interval of 200 m between every altitudinal zone was decided to assess the distribution pattern of species (Bhat et al., 2012a). The size of the plots was one-meter square used to enumerate herbs, and percent cover of herb layer species was estimated by plotting eighty (80) of 1 m^2 quadrats covering an area of 80 m^2 at every altitudinal zone.

From each sampling site, the voucher specimen of plant species was observed, collected, and identified on regional floras. Some plants in the field were also identified by the matching of plants with the pictorial field guide. Voucher specimens collected during the surveys were mounted on herbarium sheets after processing (pressing, drying, and poisoning) in the laboratory the collect specimens for all species were deposited in Botanical Survey of India (BSD, BSI-North circle, Dehradun) and Garhwal University Herbarium (GUH, HNBGU Garhwal Srinagar) and the specimens were registered with the index Herbariorum with acronym BSD and GUH. A unique identity number was generated and the identity of species cross-checked and confirmed.

Standard methods (Curtis, 1950; Phillips, 1959; Whittaker, 1972) were adopted for the analysis of vegetation parameters such as species richness (SR), density, and diversity at each altitudinal zone. The importance value index (IVI) reflects the relative importance of a species in a plant community and calculated sum of the relative frequency, relative density, and relative dominance (Phillips, 1959). Shannon-Wiener diversity (H) was calculated as methods described by Shannon and Weaver (1963) and concentration of dominance (CD) using method of Simpson (1949), Pielou equitability by Pielou (1966), Margalef diversity as describe by Margalef (1958), the indices of SR and beta diversity (Whittaker, 1972).

6.3 RESULTS

A total of 243 herb species belonging to 161 genera and 63 families were reported during the present study. The dominant family was Asteraceae with 58 herb species followed by Lamiaceae (15 species). The herb density was highest in the AZ-I (1550–1750 m) and lowest in the AZ-IV of 2900–3100 m (Table 6.1). In herbaceous layer the highest values for Shannon index (H) and Simpson index (Cd) were recorded in the AZ-V (3350–3550 m) and in the AZ-IV (2900–3100 m) respectively, while the lowest values were observed in the AZ-V (2900–3100 m) and in AZ-V (3350–3550 m) for Shannon index (H) and Simpson index (Cd) respectively (Table 6.1).

TABLE 6.1 Density and Different Diversity Indices of Herbs in All Altitudinal Zones

Parameters	AZ-I (1550– 1750 m)	AZ-II (2000– 2200 m)	AZ-III (2450– 2650 m)	AZ-IV (2900– 3100 m)	AZ-V (3350– 3550 m)
Herb density/m²	19.74	15.69	11.98	11.54	16.00
Shannon index (H′)	5.608	5.622	5.500	5.071	5.846
Simpson index (Cd)	0.032	0.031	0.026	0.039	0.024
Equitability (EC)	29.186	32.640	29.786	21.064	34.776
Beta-diversity (βD)	11.152	11.670	10.360	9.148	9.737
Margalef index (SR)	9.233	9.531	7.720	6.151	10.203

6.3.1 ALTITUDINAL ZONE-I (1550–1750 M ASL)

The herb (*Hackelochloa granularis*) was found dominant with maximum values of IVI (16.01) and density (2.36 plants/m²) followed by *Pogonatherum paniceum* (IVI, 7.46 and density, 0.88 plants/m²). The least dominant herb was *Valeriana hardwickii* (IVI, 0.46). The lowest density (0.05 plants/m²) was found for both *Pteris pseudo-quadrianrita* and *Valeriana hardwickii*. The highest (25%) frequency was for *Hackelochloa granularis* followed by *Cyanotis vaga* (20%) and the lowest for *Persicaria capitata* and *Valeriana hardwickii* with the value of 1.25% for each (Table 6.2). The different diversity indices recorded for herb layer were: Shannon (5.608), Simpson (0.032), Equitability (29.186), Beta-diversity (11.152), and Margalef (9.233) (Table 6.1).

TABLE 6.2 Phytosociological Data of Herb Species at AZ-I (1550–1750 m asl)

Species	Frequency (%)	Density (m⁻²)	Importance Value Index
Anaphalis triplinervis (Sims.) C.B. Clarke	6.25	0.21	2.09
Anemone narcissiflora L.	8.75	0.23	2.55
Apluda mutica L.	3.75	0.44	2.82
Artemisia vestita Wall. ex DC.	6.25	0.38	2.91
Arundinella setosa Trin.	12.50	0.51	4.62
Aster peduncularis Wallich	15.00	0.30	3.94
Athyrium pectinatum Wall. ex Mett.	12.50	0.45	4.30
Athyrium schimperii Moug. ex Fee.	13.75	0.53	4.88
Barleria cristata L.	10.00	0.25	2.88
Berberis vulgaris L.	8.75	0.23	2.55
Bidens biternata (Lour.) Merr. and Sherff	5.00	0.20	1.82
Bidens chinensis L.	11.25	0.24	3.02
Blainvillea acmella (L.f.) Philipson	13.75	0.38	4.12
Blumea hieraciifolia (D.Don) DC	8.75	0.15	2.17
Blumea lanceolaria (Roxb.) Druce	7.50	0.20	2.23
Chrysopogon serrulatus Trin.	12.50	0.29	3.48
Clematis montana Buch.-Ham. ex DC.	11.25	0.33	3.46
Clinopodium umbrosum (M.Bieb.) C. Koch	3.75	0.09	1.05
Cyanotis vaga (Lour.) J.A. and J.H. Schult.	20.00	0.20	4.25
Cyathula capitata Moq.	2.50	0.06	0.72
Cyathula tomentosa Moq.	15.00	0.59	5.40
Cymbopogon citratus (DC.) Stapf	6.25	0.39	2.97
Cynoglossum lanceolatum Forssk.	8.75	0.26	2.74
Diplazium frondosum Wall. ex Clarke.	13.75	0.24	3.43
Elephantopus scaber L.	7.50	0.13	1.85
Erianthus ruflinervis DC.	13.75	0.39	4.19
Eupatorium chinense L.	8.75	0.23	2.55
Euphorbia chamaesyce L.	8.75	0.14	2.11
Euphorbia pilosa Linn.	3.75	0.09	1.05
Fragaria nubicola Lindl. ex Lacaita	12.50	0.13	2.65
Gaultheria nummularioides D.Don	8.75	0.23	2.55
Gerbera gossypina (Royle) P. Beauv.	11.25	0.21	2.89

TABLE 6.2   *(Continued)*

Species	Frequency (%)	Density (m⁻²)	Importance Value Index
Girardiana diversifolia (Link) Friis	12.50	0.61	5.12
Gomphrena celosioides Mart.	7.50	0.18	2.10
Gonatanthus pumilus (D.Don) Engl. and Krause	16.25	0.93	7.31
Gonostegia hirta (Blume) Miq	2.50	0.08	0.78
Hackelochloa granularis Kuntze	25.00	2.36	16.01
Helictotrichon virescens (Nees ex Steud.) Henr.	8.75	0.16	2.24
Hemiphragma hetrophyllum Wallich	3.75	0.09	1.05
Hypolepis punctata Thunb.	10.00	0.16	2.44
Nepeta laevigata (D.Don) Hand.-Mazz.	12.50	0.41	4.11
Onychium cryptogrammoides Christ.	3.75	0.06	0.92
Paeonia emodi Wall. ex Royle	5.00	0.18	1.69
Pentanema indicum (L.) Ling	12.50	0.19	2.97
Persicaria capitata (Buch.-Ham. ex D. Don) H. Gross	1.25	0.09	0.65
Phalaris minor Retz.	6.25	0.33	2.66
Pogonatherum paniceum (Lam.) Hack.	18.75	0.88	7.46
Polygonum recumbens Royle ex Babcock	11.25	0.15	2.58
Polystichum discretum D.Don	7.50	0.11	1.78
Pteridium revolutum (Bl.) Nakai.	6.25	0.31	2.59
Pteris cretica L.	7.50	0.24	2.42
Pteris pseudo-quadrianrita Khuller.	3.75	0.05	0.86
Ranunculus membranaceous Royle.	8.75	0.13	2.05
Ranunculus trichophyllus Chaix.	3.75	0.14	1.30
Rumex hastatus D. Don	12.50	0.69	5.50
Solanum suratteuse Burm.	8.75	0.16	2.24
Sphaeranthus indicus L.	7.50	0.14	1.91
Teucrium royleanum Wall. ex Benth.	8.75	0.14	2.11
Thalictrum virgatum Hook.f. and Thomson	8.75	0.20	2.43
Triumfetta annua L.	6.25	0.08	1.39
Triumfetta rhomboidea Jacq.	8.75	0.21	2.49
Urena lobata L.	6.25	0.15	1.77
Urtica ardens Link.	2.50	0.20	1.42
Urtica dioica L.	7.50	0.49	3.68
Valeriana hardwickii Wallich	1.25	0.05	0.46
Verbascum thapsus L.	2.50	0.10	0.92
Veronica anagallis-aquatica Linn.	2.50	0.10	0.91
Xanthium pungens Wallr.	16.25	0.16	3.45
Zingiber roseum (Roxb.) Rosc.	11.25	0.43	3.97
		19.74	

6.3.2   ALTITUDINAL ZONE-II (2000–2200 M ASL)

Ainsliaea latifolia was found dominant species (IVI, 13.78) followed by *Pogonatherum paniceum* (IVI, 12.88) and *Rumex nepalensis* (IVI, 11.61). The least dominant herb was *Phytolacca acinosa* (IVI, 0.37). The highest density (1.33 plants/m²) was found for *Pogonatherum paniceum* followed by *Ainsliaea latifolia* (1.30 plants/m²) and *Rumex nepalensis* (1.13 plants/m²) while lowest density (0.03 plants/m²) was observed for *Phytolacca acinosa*. The highest (32.50%) frequency was again observed for *Ainsliaea latifolia* followed by *Pogonatherum paniceum* and *Rumex nepalensis* with a frequency value of 26.25% for each species, while the lowest (1.25%) frequency was observed for *Nepeta laevigata, Paeonia emodi,* and *Phytolacca acinosaies* (Table 6.3). The different diversity indices recorded for herb layer were: Shannon as (5.622), Simpson (0.031), Equitability (32.640), Beta-diversity (11.670), and Margalef (9.531) (Table 6.1).

TABLE 6.3   Phytosociological Data of Herb Species at AZ-II (2000–2200 m asl)

Species	Frequency (%)	Density (m⁻²)	Importance Value Index
Ainsliaea latifolia (D. Don) Sch.-Bip.	32.50	1.30	13.78
Anaphalis triplinervis (Sims.) C.B. Clarke	8.75	0.20	2.75
Anemone rivularis Buch.-Ham. ex DC	3.75	0.10	1.27
Artemisia japonica Thunb.	13.75	0.34	4.48
Artemisia vestita Wall. ex DC.	7.50	0.21	2.62
Asparagus filicinus Buch.-Ham. ex D. Don	3.75	0.05	0.95
Athyrium pectinatum Wall. ex Mett.	16.25	0.49	5.86
Athyrium schimperii Moug. ex Fee.	5.00	0.15	1.80
Berberis vulgaris L.	8.75	0.18	2.60
Bergenia ciliata (Haw.) Sternb.	3.75	0.08	1.11
Bidens bipinnata L.	6.25	0.15	2.01
Bidens pilosa L.	12.50	0.24	3.63
Bupleurum falcatum L.	8.75	0.18	2.60
Campanula pallida Wallich	13.75	0.30	4.24
Cannabis sativa L.	6.25	0.25	2.65
Cautleya spicata Baker	11.25	0.36	4.21
Clematis grata Wallich	5.00	0.26	2.52
Clinopodium umbrosum (M. Bieb.) C. Koch	8.75	0.16	2.52

TABLE 6.3 *(Continued)*

Species	Frequency (%)	Density (m⁻²)	Importance Value Index
Conyza canadensis (L.) Cronquist	7.50	0.13	2.07
Craniotome furcata (Link) O. Kuntze	8.75	0.19	2.67
Cyathula capitata Moq.	7.50	0.24	2.78
Cymbopogon citratus (DC.) Stapf	6.25	0.25	2.65
Cynoglossum glochidiatum Wall. ex Benth.	8.75	0.21	2.83
Dicliptera bupleuroides Nees	12.50	0.20	3.39
Dipteracanthus prostratus (Poir.) Nees	5.00	0.24	2.36
Dryopteris caroli-hopei Fraser-Jenkins.	10.00	0.25	3.28
Dryopteris chrysocoma (H. Christ.) C. Chr.	5.00	0.18	1.96
Elsholtzia ciliata (Thunb.) Hylander	7.50	0.24	2.78
Epilobum royleanum Haussk.	8.75	0.15	2.44
Erigeron floribundus (H.B.K.) Sch.-Bip.	7.50	0.16	2.30
Eupatorium chinense L.	5.00	0.11	1.56
Fagopyrum dibotrys (D. Don) Hara	7.50	0.23	2.70
Fragaria nubicola Lindl. ex Lacaita	7.50	0.18	2.38
Galium asperuloides Edgew.	6.25	0.14	1.93
Gentianella maddeni (C. B Clarke) Airy-Shaw	8.75	0.24	2.99
Geranium rotundifolium L.	2.50	0.08	0.90
Girardiana diversifolia (Link) Friis	6.25	0.25	2.65
Gnaphalium affine D. Don	7.50	0.15	2.22
Gnaphalium pensylvanicum Willd.	8.75	0.15	2.44
Gnaphalium purpureum L.	11.25	0.20	3.18
Hackelochloa granularis Kuntze	11.25	0.64	5.97
Hemiphragma hetrophyllum Wallich	3.75	0.05	0.95
Hypericum elodeoides Choisy	7.50	0.14	2.14
Hypodematum crenatum Forsk.	8.75	0.13	2.28
Lamium album L.	6.25	0.11	1.77
Leucostegia immersa Wallich	10.00	0.21	3.05
Morina longifolia Wall. ex DC.	3.75	0.13	1.43
Nepeta ciliaris Benth.	11.25	0.16	2.94
Nepeta laevigata (D. Don) Hand.-Mazz.	1.25	0.06	0.61
Onychium Cryptogrammoides Christ.	12.50	0.20	3.39

TABLE 6.3   *(Continued)*

Species	Frequency (%)	Density (m⁻²)	Importance Value Index
Paeonia emodi Wall. ex Royle	1.25	0.05	0.53
Persicaria glabra (Willd.) Gomes	8.75	0.14	2.36
Persicaria hydropiper (L.) Spach	8.75	0.31	3.47
Phytolacca acinosa Roxb.	1.25	0.03	0.37
Pilea umbrosa Wedd.	7.50	0.15	2.22
Pimpinella acuminata (Edgew.) C.B. Clarke	12.50	0.15	3.07
Pimpinella diversifolia DC.	7.50	0.13	2.07
Pogonatherum paniceum (Lam.) Hack.	26.25	1.33	12.88
Polystichum squarrosum D. Don	8.75	0.16	2.52
Pteracanthus angustifrons (C.B. Clarke) Bremek.	11.25	0.15	2.86
Ranunculus laetus Wall. ex D. Don	11.25	0.18	3.02
Reinwardtia indica Dumort.	5.00	0.08	1.32
Rumex nepalensis Spreng.	26.25	1.13	11.61
Saussurea albescens (DC.) Sch.-Bip.	6.25	0.09	1.61
Scrophularia polyantha Royle ex Benth.	10.00	0.15	2.65
Urtica dioica L.	6.25	0.25	2.65
Verbascum thapsus L.	3.75	0.05	0.96
Vernonia cinerea (L.) Less.	7.50	0.15	2.22
Zingiber roseum (Roxb.) Rosc.	2.50	0.09	0.98
		15.69	

6.3.4   *ALTITUDINAL ZONE-III (2450–2650 M ASL)*

Pilea umbrosa was found dominant species with highest values of IVI (10.32) and frequency (28.75%) followed by *Rumex nepalensis* with IVI (10.09) and frequency (23.75%). The least dominant herb was *Campanula pallida* (IVI, 1.35) followed by *Pimpinella acuminata* (IVI, 1.48). The highest density (0.71 plants/m²) was found in *Saussurea albescens* followed by *Rumex nepalensis* (0.66 plants/m²) while the lowest density 0.06 plants/m² was observed for *Pimpinella acuminate* (Table 6.4). The different diversity indices recorded for herb layer were: Shannon (5.500), Simpson (0.026), Equitability (29.786), Beta-diversity (10.360), and Margalef (7.720) (Table 6.1).

TABLE 6.4 Phytosociological Data of Herb Species at Altitudinal Zone-III (2450–2650 m asl)

Species	Frequency (%)	Density (m^{-2})	Importance Value Index
Ainsliaea latifolia (D. Don) Sch.-Bip.	11.25	0.16	3.51
Anaphalis contorta (D. Don) Hook.f.	10.00	0.20	3.59
Anaphalis margaritaceae (L.) Benth	6.25	0.16	2.60
Anemone rivularis Buch.-Ham. ex DC	11.25	0.26	4.35
Arisaema flavum (Forssk.) Schott	11.25	0.30	4.66
Arisaema jacquemontii Blume	8.75	0.23	3.56
Artemisia indica willd.	16.25	0.43	6.67
Artemisia roxburghiana Bess.	7.50	0.26	3.63
Artemisia vestita Wall. ex DC.	5.00	0.09	1.69
Asparagus filicinus Buch.-Ham. ex D. Don	10.00	0.33	4.63
Bergenia ciliata (Haw.) Sternb.	10.00	0.18	3.38
Bistorta amplexicaulis (D. Don) Greene	10.00	0.36	4.94
Campanula pallida Wallich	3.75	0.08	1.35
Chaerophyllum villosum DC.	7.50	0.18	2.90
Corallodiscus lanuginosus (Wall. ex DC.) B.L. Burtt	12.50	0.16	3.75
Dipsacus inermis Wallich	5.00	0.09	1.69
Dryopteris nigropaleacea Fraser-Jenkins.	8.75	0.20	3.35
Epilobum royleanum Haussk.	5.00	0.10	1.79
Erigeron karvinskianus DC.	12.50	0.44	6.05
Euphorbia pilosa Linn.	8.75	0.25	3.77
Fagopyrum dibotrys (D. Don) Hara	7.50	0.14	2.59
Filipendula vestita (Wall. ex G. Gon) Maxim.	3.75	0.10	1.55
Fragaria nubicola Lindl. ex Lacaita	10.00	0.20	3.59
Galinsoga parviflora Cav.	8.75	0.18	3.14
Galium asperuloides Edgew.	6.25	0.15	2.45
Geranium wallichianum D. Don ex Sweet	8.75	0.26	3.87
Gnaphalium affine Var. royleana D. Don	11.25	0.23	4.04
Impatiens brachycentra Kar. and Kir.	3.75	0.10	1.55
Lactuca sativa L.	13.75	0.29	5.04
Leucas lanata Benth.	7.50	0.18	2.90
Lindenbergia indica (L.) Vatke	12.50	0.24	4.38
Morina longifolia Wall. ex DC.	3.75	0.15	1.97
Onychium contiguum Wall. ex lope.	8.75	0.26	3.87
Pilea umbrosa Wedd.	28.75	0.58	10.32

TABLE 6.4 *(Continued)*

Species	Frequency (%)	Density (m^{-2})	Importance Value Index
Pimpinella acuminata (Edgew.) C.B. Clarke	5.00	0.06	1.48
Polygonatum verticillatum (L.) All.	8.75	0.11	2.62
Pteris cretica L.	5.00	0.13	2.00
Pteris wallichiana Agasdh.	17.50	0.28	5.65
Reinwardtia indica Dumort.	7.50	0.10	2.27
Rhodiala sinuata (Royle ex Edgew.) Fu.	11.25	0.19	3.72
Rumex nepalensis Spreng.	23.75	0.66	10.09
Saussurea albescens (DC.) Sch.-Bip.	16.25	0.71	9.06
Selinum wallichianum (DC.) Raizada and Saxena	10.00	0.15	3.17
Senecio rufinervis DC.	5.00	0.18	2.42
Silene edgeworthii Bocquet.	12.50	0.23	4.28
Slleggenella jacquemontii Spring.	8.75	0.13	2.72
Solidago virgaurea L.	8.75	0.13	2.72
Stachys sericea Wall. ex Benth	7.50	0.15	2.69
Synotis alatus (Wall. ex DC.) C. Jeffrey and Chen.	7.50	0.13	2.48
Tagetus minuta L.	8.75	0.14	2.83
Thalictrum reniforme Wallich	10.00	0.20	3.59
Verbascum thapsus L.	5.00	0.08	1.59
Vernonia anthelmintica (L.) Willd.	7.50	0.16	2.80
Veronica anagallis-aquatica Linn.	18.75	0.61	8.71
		11.98	

6.3.5 ALTITUDINAL ZONE-IV (2900–3100 M ASL)

Rubus nepalensis was the dominant species with the highest IVI (18.21), density (1.15 plants/m^2) and frequency (38.75%) followed by *Ainsliaea apetra* with IVI (14.88), density (0.95 plants/m^2) and frequency (31.25%). The least dominant herb species was *Impatiens scabrida* (IVI, 1.56) followed by *Bergenia ciliata* (IVI, 1.58). The lowest density (0.05 plants/m^2) and lowest frequency (2.50%) were observed for *Artemisia roxburghiana* followed by *Bergenia ciliata* and *Impatiens scabrida* with density (0.09 plants/m^2) and frequency (3.75%) values for each of them (Table 6.5). The different diversity indices recorded for herb layer are Shannon (5.071), Simpson (0.039), Equitability (21.064), Beta-diversity (9.148), and Margalef (6.151) (Table 6.1).

TABLE 6.5 Phytosociological Data of Herb Species at AZ-IV (2900–3100 m asl)

Species	Frequency (%)	Density (m⁻²)	Importance Value Index
Acomastylis elata (Wall. ex Royle) F. Bolle	5.00	0.15	2.36
Ainsliaea apetra DC.	31.25	0.95	14.88
Anaphalis contorta (D. Don) Hook.f.	23.75	0.69	11.01
Anaphalis cuneifolia (DC.) Hook.f.	20.00	0.29	6.75
Anaphalis margaritaceae (L.) Benth	10.00	0.16	3.54
Anaphalis nubigena DC.	8.75	0.44	5.65
Anaphalis royleana DC.	12.50	0.28	5.04
Androsace lanuginosa Wallich	5.00	0.25	3.23
Anemone polyanthes D. Don	12.50	0.29	5.15
Arisaema jacquemontii Blume	11.25	0.29	4.88
Artemisia roxburghiana Bess.	2.50	0.05	0.97
Begonia picta Smith	8.75	0.21	3.70
Bergenia ciliata (Haw.) Sternb.	3.75	0.09	1.58
Bistorta amplexicaulis (D. Don) Greene	21.25	0.61	9.83
Caltha palustris L.	10.00	0.24	4.19
Chenopodium album L.	6.25	0.26	3.60
Cyananthus lobatus Wall. ex Benth.	13.75	0.16	4.33
Drynaria mollis Bedd.	7.50	0.21	3.44
Elsholtzia strobilifera Benth.	10.00	0.25	4.29
Euphorbia pilosa Linn.	12.50	0.38	5.91
Fagopyrum esulentum Moench	11.25	0.29	4.88
Filipendula vestita (Wall. ex G. Gon) Maxim.	12.50	0.48	6.78
Galium aparine L.	8.75	0.20	3.59
Geranium wallichianum D. Don ex Sweet	5.00	0.15	2.36
Herminium monorchis (L.) R. Br.	7.50	0.10	2.46
Impatiens brachycentra Kar. and Kir.	10.00	0.35	5.16
Impatiens scabrida DC.	3.75	0.09	1.56
Lactuca graciliflora DC.	7.50	0.13	2.68
Morina longifolia Wall. ex DC.	6.25	0.11	2.30
Osmunda claytoniana L.	5.00	0.25	3.23
Parnassia nubicola Wall. ex Royle	15.00	0.21	5.03
Pennisetum flaccidum Griseb.	7.50	0.11	2.57
Pilea umbrosa Wedd.	7.50	0.10	2.46
Polypodium lachnopus Wall. ex Hook.	7.50	0.10	2.46

TABLE 6.5   *(Continued)*

Species	Frequency (%)	Density (m⁻²)	Importance Value Index
Roscoea alpina Royle	13.75	0.19	4.55
Rubus nepalensis (Hook.f.) Kuntze	38.75	1.15	18.21
Saxifraga diversifolia Wall. ex Ser.	10.00	0.15	3.43
Silene edgeworthii Bocquet.	7.50	0.14	2.79
Swertia chirayita (Roxb. ex Fleming) Karsten	12.50	0.21	4.50
Thermopsis barbata Royle	8.75	0.18	3.38
Verbascum thapsus L.	11.25	0.30	4.99
Viola canescens Wallich	10.00	0.19	3.75
Woodsia elongata Hook.	6.25	0.14	2.52
			11.54

6.3.6   *ALTITUDINAL ZONE-V (3350–3550 M ASL)*

Juncus himalensis was found dominant species with highest values of IVI (11.47) and density (1.39 plants/m²) followed by *Rumex nepalensis* having IVI (8.07) and density (0.71 plants/m²). The least dominant herb species was *Stellaria alsine* with IVI (0.56) followed by *Bistorta amplexicaulis*. The lowest density (0.04 plants/m²) was found for *Lindelofia stylosa* and *Stellaria alsine*. The highest frequency (27.50%) was observed in *Rumex nepalensis* followed by *Senecio chrysanthemoides* (23.75%) while the lowest (2.50%) frequency was recorded in *Bistorta amplexicaulis, Hackelia uncinata, Impatiens brachycentra,* and *Stellaria alsine* (Table 6.6). The different diversity indices recorded for herb layer were Shannon (5.846), Simpson (0.024), Equitability (34.776), Beta-diversity (9.737), and Margalef (10.203) (Table 6.1).

TABLE 6.6   Phytosociological Data of Herb Species at AZ-V (3350–3550 m asl)

Species	Frequency (%)	Density (m⁻²)	Importance Value Index
Acomastylis elata (Wall. ex Royle) F. Bolle	6.25	0.13	1.60
Aconitum hetrophyllum Wallich	18.75	0.33	4.50
Anaphalis nepalensis (Spreng.) Hand.-Mazz.	8.75	0.24	2.64
Anaphalis nubigena DC.	10.00	0.29	3.11
Anemone obtusiloba D. Don	15.00	0.26	3.61

TABLE 6.6 *(Continued)*

Species	Frequency (%)	Density (m⁻²)	Importance Value Index
Artemisia criocephala Pamp.	20.00	0.48	5.60
Bistorta amplexicaulis (D. Don) Greene	2.50	0.05	0.64
Bistorta macrophylla (D. Don) Sojak	12.50	0.21	2.97
Bistorta vaccinifolia (Wall. ex Meisn.) Greene.	7.50	0.13	1.77
Botrychium virginianum (Linn.) Swartz	6.25	0.19	1.99
Calanthe tricarinata Lindl.	11.25	0.20	2.73
Cremanthodium arnicoides (DC.) Good	11.25	0.21	2.81
Cynoglossum wallichii G. Don.	11.25	0.19	2.65
Cystopteris dickieana Sim.	8.75	0.14	2.01
Danthonia cachemyriana Jaub. and Spach.	21.25	0.34	4.91
Delphinium vestitum Wall. ex Royle	8.75	0.16	2.17
Dryopteris odonotoloma (Moore.) C. Chr.	12.50	0.43	4.30
Dryopteris wallichiana Fraser-Jenkins.	6.25	0.25	2.38
Dubyaea hispida (D. Don) DC.	3.75	0.16	1.51
Elsholtzia strobilifera Benth.	3.75	0.10	1.12
Epilobium cylindricum D. Don.	7.50	0.15	1.92
Euphorbia hypericifolia L.	6.25	0.16	1.84
Filipendula vestita (Wall. ex G. Gon) Maxim.	7.50	0.19	2.16
Galium aparine L.	6.25	0.13	1.60
Galium asperifolium Wallich.	3.75	0.10	1.12
Hackelia uncinata (Royle ex Benth.) Fisch	2.50	0.06	0.72
Impatiens brachycentra Kar. and Kir.	2.50	0.09	0.88
Juncus himalensis Klotz.	21.25	1.39	11.47
Jurinea dolomiaea Boiss.	5.00	0.08	1.13
Ligularia amplexicaulis DC.	13.75	0.25	3.37
Ligularia fischeri (Ledeb.) Turcz.	3.75	0.10	1.12
Lindelofia longiflora (Benth.) Baill.	6.25	0.11	1.53
Lindelofia stylosa (Kar. and Kir.) Brand	3.75	0.04	0.73
Lysimachia prolifera Klatt.	5.00	0.08	1.13
Maianthemum purpureum (Wall.) La Frankie	3.75	0.05	0.81
Nepeta govaniana (Benth.) Benth.	13.75	0.20	3.06
Nomocharis oxypetala (Royle.) E.H. Wilson.	11.25	0.13	2.26
Origanum vulgare L.	3.75	0.09	1.04
Pedicularis hoffmeisteri Klotz.	12.50	0.21	2.97
Pedicularis pectinata Wall. ex Benth.	10.00	0.16	2.33

TABLE 6.6   *(Continued)*

Species	Frequency (%)	Density (m^{-2})	Importance Value Index
Pennisetum flaccidum Griseb.	8.75	0.26	2.79
Phlomis bracteosa Royle ex Benth.	7.50	0.18	2.08
Picrorhiza kurrooa Royle ex Benth.	12.50	0.23	3.05
Pimpinella diversifolia DC.	8.75	0.19	2.32
Plantago depressa Willd.	10.00	0.16	2.33
Plantago himalaica Pilger.	10.00	0.20	2.57
Podophyllum hexandrum Royle.	11.25	0.19	2.65
Polygonum tortuosum D. Don	10.00	0.20	2.57
Potentilla argyrophylla Wall. ex Lehm.	16.25	0.28	3.86
Potentilla atrisanguinea Lodd. ex Lehm	20.00	0.36	4.90
Potentilla sericea L.	10.00	0.14	2.18
Primula denticulata Sm.	18.75	0.31	4.42
Ranunculus hirtellus Royle.	20.00	0.38	4.98
Rhodiala himalensis (D. Don) Fu	15.00	0.28	3.69
Rubus nepalensis (Hook.f.) Kuntze	5.00	0.11	1.36
Rumex nepalensis Spreng.	27.50	0.71	8.07
Salvia hians Royle ex Benth.	12.50	0.21	2.97
Salvia nubicola Wall. ex Sw.	3.75	0.06	0.88
Saussurea auriculata (Spreng. ex DC.) Sch.-Bip.	13.75	0.19	2.98
Saussurea piptathera Edgew.	12.50	0.19	2.82
Saussurea roylei (DC.) Sch.-Bip.	15.00	0.24	3.46
Saxifraga heterophylla Sternb.	7.50	0.13	1.77
Selinum candollii DC.	8.75	0.20	2.40
Senecio chrysanthemoides DC.	23.75	0.64	7.11
Senecio graciliflorus DC.	8.75	0.15	2.09
Senecio laetus Edgew.	13.75	0.31	3.76
Sibbaldia parviflora Willd.	10.00	0.18	2.41
Stellaria alsine Grimm.	2.50	0.04	0.56
Stellaria decumbens Edgew.	5.00	0.05	0.97
Swertia alternifolia Royle	17.50	0.48	5.27
Swertia ciliata (G. Don) Burtt.	18.75	0.30	4.34
Taraxacum officinale Weber.	7.50	0.13	1.77
Trachydium roylei Lindl.	5.00	0.05	0.97
Woodsia elongata Hook.	6.25	0.10	1.45
		16.00	

6.4 DISCUSSION

Understanding the floristic composition is essential to comprehend the structure and function of any ecosystem (Bhat, 2012a; Malik and Bhatt, 2015). The number of species (often called SR) is a very simple way to describe the plant communities (Magurran, 1988). The main mechanism responsible for the altitudinal changes in SR is attributed to the altitudinal dependence of productivity. An increase in productivity may cause a gradual increase in SR (MacArthur, 1965) or may result in a decline after reaching a point where productivity exceeds its mean value (Rosenzweig, 1995).

In the present study, a total of 243 herb species were recorded from five (5) zones along an altitudinal gradient. Pokhriyal et al. (2009) reported a total of 179 species of which were 27 trees, 21 shrubs and 39 herb species in Pathri Rao and 24 trees, 23 shrubs and 45 herbs in Phakot watershed of Garhwal Himalaya. Uniyal et al. (2010) reported a total of 182 species belonging to 144 genera in which 24 were trees, 55 shrubs, and 103 herbs in Oak forest of Dewalgarh watershed of Garhwal Himalaya. Pant and Samant (2007) reported 289 species in which 37 were trees, 37 shrubs, and 215 herbs in Mornaula Reserve forest in western Himalaya between 1500–2200 m altitudes. Chandra et al. (2010) reported 209 plant species, out of which 29 were trees, 50 shrubs, 102 herbs, 11 climbers, 7 epiphytes, 4 pteridophytes, 3 bryophytes, and 3 parasites along an altitudinal gradient of Garhwal Himalaya. The differences in a number of species from previous studies with present study may be due to difference in locality factors like temperature, altitude, humidity, rainfall, aspect, and interaction of other site factors. The variation in a number of species may be also due to the size and type of the forest as well as the size of the altitudinal gradient. The altitudinal gradient in the present study was broad and large as compared to the previous studies.

The study of Chandra et al. (2010) along altitudinal gradient in Garhwal Himalaya reported that there are a large number of environmental factors that may influence the SR and composition, such as elevation and habitat. The study of SR patterns is very common in ecological and biogeographical investigation, and in general, the view held is that SR decreases with an increase in altitude (MacArthur, 1972; McCoy, 1990). The SR in the present study for trees, shrubs, and herbs also decreased with increasing altitude. According to Grabherr et al. (1995), who worked on southern Norway along an elevational gradient, upper limits are obviously set by climatic severity and resource restrictions, which have been demonstrated for many plant species along gradients approaching to nival zone. Grytnes and Vetaas (2002) in Nepal Himalaya have reported in decreasing trend of SR with altitude. The linearly decreasing in SR from 85 to 30 species with increasing altitude

(Aubry et al., 2005). Sklenar (2006) working in the Volcan Iliniza of Ecuador also observed a decline in SR with increasing altitudinal levels.

In terrestrial environments, SR decreases with altitude (Rahbek, 1995; Theurillat et al., 2003, 2011; Stanisci et al., 2005; Grytnes et al., 2006). Studies have shown that when complete gradient is considered, then the species-richness curve is hump-shaped with a maximum in the lower half of the gradient, but it shows a monotonic decrease with latitude (Rahbek, 1995, 2005). The assessment with altitudes is also complicated due to numerous of intercorrelated environmental gradients, i.e., temperature, rainfall, solar radiation, wind, season length, atmospheric pressure, steepness, and human land use (Korner, 2007; Rahbek, 2005). However, several other plausible explanations for the linear relationship of SR and altitudes have been given (Givnish, 1999).

The plant species distribution patterns in phyto-geographical locations have also determined by solar radiation, aspect, and angle of slope (Bormann et al., 1970; Ebermayer, 1976). Species diversity of individual generally considered as a function of relative distribution among species. Bhat et al. (2020) has reported that the diversity of woody plants declined as altitude increased, indicating that environmental filtering acts as a key factor in shaping the woody vegetation at high altitudes. Species diversity is regulated by long-term factors like community stability and evolutionary time as heterogeneity of both micro and macroclimate affects the diversification among different communities (Verma et al., 2004).

The Shannon diversity (H') of herbs in the present study ranged from 5.500 (AZ-III 2450–2650 m) to 5.846 (AZ-V 3350–3550 m) (Table 6.1). Similar diversity values had also been reported by many workers in Himalayas. Uniyal et al. (2010) reported diversity values of 3.30 to 3.86 for herbs in Dewalgarh watershed forests of Garhwal Himalaya. Pande et al. (2001) reported diversity values ranging from 2.54 to 2.99 for herbs in certain forests of Garhwal Himalaya. Khumbongmayum et al. (2005) carried out a study in Manipur forests (sacred groves) and observed herb diversity range from 2.77 to 3.13. However, the present diversity values were observed higher than the values reported by Pokhriyal et al. (2009) in the forests of Garhwal Himalaya, their diversity values ranged between 0.76 to 0.81 for herbs, the reason behind higher diversity values of the present study might be because the present study was carried out in a protected area. The CD in the present study for herbs ranged from 0.024 (AZ-V 3350–3550 m) to 0.039 (AZ-IV 2900–3100 m) (Table 6.1). CD showed opposite trend as compared to species diversity, which is supported by the studies of Khumbongmayum et al. (2005) and Kumar et al. (2009). In general species diversity and CD are inversely related (Khumbongmayum et al., 2005). The present values of CD are comparable with the reported values of various workers. Pant and Samant (2007) reported Cd values of herbs from 0.01 to 0.52 in Mornaula Reserve Forests of Central Himalaya. Pokhriyal et al. (2009) reported Simpson's index value of 0.07 to 0.10 for herbs in the forests of Garhwal Himalaya. Uniyal et al. (2010) reported Cd values in the range of 0.28 to 0.050 for herbs in two forest types along the disturbance gradient in Dewalgarh watershed of Garhwal Himalaya.

The equitability in the present study for herbs ranged from 21.064 (AZ-IV 2900–3100 m) to 34.776 (AZ-V 3350–3550 m) (Table 6.1). The equitability values found by various workers are comparable to the present study. The βD for herbs in the present study ranged from 9.148 (AZ-IV 2900–3100 m) to 11.670 (AZ-II 2000–2200 m) (Table 6.1). Communities under different environmental conditions differ in the number of species they contain. The number of species in the community is usually referred to as richness. The concept of diversity relates simply to the richness of a community. The present investigation reveals that, the Margalef index (SR) 'SR' for trees ranged from 1.176 (AZ-V 3350–3550 m) to 3.430 (AZ-I 1550–1750 m), for shrubs 1.192 (AZ-V 3350–3550 m) to 3.713 (Zone-III 2450–2650 m) and for herbs 6.151 (Zone-2900–3100 m) to 10.203 (AZ-V 3350–3550 m) (Table 6.1). Uniyal et al. (2010) reported SR in the range of 5.34 to 9.9 for herbs in two forest types along the disturbance gradient in Dewalgarh watershed of Garhwal Himalaya. Ram et al. (2004) also reported herb richness as 2.0 for six forest types in Central Himalayas. Khumbongmayum et al. (2005) reported SR from 0.53 to 0.79 for herbs in Manipur forests of northeast India, and these values are lower than the values of the present study. The diversity in the present study decreased with increase in altitude, which might be due to site quality.

KEYWORDS

- **altitudinal zone**
- **Convention on biological diversity**
- **flora**
- **herbaceous**
- **importance value index**
- **protected area**
- **protected area network**
- **species richness**

REFERENCES

Aubry, S., Magnin, F., Bonnet, V., & Preece, R. C., (2005). Multi-scale altitudinal patterns in species richness of land snail communities in south-eastern France. *J. Biogeogra, 32*, 985–998.

Bhat, J. A., (2012a). *Diversity of Flora along an Altitudinal Gradient in Kedarnath Wildlife Sanctuary.* PhD thesis, HNB Garhwal University Srinagar, Garhwal.

Bhat, J. A., Iqbal, K., Kumar, M., Negi, A. K., & Todaria, N. P., (2013b). Carbon stock of trees along an elevational gradient in temperate forests of Kedarnath Wildlife Sanctuary. *For. Sci. Pract., 15*(2), 137–143.

Bhat, J. A., Kumar, M., & Bussmann, R. W., (2013a). Ecological status and traditional knowledge of medicinal plants in Kedarnath Wildlife Sanctuary of Garhwal Himalaya, India. *J. Ethnobiol. Ethnomed., 9*(1), 1.

Bhat, J. A., Negi, K. M., Ajeet, K., & Todaria, N. P., (2012). Anthropogenic pressure along altitudinal gradient in a protected area of Garhwal Himalaya, India. *J. Environ. Res. Develop., 7*(1), 62–65.

Bhat, J. A., Kumar, M., Negi, A. K., Todaria, N. P., Malik, Z. A., Pala, N. A., Kumar, A., & Shukla, G., (2020). Species diversity of woody vegetation along altitudinal gradient of the Western Himalayas. Glob. Ecol. Conserv. 24, e01302. doi: 10.1016/j.gecco.2020.e01302.

Bliss, L. C., (1962). In: Leith, H., (ed.), *Net Primary Production of Tundra Ecosystems* (pp. 35–48). Die Stoffproduktion der Pfanzendecke. Gustav Fischer Velag Stuttgart.

Bormann, F. H., Siccama, T. G., Likens, G. E., & Whittaker, R. H., (1970). The Hubbard Brook ecosystem study: Composition and dynamics of the tree stratum. *Ecol. Monog., 40*, 377–388.

Brandon, K., Redford, K. H., & Sanderson, S. E., (1998). *Parks in Peril: People, Politics, and Protected Areas.* Island Press: Washington, DC, USA.

Champion, H. G., & Seth, S. K., (1968). *A Revised Survey of Forest Types in India.* Government of India Publication, New Delhi.

Chandra, J., Rawat, V. S., Rawat, Y. S., & Ram, J., (2010). Vegetational diversity along altitudinal range in Garhwal Himalaya. *Int. J. Biodiver. Conser., 2*(1), 014–018.

Curtis, J. T., & McIntosh, R. P., (1950). The interrelations of certain analytic and synthetic phytosociological characters. *Ecol., 31*, 434–455.

Ebermayer, E., (1976). *The Gesamate Lehe Dev Waldstrenmit Rucksicht any Dis Chemische Static Des Waldbane* (p. 11). Berlin Julius Springer.

Givnish, T. J., (1999). On the causes of gradients in tropical tree diversity. *J. Ecol., 87*, 193–210.

Grabherr, G., Gottfried, M., Gruber, A., & Pauli, H., (1995). Patterns and current changes in alpine plant diversity. In: Chapin, et al., (eds.), *Artic and Alpine Biodiversity: Patterns, Causes and Ecosystem Consequences* (pp. 167–181). Springer-Verlag, Berlin Heidelberg.

Grytnes, J. A., & Vetaas, O. R., (2002). Species richness and altitude: A comparison between null models and interpolated plant species richness along the Himalayan altitudinal gradient, Nepal. *The Am. Nat., 159*(3), 294–304.

Grytnes, J. A., Heegaard, E., & Ihlen, P. G., (2006). Species richness of vascular plants, bryophytes, and lichens along an altitudinal gradient in western Norway. *Acta Oecol., 29*, 241–246.

Hanson, H. C., & Churchill, E. D., (1961). *The Plant Community.* Reinhold Publishing Corporation, New York.

Hough, J. L., (1988). Obstacles to effective management of conflicts between national parks and surrounding human communities in developing countries. *Environ. Conser., 15*(2), 129–136.

Ite, U. E., (1996). Community perceptions of the cross-river National Park, Nigeria. *Environ. Conser., 23*(4), 351–357.

Khumbongmayum, A. D., Khan, M. L., & Tripathi, R. S., (2005). Sacred groves of Manipur, northeast India: Biodiversity value, status and strategies for their conservation. *Biodiv. Conser., 14*, 1541–1582.

Korner, C., (2007). The use of 'altitude' in ecological research. *Tren. Ecol. Evol., 22*, 569–574.

Kramer, R., Van-Schaik, C., & Johnson, J., (1997). *Last Stand: Protected Areas and the Defense of Tropical Biodiversity.* Oxford University Press: New York, USA.

Kumar, M., Sharma, C. M., & Rajwar, G. S., (2009). The effects of disturbance on forest structure and diversity at different altitudes in Garhwal Himalaya. *Chin. J. Ecol., 28*(3), 424–432.

Liu, X., & Chen, B., (2000). Climatic warming in Tibetan Plateau during recent decades. *Int. J. Climatology, 20*, 1729–1742.

MacArthur, R. H., (1965). Patterns of species diversity. *Biol. Rev., 40*, 510.

MacArthur, R. H., (1972). *Geographical Ecology.* Harper and Row, New York.

Magurran, A. E., (1988). *Ecological Diversity and its Measurements.* Princeton University Press, Princeton.

Malik, Z. A., & Bhatt, A. B., (2015). Phytosociological analysis of woody species in Kedarnath Wildlife Sanctuary and its adjoining areas in Western Himalaya, India. *J. For. Environ. Sci., 31*(3), 149–163.

Malik, Z. A., Bhat, J. A., & Bhatt, A. B., (2014). Forest resource use pattern in Kedarnath wildlife sanctuary and its fringe areas (a case study from Western Himalaya, India). *En. Pol., 67*, 138–145.

Margalef, R., (1958). Information theory in ecology. *Gen. Syst., 3*, 36–71.

Mathur, V. B., Kathyat, J. S., & Rath, D. P., (2000). *Envis Bulletin: Wildlife and Protected Areas,* (Vol. 3, p. 1). Wildlife Institute of India, Dehradun.

McCoy, E. D., (1990). The distribution of insects along elevational gradients. *Oikos, 58*, 313–322.

MEA (Millennium Ecosystem Assessment), (2005). *Ecosystems and Human Wellbeing: Synthesis* (p. 155). Island Press: Washington DC.

Nogues-Bravo, D., Araujo, M. B., Errea, M. P., & Martinez-Rica, J. P., (2007). Exposure of global mountain systems to climate warming during the 21st century. *Glob. Environ. Chang., 17*, 420–428.

Odum, E. P., (1971). *Fundamentals of Ecology.* W.B. Saunders Co., Philadelphia.

Pande, P. K., Negi, J. D. S., & Sharma, S. C., (2001). Plant species diversity and vegetation analysis in moist temperate Himalayan forest. *Ind. J. For., 24*(4), 456–470.

Pant, S., & Samant, S. S., (2007). Assessment of plant diversity and prioritization of communities for conservation in Mornaula reserve forest. *Appl. Ecol. Environ. Res., 5*(2), 123–138.

Peters, J., (1999). Understanding conflicts between people and parks at Ranomafana, Madagascar. *Agr. Hum. Val., 16*, 65–74.

Phillips, E. A., (1959). *Methods of Vegetation Study.* Henry Hill & Co. Inc., U.S.A.

Pielou, E. C., (1966). The measurement of diversity in different types of biological collections. *J. Theor. Biol., 13*, 131–144.

Pimbert, M. P., & Pretty, J. N., (1995). *Parks, People and Professionals: Putting 'Participation' into Protected Area Management.* United Nations Research Institute for Social Development, Geneva, Switzerland.

Pokhriyal, P., Naithani, V., Dasgupta, S., & Todaria, N. P., (2009). Comparative studies on species, diversity and composition of *Anogeissus latifolius* mixed forests in Phakot and Pathri Rao Watersheds of Garhwal Himalaya. *Curr. Sci., 97*(9), 1349–1355.

Diversity and Dynamics in Forest Ecosystems

Rahbek, C., (1995). The elevational gradient of species richness: A uniform pattern? *Ecography, 18*, 200–205.

Rahbek, C., (2005). The role of spatial scale and the perception of large-scale species-richness patterns. *Ecol. Lett., 8*, 224–239.

Ram, J., Kumar, A., & Bhatt, J., (2004). Plant diversity in six forest types of Uttaranchal, Central Himalaya, India. *Curr. Sci., 86*, 975–978.

Rosenzweig, M. L., (1995). *Species Diversity in Space and Time* (p. 436). Cambridge University Press: Cambridge.

Shannon, C. E., & Wiener, W., (1963). *The Mathematical Theory of Communication*. University of Illinois Press, Urbana.

Shrestha, A. B., Wake, C. P., Mayewski, P. A., & Dibb, J. E., (1999). Maximum temperature trend in the Himalaya and its vicinity: An analysis based on temperature records from Nepal for the period 1971–94. *J. Climate, 12*, 2775–2786.

Simpson, E. H., (1949). Measurement of diversity. *Nature, 163*, 688.

Sklenar, P., (2006). Searching for altitudinal zonation: Distribution and vegetation composition in the Super paramo of Volcan Iliniza, Eucador. *Pl. Ecol., 184*, 337–350.

Stanisci, A., Pelino, G., & Blasi, C., (2005). Vascular plant diversity and climate change in the alpine belt of the central Apennines (Italy). *Biodiv. Conser., 14*, 1301–1318.

Stevens, S., (1997). *Conservation through Cultural Survival: Indigenous Peoples and Protected Areas*. Island Press: Washington, DC, USA.

Tello, B., Fiallo, E. A., & Naughton-Treves, L., (1998). Ecuador: Podocarpus National Park. In: Brandon, K., et al., (eds.), *Parks in Peril: People, Politics, and Protected Areas* (pp. 287–322). Sanderson, Island Press: Washington, DC, USA.

Theurillat, J. P., Iocchi, M., Cutini, M., & De, M. G., (2011). Vascular plant richness along an elevation gradient at Monte Velino (Central Apennines, Italy). *Biogeogra., 28*, 149–160.

Theurillat, J. P., Schlussel, A., Geissler, P., Guisan, A., Velluti, C., & Wiget, L., (2003). Vascular plant and bryophyte diversity along elevation gradients in the Alps. In: Nagy, L., et al., (eds.), *Alpine Biodiversity in Europe* (Vol. 167, pp. 185–193). Springer, Heidelberg.

Uniyal, P., Pokhriyal, P., Dasgupta, S., Bhatt, D., & Todaria, N. P., (2010). Plant diversity in two forest types along the disturbance gradient in Dewalgarh Watersheds Garhwal Himalaya. *Curr. Sci., 98*(7), 938–943.

Verma, R. K., Kapoor, K. S., Subramani, S. P., & Rawat, R. S., (2004). Evaluation of plant diversity and soil quality under plantation raised in surface-mined areas. *Ind. J. For., 27*(2), 227–233.

Wells, M., & Brandon, K., (1992). *People and Parks: Linking Protected Area Management with Local Communities*. The World Bank: Washington, DC, USA.

West, P. C., & Brechin, S. R., (1991). *Resident Peoples and National Parks: Social Dilemmas and Strategies in International Conservation*. The University of Arizona Press, Tucson, Arizona.

Western, D., & Wright, R. M., (1994). The background to community-based conservation. In: Western, D., (ed.), *Natural Connections: Perspectives in Community-Based Conservation* (pp. 1–12). Island Press: Washington, D. C., USA.

Whittaker, R. H., (1972). Evolution and measurement of species diversity. *Taxon, 21*, 213–251.

CHAPTER 7

Forage Crop Genetic Resources of North-Western Himalayas: An Underutilized Treasure

SHEERAZ SALEEM BHAT,[1] SUHEEL AHMAD,[1] NAZIM HAMID MIR,[1] SHEIKH MOHAMMAD SULTAN,[2] and SUSHEEL KUMAR RAINA[2]

[1]*Regional Research Station, ICAR-Indian Grassland and Fodder Research Institute, CITH Campus, Rangreth, Srinagar–191132, Jammu and Kashmir, India, E-mail: shrzbhat@gmail.com (S. S. Bhat)*

[2]*Regional Research Station, ICAR-National Bureau of Plant Genetic Resources, CITH Campus, Rangreth, Srinagar–191132, Jammu and Kashmir, India*

ABSTRACT

Forage crop genetic resources (FCGRs) of the North-Western Himalayan region are extremely important for poverty alleviation and food security for the local inhabitants for their role in animal production systems, besides their environmental significance. Major forage legumes of the region include red clover (*Trifolium pratense*), alfalfa (*Medicago sativa*), white clover (*Trifolium repens*), Egyptian clover/berseem (*Trifolium alexandrinum*), hairy vetch (*Vicia villosa*), crown vetch (*Coronilla varia*), and sainfoin (*Onobrychis viciifolia*), while the prevalent forage grasses include ryegrass (*Lolium*s pecies), tall fescue (*Festuca arundinaceae*), orchardgrass (*Dactylis glomerata*), bromegrass (*Bromus* species), foxtail millet (*Setariai talica*) and phalaris (*Phalaris* species), besides other forage crops like oats and maize. There exists a huge gap between the demand and the supply of both green as well as the dry fodder in the region, which affects the livestock productivity badly. On the other hand, the FCGRs in the region are under threat due to genetic erosion and narrowing of genetic base as a result of conversion of grasslands and pastures to commercial agriculture, mismanagement, and degradation of grasslands and pastures, consideration of FCGRs as orphan crops, ignoring the traditional knowledge of local farmers and pastorals regarding FCGRs, and the small number of species and cultivars under selection, use, maintenance, and improvement. The need of the hour is to devise sustainable protocols for sustainably managing, conserving, and maintaining the diversity of FCGRs through promoting *in-situ* conservation involving local farmers and the pastorals and recognize properly the different functions of FCGRs. Broadening of genetic base and improvement programs of FCGRs are required. The different agencies and institutions working on FCGRs in the region should collaborate and share knowledge and germplasm, explore new populations, evaluate, and improve the existing FCGRs and develop a database of the same for end-users. Besides this, an International treaty on their conservation and germplasm exchange is needed for sustainable fodder production and livestock productivity in the region.

7.1   INTRODUCTION

Plant genetic resources (PGRs) consist of the diversity of genetic material contained both traditional varieties and modern cultivars grown by farmers as well as crop wild relatives and other wild plant species used for food, feed for domestic animals, fiber, clothing, shelter, wood, timber, energy, etc., (FAO, 1997). These are of tremendous use from meeting out of daily requirements and the sustainability of life on the earth, and have classified based on their utilization pattern. Despite all the tangible and intangible benefits provided by the PGRs, many of them have been facing threats of extinction, while others have been well domesticated and improved for better livelihood and conservation. Forage crop genetic resources (FCGRs) consist of all those crop species which provide fodder and forage to the livestock, and hence play a vital role in the socio-economic upliftment of the agrarian population of the country. There are different sources of fodder available in the different regions of the state, but still, there is a shortage of fodder availability for sustaining the current livestock population. However, there are still some forage crop species which are underutilized in one region or the other, and are called underutilized FCGRs.

Currently, there is a lack of clear definition for the so-called underutilized or neglected crops. They are sometimes also called orphan crops, forgotten crops or minor crops also. However, one must keep in mind that their crops may be underutilized in one region but not in the other. The underutilized crop genetic resources may be regarded as those crops which have not been previously considered as major crops, under-researched, under-exploited crops which are currently not much utilized and mainly restricted to small landholder farmers. It is pertinent to mention that such crops have been playing an important role in ensuring food security of livestock and human beings from time immortal, especially when the main crops have failed. Such crops serve as a vast resource in terms of being rich gene pool and augmenting the major crop genetic resources, besides their ecological, social, cultural, and economic significance. For many cultivated crop species, their wild forms have got extinct since their breeding for domestication has resulted in plant species being unable to reproduce without the helpful hand of humans. The problem with forage crops gets aggregated as they are less domesticated as compared to major crop species (Harlan, 1983), and as a result, many wild forms of common forage species still exist, as well as their feral forms. We call such wild populations as "semi-natural" because they have persistent in the nature without conscious selection and can be regarded similar to crop wild relatives. Therefore, the closeness of cultivated forage crop species and their wild ones makes a wealth of natural genetic variation readily accessible for use in their breeding and improvement programs. The use of adapted genetic material collected from permanent grasslands and pastures has been of great benefit to early fodder crop breeding and has dramatically improved the persistency of fodder grasses.

7.2   LIVESTOCK AND FORAGE STATUS

India has the highest livestock population in the world with 512.05 million, which includes 190.90 million cattle, 108.70 million buffaloes, 135.173 million goats, and 65.1 million sheep population and others. The gap between the current production and requirement of fodder and other forages to sustain such a huge population is increasing. Presently, 2/3rd of the feed requirement is met from crop residues. Presently, India requires 883.95 Mt green fodders and 583.66 Mt dry fodders, against which the fodder production is only 664.73 Mt green fodders and 355.93 Mt dry fodders. The existing gap of 218.22 Mt green fodders and 227.73 Mt dry fodders has a substantial effect on the livestock productivity and the socioeconomic status of the farming community. It is pertinent to mention that we have only less than 5% of the gross cropped area of the land allocated to green fodder production (Earagariyanna et al., 2017). Surplus forage resources are available in cultivated fields, forestlands and other common property resources, harvested for haymaking for livestock feeding during lean months. Concentrate feeding, although limited, is also in vogue and mostly economically sound farmers purchase feed pellets, wheat straw, bran, and cakes for supplementation during the winter season. Even the fortification of fodders, feeding of mineral mixtures, silages is not much prevalent in the country. Sedentary, semi-sedentary, and migratory livestock rearing systems are commonly observed in livestock farming systems.

When we have a glance at the North-Western Himalayan livestock sector, the aggregate livestock population in Jammu and Kashmir, Himachal Pradesh and Uttarakhand in 2012 (19th Livestock census, 2012) was 9200842, 4844431, and 4794730, i.e., 3.67% of the country (512057301). Due to acute shortage of fodder, especially the green nutritious fodder, we have low livestock productivity, which a major concern of the day. A limited area (1%) of the entire Himalayan region is under forage cultivation (4% in Jammu and Kashmir). In J&K, deficit in green and dry fodder is 67% and 27.31%, respectively. In Himachal Pradesh, the deficit of annual requirement of green and dry fodder has been calculated at 54 and 34%, respectively (Radotra, 2015).

In Jammu and Kashmir, including Ladakh, the number of livestock units per 1000 human population is 736 animals against the national average of 409 animals/1000 human populations. The estimated livestock population (excluding poultry) of the both union territories is 9200842, which shares about 1.78% of the country's total livestock population. The density of livestock per sq. km of area is 98 animals against 90 animals recorded by 19th Livestock census, conducted in 2012, with sheep constituting the major share (36.84%), followed by cattle (30.41%) and goat (21.93%). There has been a significant decline by 16.25% from the previous census. The total fodder production of Jammu and Kashmir is 86.5 lakh tons of which green fodder contributes 61.4 and dry fodder 25.1 lakh tons. The union territories have a deficit of is 67% green fodder and 27.31% dry fodder (Ahmad et al., 2017). Decreasing trend in the area of grasslands and pasturelands because of encroachment, conversions, developmental projects, and other factors is also a cause of concern for the livestock sector of the union territories.

7.2.1 GRASSLANDS OF THE REGION

For semi-sedentary and migratory livestock, grasslands, and pasturelands play a major role in their sustenance. We have five types of grasslands in our country (Dabadghao and Shankarnarayan, 1973). These are *Sehima-Dichanthium* type, *Dichanthium-Cenchurus-Lasiurus* type, *Phragmites-Saccharam-Imperata* type, *Themada-Arundinella* type and the temperate-alpine type. Major grassland in the region is *Themada-Arundinella* type and the temperate-alpine type.

The *Themada-Arundinella* type of grass cover occurs in major tracts of the northwestern region down the plains up to 1800 m asl. Snowfalls are frequent. The prominent grasses in the region are *Arundinella nepalensis, Apluda mutica, Cynodon dactylon, Heteropogan contortus, Chrysopogon fulvus,* and *Cymbopogon jwarancus.* This type of grassland should be subjected to minimum biotic interference and stall-feeding should be preferred in the region. Adequate soil conservation measure and grazing management is required for the better management of such grasslands (Mukherjee and Maiti, 2008). The temperate-alpine type grassland occurs at higher elevations, varying from 1500 m asl. Snowfall during winter is quiet common in these grasslands. The prominent species are *Poaannua, Poapratensis, Pheliumalpinum, Phelium pratense, Stipaconcinna, and* different species of *Festuca, Lolium, and Agrotis.* These grasslands need improvement for meeting the fodder requirement of the livestock of the region (Mukherjee and Maiti, 2008). Our sedentary livestock sector majorly depends on the cultivated oats, fodder maize, sorghum, bajra, Berseem, and lucerne for feed and fodder. These crops are majorly grown as sole crops.

7.3 UNDERUTILIZED TEMPERATE FORAGE CROP GENETIC RESOURCES (FCGRS)

The temperate region of the country is blessed with huge diversity of temperate grasses and legumes, but most of them are underutilized in practice. Major fodder available in the region is from oats and fodder maize, either green during the spring and summer or in the dried form during the winter. Major legumes of the region include red clover (*Trifolium pratense*), alfalfa (*Medicago sativa*), white clover (*Trifolium repens*), sainfoin (*Onobrychis viciifolia*), cowpea (*Vignaun guiculata*), crown vetch (*Coronilla varia*), hairy vetch (*Vicia villosa*), and the Egyptian clover/berseem (*Trifolium alexandrinum*). The prevalent grasses include perennial rye grass (*Lolium perenne*), tall fescue (*Festuca arundinaceae*), orchardgrass (*Dactylis glomerata*), phalaris hybrid (*Phalaris arundinacea × P. tuberosa*), bromegrass (*Bromus unioloides*), and foxtail millet (*Setaria italica*). Though red clover, white clover, and alfalfa have been taken up by the farmers to some extent, still all these grasses and legumes are highly underutilized, despite huge scope and potential as sole crops or under integrated farming systems in the northwestern Himalayan region of India.

Besides these grasses and legumes, many fodder trees are also underutilized in the region. These include *Robinia pseudoacacia, Morus alba, Celtis australis, Ailanthus* species, poplars, and willows. Few underutilized forage legumes and grasses are briefly discussed in subsections (Figures 7.1 and 7.2).

(a) Sainfoin plantation	(b) Crown vetch
(c) Perennial ryegrass	(d) Tall fescue
(e) Orchardgrass	(f) Harding grass

FIGURE 7.1   Different temperate forage grasses and legumes.

7.3.1   RED CLOVER (TRIFOLIUM PRATENSE)

Red clover is an herbaceous, short-lived temperate perennial legume, thrives well for 2–4 years, variable in size, growing 20–80 cm tall. Prevalent root is a deep taproot, up to 1 m, with secondary adventitious roots up to 30 cm, which makes drought tolerant and checks soil erosion. It is native to Europe, northwest Africa, and Western Asia and has been growing well in the temperate Himalaya. It has a good N-fixing ability and has several medicinal uses.

FIGURE 7.2   Seeds of different temperate forage crops.

The leaves in red clover are alternate, trifoliate in nature, each leaflet measures 15–30 mm in length and 8–15 mm in breadth, green with a characteristic pale crescent in the outer half of the leaf. The petiole is up to 44 cm in length, with two basal stipules which abruptly narrow down to a bristle-like point. The flowers produced in a dense inflorescence are dark pink in color, with a paler base, 12–15 mm in length. Fodder harvesting is done at the early-flowering stage for better quality, with crude protein of up to 25% and more quantity. 2–3 cuts/year are possible for fodder harvesting, provided that the sward receives sufficient fertilizers (P, K).

Red clover seed can be planted in autumn or early spring. Being an N-fixing legume, it can be used as green manure also and it can provide about 4.2 t DM/ha and 93 kg N/ha to the soil. When used in orchard floor management, it smothers spring weeds and improves soil tilth, besides having a beneficial effect on the fruit trees.

7.3.2   WHITE CLOVER (TRIFOLIUM REPENS)

White clover (*Trifolium repens* L.) is an herbaceous perennial legume 10–40 cm high with fairly deep roots with adventitious roots, leafy branches with petiolated trifoliate leaflets and inflorescence stalks. Leaves can vary widely in form and size, depending on cultivar or type (ovate, broad, solid dull green). It's probable origin is an eastern Mediterranean region of Asia Minor and is indigenous to the whole of Europe, Central Asia west of Lake Baikal, and to small areas in Morocco and Tunisia.

White clover is a widely used grown alone or mixed with grasses under rainfed or irrigated conditions. It is highly palatable and has a high nutritive value. However, it can be grown on a wide range of soils, but higher productivity is achieved in clayey or loamy soils than on sandy soils. The species has moderate drought tolerance, which can be improved by growing the same along with a companion grass for shade and lowering the temperature at the ground level.

Fodder productivity of the legume sown in pure stands yields 12–15 t DM/ha under irrigated conditions and up to 10 t DM/ha under rainfed conditions. The species also has a capacity to fix up to 400 kg N/ha/year. The fodder has high crude protein (20–25%), which decreases with maturity and high digestibility but causes bloating when fed alone or >30% fodder mixture.

7.3.3 ALFALFA (MEDICAGO SATIVA)

In Arabic, Alfalfa means the best. It is a highly palatable legume and is known for its excellent nutritional quality, besides being a rich source of essential vitamins, minerals, and amino acids with wide adaptability and distribution, but a native of South-Western Asia. It performs better in cooler and drier conditions, having well-drained deep loamy soils rich in calcium, phosphorus, and potassium with pH 5.5–8.8.

It is a perennial herb that can grow up to 1.6 m wit upright stem. Leaves are alternate, pinnately trifoliate and born on short petioles with stipules adenate to petioles. Small yellow or violet papilionaceous flowers are borne on short axillary racemes. Pods are dehiscent, several seeded and spirally twisted. Deep taproots up to 100 cm are present. Lucerne regenerates with basal buds, thus form branched crowns at or near ground level.

Lucerne is an excellent fodder legume with around 20% crude protein and 30% crude fiber even in the dry matter. Green herbage productivity varies from 953 to 1243 quintal/ha. Lucerne is usually fed as green forage to the livestock and even protein-rich concentrates are made from it for lean season.

7.3.4 SAINFOIN (ONOBRYCHIS VICIIFOLIA)

The *Onobrychis* genus (Fabaceae family) originated in arid regions of Eurasia, first domesticated in Europe. In Old French, the translation of sainfoin is *"healthy hay."* Amongst the different species of the *Onobrychis* genus, sainfoin (*Onobrychis viciifolia*) is the most widespread species, being tolerant to drought, cold, and low nutrient status. It grows from 40 to 100 cm in height. Many hollow stems from basal buds form a branched crown. Each stem has pinnate leaves formed with 10 to 28 leaflets grouped in pairs on long petioles and with a terminal leaflet. The inflorescences develop on axillary tillers with about 80 pinkish-red melliferous flowers, which present an esthetic view also. Each flower can produce a kidney-shaped seed contained in a brown pod. The fruit is either spiny or spineless.

Although sainfoin has low productivity (14–16 tons DM ha^{-1}) and is more difficult to maintain than other legumes, it is known to have valuable characteristics such as palatability, higher protein digestibility, and anthelminthic properties due to its unique tannin and polyphenol composition that prevent bloating. Thus it has the potential to reduce greenhouse gas (GHG) emissions, which in turns acts as a measure for climate change mitigation. Being drought-tolerant, it is a climate change smart forage legume. Positive effects on wildlife and honey production from the pinkish-red flowers could also be advantageous in the context of sustainable farming. It has been now an orphan crop, and modern breeding programs have not been a priority for its improvement, leading to a lack of genetic knowledge in comparison to extensively used forage legumes. Breeding priorities for the better domestication of Sainfoin should be focused on enhancing seed germination, early establishment, improvement of forage productivity, quality fodder traits, and some specific traits like anthelmintic properties observed in some varieties (Carbonero et al., 2011).

Sainfoin is self-incompatible legume species, mostly pollinated by insects. Propagation is done through seed. Seed weight is about 20g/1000 milled seed. Seed size varies and it ranges from 2.5 to 4.5 mm in length, 2 to 3.5 mm in breadth, and 1.5 to 2 mm in thickness. A seed rate of 7 kg/ha and a row spacing of 60 cm are recommended for seed production (or 40 kg/ha @ 15 cm for hay production) (Goplen et al., 1991).

7.3.5 CROWN VETCH (CORONILLA VARIA)

Coronilla varia, commonly known as crown vetch, is a low-growing legume-vine native to Asia, Africa, and Europe. Besides forage production, the species is suitable for soil and water conservation aspects. The vine is a tough, aggressive spreading fodder legume plant that will crowd out its neighbors. Its deep, tenacious, complex root system and thick, fern-like leaves provide excellent erosion control.

Crown vetch was earlier labeled as poor forage for farm animals, but more recently, it has been deemed as good forage when fed as hay to or grazed by ruminants. It contains 21.7% crude protein and 22.2% crude fiber (Reynolds et al., 1967). Digestibility of crown vetch in sheep has revealed that the crude protein is 65.6% and the crude fiber is 46.2% digestible, which is comparable to that of alfalfa forage. This fodder species only suits to ruminant animals as it is toxic to non-ruminants because of the presence of nitroglycosides, which are degraded in the ruminant digestive tract (Moyer and Gustine, 1984). A rainfed crop has the potential to yield 200 quintals/ hectare green fodder.

7.3.6   PERSIAN CLOVER (TRIFOLIUM RESUPINATUM)

This legume is native to central and southern Europe, all the Mediterranean region and southwest Asia. It is cultivated as a winter annual, sown in autumn and provides forage during the spring season. It can be grown as summer forage also. It is locally known as shaftal, it is an erect decumbent or prostrate, coarse herbaceous legume branching from the base and reaches up to height of 30–100 cm. it has pink mauve, sweet-smelling flowers, ovoid pods and pale brown seeds. Prominent varieties are SH-48 and SH-69 (Bhat et al., 2019). The nutritive value of the dry matter produced from the above-ground parts including the leaves and tender stem is crude fiber 16.9%, crude protein 21.5%, total ash content 17.7%, ether extract 1.9%, and nitrogen-free extract of 42%.

7.3.7   PERENNIAL RYEGRASS (LOLIUM PERENNE)

Perennial ryegrass is native of Eurasia and North Africa. But it is presently one of the most commonly sown grasses for forage production, turf, and soil stabilization, because of which it has been introduced all over the world. It is a tufted annual or short-lived perennial grass with feather-like seed heads, which lacks awns and is commonly found in lawns.

It grows to up to 100 cm in height and is erect or spreading. The root system consists of fibrous roots and rhizomes. The entire plant is smooth and hairless. There are numerous narrow, long, and stiff leaves near the base of the plant. However, *Lolium* has a wide range of adaptability to edaphic factors, but it thrives best on dark rich soils in regions with mild climates. A fine, firm seedbed gives the best grassland establishment. Mulched seed-ings on graded soil germinate readily during the spring. It responds well to different management practices, such as intensive rotational grazing and fertilizer applications. Perennial ryegrass is a valuable forage and soil stabilization plant. Usually, the tetraploid cultivars are used for forage production while as the diploid ones are used for lawns and conservation plantings (Anonymous, 2002a). Crude protein of fresh vegetative parts is around 19%.

7.3.8 TALL FESCUE (FESTUCA ARUNDINACEA)

Tall fescue (*Festuca arundinacea*) is a robust, long-lived, deep-rooted bunchgrass, with huge potential for forage production, especially in the Northwestern Himalayan region. The stems are up to 4 feet in height, supporting a nodding panicle which is 4–12 inches in length. The leaves are flat, broad, smooth, and shiny on the underside, with clearly visible ribs on the upper surface. The species is well adapted to cool and humid climates and moist soils with a pH 5.5–7. Although it grows fairly well on soils with low fertility status, but it is better adapted to fertile conditions. Grows vigorously in spring, summer, and early autumn, but has slow winter growth. Tall fescue is easy to establish due to its speedy germination and good seedling vigor (Anonymous, 2002b).

The grass is propagated through seeds and rooted slips. Seeds have awns (bristles) unlike perennial ryegrass. The best time to sow tall fescue seeds in autumn and early winter when soil moisture is becoming adequate. Approximately, 420000 tall fescue seeds weight per kilogram. Recommended seeding rate @10–15 kg/ha. Tall fescue has poor seedling vigor with the roots and crown developing slowly, so sufficient seed should be sown to promote good ground cover. Temperate cultivars of tall fescue are highly productive with greatest growth over spring and early summer (40–60 kg dry matter/ha/day). Nutritive value of the species is better when compared to perennial ryegrass, *Phalaris*, and cocksfoot, with higher digestibility. Prominent varieties are Demeter, with dry matter yield of 80–90 q/ha/year in 2–3 cuttings; HIMA-1 with dry matter yield of 5–5.2 t/ha/year in 2–3 cuttings and HIMA-4 (Ahmad et al., 2019).

7.3.9 ORCHARD GRASS/COCK'S FOOT (DACTYLIS GLOMERATA)

Orchardgrass (*Dactylis glomerata*), due to its high productivity, disease resistance under different climatic conditions considered the most important forage grasses in the world, this grass is suitable for orchard floor management, pasture management, hay, and silage making. The main advantage of cocksfoot is its higher forage production during the summer compared to other species of grasses. The grass stays green after most of the other grasses have dried. The grass is strongly tufted, deep-rooted, long-lived perennial grass reaching a height of 60–150 cm. For optimal growth annual day temperatures ranging from 4.3°C to 23.8°C, annual rainfall from 480 mm to 750 mm, on normally drained to dry soils, rich soils of heavy types such as clays and loams, with a pH ranging from 4.5 to 8.2 (Mir et al., 2018).

Due to its better forage quality shade tolerance and persistence, it is more suitable for cultivation than many other cool-season perennial grasses. As a cool-season perennial, cocksfoot may be harvested 3–4 times a year and remains productive for up to 8 years. The green fodder yield of 22.32 t/ha and dry fodder yield of 7.39 t/ha in cocksfoot was obtained in an apple based hortipastoral system under temperate conditions of Kashmir (Ahmed et al., 2017). Crude protein in the species varies around 18%.

7.3.10   RED FESCUE (FESTUCA RUBRA)

Festuca rubra, locally known as red fescue, is a widespread perennial grass across much of the Northern Hemisphere. Though the species has wider adaptability, it can thrive well in many habitats and climates but is best adapted to well drained soils in cool, temperate climates and prefers shadier areas. It can grow up to 50 cm in height. Like all fescues, the leaves are narrow and needle-like. It is cultivated as an ornamental plant for use as turfgrass and ground covering. It can be left completely unmowed, or occasionally trimmed for a lush green meadow development.

Sowing can be either done in autumn) or spring. For autumn, sowing should be done from mid-September to November. A seed rate of 8–10 kg/ha (as sole) should be sown in lines 30 or 50 cm apart. Seed should be sown @3–5 kg/ha, when sown in mixture with legume.

7.3.11   TIMOTHY (PHELIUM PRATENSE)

Timothy, also known as meadow cat's-tail or common cat's tail, is a perennial grass, native to most of Europe except for the Mediterranean region. It grows 80–100 cm in height, with hairless leaves up to 43 cm long and 1.3 cm broad, leaves are rolled rather than folded with their lower sheaths turn dark brown.

The flower head is 70–152 mm long and 6.4–12.7 mm broad, with densely packed spikelets. It flowers during June-July, with pink stamens and short and blunt ligules.

It grows well in heavy soils, and the species is cold and drought tolerant, which enables it to grow in dry upland or poor sandy soils. In pasturelands, it tends to be overwhelmed by other more competitive grasses. After cutting, it grows slowly. Like other temperate grasses, it can be sown in either autumn (October-November) or early spring (February-March). Its seed is smaller in size than *Dactylis*, tall fescue and Bromus. A well-prepared, moist seedbed good for its germination and subsequent growth. A seed rate of 4–5 kg/ha is enough for raising a good crop of timothy.

7.3.12   BROME GRASS (BROMUS SPECIES)

The genus *Bromus* consists of different species like *B. inermis, B. uniloides*. They are annual to short-term perennial grasses, but pasture life can be extended by better management practices. It provides quality forage up to autumn, when most of the other grasses are not green and much productive. It has a similar growth habit to perennial ryegrass, but grows more into summer and is more heat-tolerant. It has a good ability to recover from hard grazing. It performs well when grown with legumes, but best productivity is achieved when grown with red and white clover in pure swards with legumes. It can even be used as a permanent pasture if well managed. The common name rescue grass refers to the ability of the grass to provide forage after harsh droughts or severe winters.

It grows 0.2–1 m (7.9 in–3 ft in) in height. The culms of the grass are glabrous and 2–4 mm (0.079–0.157 in) thick. This grass grows well in areas receiving around 850 mm annual rainfall. It requires highly fertile, well-drained soils, with best growth on soils with pH 5.5 or above. It performs poorly on waterlogged soils. A seed rate up to 20 kg/ha in mixtures with other perennial grasses is optimum, but it can be better managed as pure swards.

7.3.13   HARDING GRASS (PHALARIS HYBRID)

It is also known as white-crested grass, referring to the appearance of the seed head. It is a cross between *Phalaris arundinacea E.* and *Phalaris tuberose*. It is palatable to animals, and is good grass for high nutrient hay or silage making. It should preferably be cut before reaching the heading stage as the green fodder turns harder, which is not preferred by the animals. It is triploid in nature and cannot be propagated through seed. It is propagated through rooted-slips, which should be transplanted in rows 40 m apart in March-April or November-December. About 60,000 slips required per hectare.

Though there are not any specific varieties available but we have six different selections at our station, which have been named as IGFRIRS-Phalaris-1, IGFRIRS-Phalaris-2, IGFRIRS-Phalaris-3, IGFRIRS-Phalaris-4, IGFRIRS-Phalaris-5 and IGFRIRS-Phalaris-6. We have observed an average green fodder yield of 602 qtl/ha/yr from these collections. It is pertinent to mention that IGFRIRS-Phalaris-2 yields about 750 qtl green fodder/ha/yr.

7.4   BENEFITS FROM FCGRS

FCGRs form the major part of our natural grasslands and pastures. The major benefits from FCGRs include:

7.4.1   *PRODUCTIVE BENEFITS*

They provide fodder for the livestock in pastures. Grasslands and the fields of the local farmers. Forage legumes also fix nitrogen, which increases the productivity of our land. Some grasses and legumes have medicinal and aromatic properties also, e.g., lemongrass. Tea is also made from different grasses and legumes, e.g., red clover tea.

7.4.2   *ENVIRONMENTAL BENEFITS*

These include soil and water conservation, carbon sequestration and mitigation of climate change, phytoremediation, and check pollution. Rehabilitation of degraded areas from mining, industries, deforestation, and other such causes can be done using different FCGRs. As per Batello et al. (2008), grass-legume mixtures can also reduce nitrate leaching and check pollution.

7.4.3   *RECREATIONAL BENEFITS*

The northwestern Himalayan states are famous globally for tourism and some natural pastures and grasslands. Prominent ones from Kashmir

Himalaya include Sonamarg, Gulmarg, Yousmarg, Pahalgam, amongst others. Eco-tourism has been a recent concept, and the same can be well organized and remunerative in the northwestern Himalayan states because of these grasslands and pastures.

7.4.4 SOCIO-ECONOMIC BENEFITS

They support livelihood and socioeconomic base of all those associated with livestock rearing, especially pastorals and local farmers at the grassroots level and the associated industries with livestock products at higher levels.

7.4.5 BIODIVERSITY AND WILDLIFE

Grasslands and pastures provide a good habitat for different animals and provide an elementary base for different food chains and food webs. They are the cheapest form of *in-situ* conservation of different FCGRs. The different vegetation from these grasslands and pastures serves as an important resource for the genetic improvement of different major and minor crops, keeping in view their nature to survive against different biotic and abiotic stresses.

7.5 MAJOR ISSUES AND THEIR SOLUTIONS

The different issues regarding grasslands, pastures, and FCGRs can be settled through the following measures:

7.5.1 CONSERVATION AND MANAGEMENT OF FCGRS

Keeping in view the different tangible and intangibles benefits, especially from the forage, environment, and biodiversity point of view, we should conserve and manage our FCGRs properly. We cannot close a grassland/pasture for grazing and expect the vegetation to remain the same. In the north-western Himalayan region, both the *in-situ* and *ex-situ* measures of conservation, germplasm explorations and the inventorization of FCGRs are still in the infancy stage, for which we must sustainable protocols for sustainably managing, conserving, and maintaining the diversity of FCGRs.

7.5.2   *FCGR GERMPLASM EXCHANGE AND IMPROVEMENT*

Broadening of genetic base and improvement programs of FCGRs are required, which can be achieved through germplasm exchange at national and international levels, explorations from different zones of the region in general and biodiversity hotspots/reserves in particular, cooperation between traditional guardians of FCGRs and research agencies for their identification, collection, and evaluation.

7.5.3   *QUALITY SEED PRODUCTION AND AVAILABILITY*

Intensive large-scale cultivar and variety development of forage crops is still in the infancy stage in our region. Lack of commercial seed production of improved varieties and their availability at a reasonable price to the local farmers is one major issue regarding FCGRs.

7.5.4   *INVOLVE THE LOCALS AND END-USERS THROUGH PARTICIPATORY MANAGEMENT*

Improved rangeland management will only happen in a non-threatening environment in which the representatives of the local guardians can make their voices heard. Heterogeneous committees with leaders from different groups representing different ethnicity, gender, age, pastoral experiences, and well-being need to be composed for management on the lines of the joint forest management (JFM).

7.5.5   *GRAZING MANAGEMENT AND FODDER COLLECTION*

Rampant grazing without taking care of regeneration and vegetative growth stage of the forage grasses and legumes in the grasslands and pastures leads to the decline of the composition of better quality forage grasses and legumes. Grazing should be done in a controlled manner, on a rotational basis till the previous block or compartment gets enough regeneration and stocking to sustain more livestock. The problem becomes more complex when a single type of animals like sheep, goats, cattle are grazing. Their selective crop species gets more pressure from the grazing, and its regeneration declines. Other less grazed species dominate and create problems more often. This can even lead to genetic erosion of the selective species grazed. To check the problem, we should graze different types of animals as mixed grazing on the same piece of pasture/grassland. Regarding fodder collection from natural pastures/grasslands, we should leave enough regeneration material or let the species seed naturally. In Ladakh, we have observed that *Cicer microphyllum* (wild chickpea) is being collected for fodder when succulent and the locals do not keep enough plants for seed dispersal and it has lead to its regeneration problem, which in turn can cause its genetic erosion.

7.5.6 IMPROVE THE CONDITION OF RANGELANDS AND GRASSLANDS

The productivity of our natural grasslands and pastures is not up to mark. It needs to be improved to sustain the livestock pressure during the grazing season. This can be done through re-seeding of different productive and good quality forage grasses and legumes. Application of fertilizers, wherever possible, should also be done.

7.5.7 THREATS TO FCGRS

Major threats to FCGRs include conversion of natural grasslands and pastures to agricultural lands, pasture degradation, and deforestation, diversion of grasslands and pastures to developmental projects, improper grazing management, and rural/urban encroachment. Policies and guidelines should be drafted in favor of natural pastures and grasslands so that they survive sustainably and we will not lose the forage crop genetic resource wealth from our northwestern Himalayan states.

7.5.8 IDENTIFY ALTERNATIVE ENERGY AND LIVELIHOOD OPTIONS

We need to reduce the pressure on our grasslands and pastures, for which we must promote alternative livelihood options for small ruminant livestock keepers to reduce the risk.

7.5.9   *RESEARCH AND EDUCATION REGARDING FCGRS*

Since the FCGRs are important from livestock and food security point of view, more research and education regarding these crops is required, for which different governmental and non-governmental agencies should come forward and plan and execute different research and development programs. More funds should be diverted towards the cause as the ecology of the region is fragile, often prone to landslides, soil, and water erosion, floods, and other such threats.

7.5.10   *EXTENSION PROGRAMS AND DEMONSTRATIONS*

Unlike major crops like cereals and commercial ones, very less extension-based activities are being conducted. There is an urgent need from research institutes and extension agencies to collaborate and demonstrate the advanced technologies, varieties, cultivars, new introductions, and other production and management technologies to the local farmers and pastorals. This will increase the area under forage crops, conserve the FCGRs on-farm, reduce pressure on natural grasslands and pastures, stabilize livestock rearing system, and improve the livestock productivity.

7.5.11   *INTER-INSTITUTIONAL COLLABORATIONS*

The different agencies and institutions working on FCGRs in the region should collaborate and share knowledge and germplasm, explore new populations, evaluate, and improve the existing FCGRs and develop a database of the same for end-users. Besides this, an International treaty on their conservation and germplasm exchange is needed for sustainable fodder production and livestock productivity in the region.

7.5.12   *ENCOURAGE MIXED FARMING*

Strategies need to be devised to increase forage productivity and encourage forage production undermixed farming systems for promotion, management, and conservation of FCGRs. On the other hand, strategies for efficient utilization of available resources like non-conventional feeds, crop residues, and underutilized fodders need to be formulated properly. Stakeholder participation and institutional support is important for their promotion.

7.6 RESEARCH INSTITUTIONS DEALING WITH FCGRS IN THE REGION

There are a number of state as well as state-level research institutions dealing with the different aspects of these FCGRs. These include state agricultural universities, institutions under the Indian Council of Agricultural Research, institutions under the Ministry of Environment and Forests, Climate Change, besides the state line Departments like Department of Agriculture, sheep husbandry, livestock, forests, watershed management, and others.

ICAR-Indian Grassland and Fodder Research Institute Jhansi is the main institute under the National Agricultural Research System working on different aspects of forage crops like grassland and pasture management, forage crop production, genetic improvement of forage crops, plant-animal relationship studies, post-harvest technology development, and transfer of modern technologies to the stakeholders. This institute is having one full-fledged Research Station in Srinagar, Jammu, and Kashmir, wherein we are working on these different aspects of grasslands/pastures and forage crops, besides one center at Palampur, Himachal Pradesh.

ICAR-Indian Grassland and Fodder Research Institute, Regional Research Station, Srinagar, and ICAR-National Bureau of Plant Genetic Resources Regional Station, Srinagar have taken initiative regarding different aspects of germplasm like collection, evaluation, documentation, and improvement; standardization of packages and practices of different FCGRs; production of forages under different cropping systems like hortipasture; forage crop seed production and distribution among different stakeholders especially farmers; extension of different technologies of forage crops and different aspects of grassland and pastures.

In total, 230 germplasm accessions consisting of 164 temperate grasses viz., tall fescue, orchard grass, Harding grass, prairie grass, timothy grass, red fescue grass and 66 temperate legumes viz., white clover, red clover, sainfoin are maintained *in situ* at germplasm block of IGFRI RRS Srinagar, J&K (Table 7.1). We are actively involved in germplasm exploration, multi-plication, improvement, and exchange programs at our station regarding our forage crops.

ICAR-All India Coordinated Project on Forage Crops Jhansi has its one center at Shere Kashmir University of Agricultural Sciences and Technology, Srinagar, and they are actively involved in research on forage crops especially oats and maize besides conducting the multi-location trials of the AICRP Forage crops in their research farm. The Agronomy and crop improvement divisions of the university are also actively involved in forage crop research. Sabzar variety of oats is one major contribution of the university.

The contribution of Himachal Pradesh Krishi Vishwavidyaliya Palampur towards forage crops is also worth mentioning. As per the Annual Report of the University (2003), they had evaluated different grasses under rangeland conditions over the years, which revealed the superiority of *Setaria anceps* var. *Setaria*-92 in terms of adaptability and productivity in the cool sub-tropical Himalayan region. The variety was released in subtropical grasslands and pastures of Himachal Pradesh and Uttaranchal. They have also identified *Lespedeza* as one of the best perennial legumes suitable for harsh dry temperate conditions in the state. They evaluated improved varieties of tall fescue and the results revealed that three strains, viz., Hima-1 (8.89 t/ha), Hima-3 (9.94) and EC-178184 (11.66 t/ha), were statistically at par with check, Hima-4 (9.94 t/ha) in term of forage yield.

TABLE 7.1   Details of Different Collections of Forage Crops Maintained at Regional Research Station, IGFRI Srinagar

SL. No.	Crops	Botanical Name	No of Germplasm Collections	IC Nos...
1.	Tall Fescue	*Festuca arundinacea* (Schreb) Hack	25	IC-0615879... IC-0615903
2.	Orchardgrass	*Dactylis glomerata* L.	26	IC-0615904... IC-0615929
3.	Harding grass	*Phalaris* hybrid	6	–
4.	Prairie grass	*Bromus unioloides* Kunth	33	IC-0615930... IC-0615962
5.	Timothy grass	*Phleum pratens* L.	30	IC-0615739... IC-0615768
6.	Red Fescue	*Festuca rubra* L.	44	IC-0615835... IC-0615878
		Temperate Grasses	**164**	
7.	White clover	*Trifolium repens* L.	28	IC-0615796... IC-0615823
8.	Red clover	*Trifolium pratense* L.	27	IC-0615769... IC-0615795
9.	Sainfoin	*Onobrychis viciifolia* Scop.	11	IC-0615824... IC-0615834
		Temperate Legumes	**66**	

7.7 PASTURE AND GRASSLAND IMPROVEMENT

The productivity of natural pastures and grasslands has been declining at they need measures to improve their health and increase stocking. The productivity of these unimproved native grasslands is low due to the nature of their species, which have been persisting despite low nutrient status and the unfavorable climatic conditions. Though the species robustly re-grow after their dormant season, they have low fodder productivity, low crude protein concentrations, and even low digestibility. Such conditions can be overcome by replacing the native species by productive and persistent legumes or grasses or by dribbling legume seeds wherever possible, use of non-N fertilizers or planting non-exacting leguminous species with good fodder quality traits (Shaw and Byron, 1976). Without persistent and effective legumes, the amount of soil nitrogen and mineral availability limits the production of forages to half (Mannetje, 1997). The chief objective should be increasing the pasture/grassland productivity and optimizing the animal production. Due care must be taken of the harvesting/grazing stage in the grassland/pasture as the fodder quality declines after the flowering stage, though the problem is not severe in temperate regions for C_3 grasses.

Efforts must be directed in an integrated way after understanding the complex fodder production system and activities must be aimed at improving the soil health, grazing management, multi-animal grazing system besides the botanical composition of these natural pastures and grasslands. The involvement and support of local farmers and pastorals is very important for the pasture and grassland improvement activity, for which awareness camps and demonstrations should be held by the line Departments and research institutions working in this sector.

7.8 LOCAL GUARDIANS AND FCGRS

Local guardians of these grasses and legumes are the local farmers and tribal pastorals, who feed their livestock by these forage crops and feed the world meat and other livestock products, though they themselves have remained poor and even hunger-stricken in different parts of the world (Batello et al., 2008). In the north-western Himalayan states, they include the Gujjars, Bakerwals, Gaddis, Chopans in specific and local farmers associated with livestock rearing in general. They play an important role in propagation, conservation, documentation, and management of these FCGRs in the following ways:

1. They are reservoirs of local knowledge about FCGRs, including their identification, use, propagation, reproduction, and the local social, cultural, and religious taboos associated with these crops.
2. They can be helpful in planning for *in-situ* as well as *ex-situ* conservation of FCGRs through sacred grooves, biosphere reserves, seed lots, and even on-farm cultivation approaches.
3. They conserve the wild relatives of major crops *in-situ,* and the same can be used in different crop improvement programs.
4. They play an immense role in grazing management and regeneration of grasslands and natural pastures. The obnoxious weeds and other unwanted dominating vegetation can be checked through grazing management in natural grasslands and pastures.
5. They play a vital role in wildlife management vis-a-vis grassland/pasture management.
6. The local guardians play an urgent role in conducting scientific germplasm explorations in unexplored areas. Having knowledge of the local vegetation, they can be helpful in accessing the local novel germplasm of crop genetic resources and widen the genetic base of available forage crop and food crop genetic resources against biotic and abiotic stresses.
7. Pasture/grassland improvement cannot be achieved without the support of their local guardians. Their cooperation is highly required in reseeding, introduction of legumes for nitrogen fixation and forage value improvement and grazing management of these pastures.
8. Their role in execution of different plans and programs of different line departments like animal husbandry, agriculture, and forestry, research agencies, national, and international organizations for pasture improvement, restoration of degraded grasslands and pastures, their maintenance, and management, should be highly acknowledged.

7.9 CONCLUSION

FCGRs of the northwestern Himalayan region are extremely important for poverty alleviation and food security of the local inhabitants for their role in animal production systems, besides their environmental significance viz., soil, and water erosion control, carbon sequestration and climate change mitigation, biodiversity conservation and wildlife habitats amongst others. There is huge scope to enhance the production and availability of green fodder through various technological interventions and the underutilized FCGRs in the region. There is a huge gap between the demand and supply of both green and dry fodder in the region, which affects the livestock productivity badly. On the other hand, the FCGRs in the region are under threat due to genetic erosion and narrowing of genetic base as a result of conversion of grasslands and pastures to commercial agriculture, mismanagement, and degradation of grasslands and pastures, consideration of FCGRs as orphan crops, ignoring the traditional knowledge of local farmers and pastorals regarding FCGRs, and the small number of species and cultivars under selection, use, maintenance, and improvement.

Broadening of genetic base and improvement programs of FCGRs are required, which can be achieved through germplasm exchange at national and international levels, explorations from different zones of the region in general and biodiversity hotspots/reserves in particular, cooperation between traditional guardians of FCGRs and research agencies for their identification, collection, and evaluation. Reclamation of degraded grazing and pasture-lands through plantation of underutilized forage crop species should be done. Strategies to increase forage production per unit area and encouraging forage production in mixed farming systems is an open option. Strategies for efficient utilization of available resources like crop residues, non-conventional feeds and fodders need to be formulated properly. Stakeholder participation and institutional support is important for their promotion. The different agencies and institutions working on FCGRs in the region should collaborate and share knowledge and germplasm, explore new populations, evaluate, and improve the existing FCGRs and develop a database of the same for end-users. Besides this, International treaty on their conservation and germ-plasm exchange is needed for sustainable fodder production and livestock productivity in the region.

KEYWORDS

- **forage crop genetic resources**
- **genetic erosion**
- **legumes**
- **livestock production**
- **plant genetic resources**
- **policy measures**

REFERENCES

Ahmad, S., Khan, P. A., Verma, D. K., Mir, N., Singh, J. P., Dev, I., & Roshetko, J., (2017). Scope and potential of hortipastoral systems for enhancing livestock productivity in Jammu and Kashmir. *Indian Journal of Agroforestry, 19*(1), 48–56.

Ahmad, S., Mir, N. H., Bhat, S. S., Verma, D. K., & Suman, M., (2019). *Temperate Perennial Forage Grasses: An Introduction* (pp. 1–8). IGFRI Grasses Folder.

Anonymous, (2002a). *Perennial Ryegrass* (p. 2). USDA NRCS Northeast Plant Materials Program (online).

Anonymous, (2002b). *Tall Fescue* (p. 2). USDA NRCS Northeast Plant Materials Program (online).

Batello, C., Brinkman, R., Mannetje, L. T., Martinez, A., & Suttie, J., (2008). *Plant Genetic Resources of Forage Crops, Pasture and Rangelands* (pp. 1–62). FAO Publication.

Bhat, S. S., Ahmad, S., Mir, N. H., & Suman, M., (2019). *Temperate Forage Legumes for Sustainable Livestock Production* (pp. 1–8). IGFRI legumes Folder.

Carbonero, C. H., Mueller-Harvey, I., Brown, T. A., & Smith, L., (2011). Sainfoin (*Onobrychis viciifolia*), a beneficial forage legume. *Plant Genetic Resources: Characterization and Utilization, 9*(1), 70–85.

Dabadghao, P. M., & Shankarnarayan, K. A., (1973). *The Grass Cover of India* (p. 713). ICAR, New Delhi.

Earagariyanna, M., Jagadeeswary, V., Kammardi, S., & Sriramaiah, M. (2017). Fodder Resource Management in India: A Critical Analysis. International Journal of Livestock Research, 7(7), 14–22.

FAO, (1997). *The State of the World's Plant Genetic Resources for Food and Agriculture* (p. 501). Food and Agriculture Organization of the United Nations, Rome.

Goplen, B. P., Richards, K. W., & Moyer, J. R., (1991). *Sainfoin for Western Canada*. Agriculture Canada Publication, 1470/E.

Harlan, J. R., (1983). The scope of collection and improvement of forage plants. In: McIvor, J. G., & Bray, R. A., (eds.), *Genetic Resources of Forage Plants* (pp. 3–14). Commonwealth Scientific and Industrial Research Organization, East Melbourne, Australia.

Mannetje, L. T., (1997). Potential and prospects of legume-based pastures in the tropics. *Tropical Grasslands, 31*, 81–94.

Mir, N. H., Ahmad, S., & Bhat, S. S., (2018). *Dactylis glomerata* L. (Cock's Foot/Orchard Grass), A potential temperate forage grass for cultivation in north-western Himalaya. *Advances in Research, 15*(5), 1–10.

Moyer, B. G., & Gustine, D. L., (1984). Regeneration of *Coronilla varia* L. (crown vetch) plants from callus culture. *Journal of Plant Biotechnology, 3*, 143–148.

Mukherjee, A. K., & Maiti, S., (2008). *Forage Crops- Production and Conservation* (p. 255). Kalyani Publishers, New Delhi.

Radotra, S., Dev, I., & Ahmad, S., (2015). Pasture and forages in north-western Himalayan region: Current status and future strategies. In: Bhar, et al., (eds.), *Current Status and Future Prospects of Animal Production System in North Western Himalayan Region* (pp. 49–57.). IVRI Palampur (H.P).

Reynolds, P. J., Jackson, C., Lindahl, I. L., & Henson, P. R., (1967). Consumption and digestibility of crown vetch (*Coronilla varia* L.) forage by sheep. *Agronomy Journal, 59*, 589–591.

Shaw, N. H., & Bryan, W. W., (1976). *Tropical Pasture Research: Principles and Methods*. Commonwealth Bureau of Pastures and Field Crops Bulletin No. 51.

CHAPTER 8

Livelihood Security and Forest Resource Extraction by Forest Fringe Communities in Indian Himalayan Region

MUNEESA BANDAY, M. A. ISLAM, NAZIR A. PALA, MEGNA RASHID, PEERZADA ISHTIYAK AHMAD, M. M. RATHER, and RAMEEZ RAJA

Faculty of Forestry, Sher-e-Kashmir University of Agricultural Sciences and Technology of Kashmir, Benhama, Ganderbal–191201, Jammu and Kashmir, India, E-mail: 13forestry08@gmail.com (M. Banday)

ABSTRACT

Unsustainable exploitation of resources from forests often leads to forest degradation. A number of studies have reported the negative impacts of overexploitation of forest resources and its influence in loss of the diversity of native species in various regions. The correlation analysis between overall forest resources utilization pattern and forest products consumption pattern and various independent variables of the respondents have exhibited positive and significant association with size of landholding, farm income, farming composition, fanning experience and herd size. Increased household size and low level of literate population greatly influence the quantity of forest products gathered. A positive relationship between incomes and fodder consumption has been reported for fodder collection resources. The correlation between income and fuelwood consumption has largely been found negative. Agricultural induced land-use change has been the leading cause of forest degradation in developing countries. Livelihood induced forest extraction like grazing, fuelwood removal, etc., has also depleted the biomass of the forests of India. Basically, in rural and urban sectors of India, most of the population primarily depends on forests to meet their requirement. Joint forest management (JFM) is conceived as an efficient strategy for sustainable development of the people. Sustainable extraction of forest resources by the forest dwellers has been emphasized as a prominent strategy for their income and employment generation. Agroforestry plantation is recognized as the most competent land-use option for ensured sustainable development, increased productivity of land, eased environmental stress, and enhanced livelihoods.

8.1 INTRODUCTION

The World Bank has estimated that 1.6 billion people of the world depend on forest for livelihood (Sraku-Lartey, 2014). Forest resources play an important role in food security, fodder/livestock security, agricultural support, bio-energy security, housing security, cottage industry, health security, socio-cultural security, income security, and employment security for local people in developing countries (Shit and Pati, 2012). Thus a country with a population like India and high poverty induced dependence or exploitation leads to the perpetual decrease in forest wealth, especially when such a huge proportion (40%) of individuals live in the vicinity of forests (Saha and Guru, 2003; MoEF, 2009). India has a forest-dwelling population of over 100 million belonging to 550 communities of 227 ethnic groups, of which 60% live in forest and depend on for sustenance (Nautiyal et al., 2002). The tribes live closely to the forest have rich traditional knowledge for the uses of natural resources (food, fodder, shelter, and healthcare) for their sustenance (Kumar et al., 2012).

8.2 FOREST RESOURCE OF INDIA AND BIOMASS CARBON OUTFLOW

Around 44%, 11% and 45% of terrestrial vegetation carbon (650 billion tons of Carbon) is stored in biomass, dead wood and soil, respectively (FAO, 2010). More than 45% of terrestrial carbon is contributed by forests with an annual exchange of 10% carbon with the atmospheric carbon (Bonan, 2008; Schimel, 1995). Quantification of terrestrial carbon poses a great challenge to the researchers owing to its high level of uncertainty in the previous estimates (Bradford et al., 2009). In case of forests, the factors that influence carbon sequestration in forests like forest extraction by fringe communities, timber felling, grazing, etc., need to be taken into account (Schulze et al., 2000; Randerson et al., 2002). In India, the total estimated carbon stock in forests varied from 3325 to 3161 Mt during the years. There was a net flux of 372 Mt of CO_2 in assessment period I (ASP-I) 2003–2005 and 288 Mt of CO_2 in

ASP-II (2005–2007). The total forest biomass stored in Western Himalayas was 1336.35 1348.52 and 1397.58 Mt for the years 2003, 2005, and 2007 respectively. The carbon stock in general has decreased continuously from 2003 onwards, despite a slight increase in forest cover. With nearly 173000 villages classified as forest fringe villages, there is obviously a large dependence of communities on forest resource and thus higher flow of carbon stock to these areas (Sheikh et al., 2011). The climate change mitigation and adaptation projects based on forests widely promoted to enable households to adapt the challenge on it (Chia et al., 2013; Rennaud et al., 2013; FAO, 2015). Hajost and Zerbock (2013) and reported that forest-based adaptation initiatives to succeed if they built on the lessons learned from community based forest management. Therefore, reliable information of those factors that influence rural people's engagement in sustainable use and management of forest resources and socio-economic characteristics of people influence them is crucial (Cardona, 2005; Chia et al., 2013).

8.3 FOREST-BASED LIVELIHOODS AND RURAL SOCIO-ECONOMY OF FOREST FRINGE COMMUNITIES

The rural households and forest-dependent communities closely interact with forest and derive economic livelihood and often their cultural and spiritual identity (Byron and Arnold, 1999). Forest resources have been identified as one of the key sources for livelihoods and food security of tribal households (Dovie, 2003) because of their free access and subsistence to the poorest of the poor receive priority for their development and management. In India, more than 50 million people dependent on NTFPs for subsistence and cash income (Sarmah and Arunachalam, 2011). NTFPs provide 50% of the household income for 20 to 30% of the rural population, particularly for the tribal (Sharma et al., 2015). About 16,000 plant species in India, nearly of total 3000 species yield 20% NTFPs, and only about 126 species (0.8%) have been developed commercially (Singh and Quli, 2011). Approximate 50% of forest revenues and 79% of forest-based export in the country is based on NTFPs. Thus, NTFPs are mainstays of income and subsistence for many tribal communities (Opaluwa et al., 2011). Chaudhury et al. (2004) suggested that the socio-economic status conditions of forest dwellers associated with the type and extent of forest. Households' socio-economic characteristics indicate both; what the forest resources are utilized and also the extent to which they are harnessed (Mamo et al., 2007; Vedeld et al., 2007; Babulo et al., 2008).

Tribal regions in Himalayas are sparsely populated in small settlements with high dependence on rain-fed agriculture and adjoining forests (Pandey and Gupta, 2013). At higher elevations of the Himalayas, little agriculture is practiced due to severe climate and shallow soils (InderDev, 2001). The rearing of livestock in mountain is one of the important occupation (Singh, 1995), contributes 20% of the household cash income (Tulachan and Neupane, 1999). Stall-feeding is frequent basically for buffaloes and cows grazed outdoors mostly in forest areas. Pandey and Gupta (2013), in a study in Uttarakhand state of Indian Himalayas, found that the average household size in such villages vary from 5.36 persons near open forests to 6.34 persons near moderate density forests. The majority of the people above 60 years of age were uneducated, whereas the literacy rate was seen higher in young age classes. The landholding of such households was less than 0.5 ha due to hilly terrain and lack of irrigation facilities. In the same study, reported that the average livestock size per household was 1.58 (cows), 0.97 (buffaloes) and 1.36 (goats). Fodder is collected from forests mostly by women, which is more frequent in summertime and less frequent in winter. The average time spent per day was found to ranging from 2–5 hours depending on the distance of households from the fodder source areas. The mean annual cash income per household ranged from ₹ 4,272 to 5,178, which is well below the poverty line as per the Government of India which is prescribed as below ₹ 12,000 in rural households (Sharma et al., 2009). In a study (Hussain et al., 2016) in Uttarakhand on Van Gujjars living near Corbet Tiger Reserve, it was reported that the household size varied from 2 to 28 persons. The overall average family size was 13.69 members. The main sources of livelihood were pastoralism and sometimes labor. The average cattle unit (ACU) holding was found to be 29.21 per household. Livestock possession was found to have a significant bearing on family income. The houses were of traditional type, i.e., mud-walled and thatched roof. Kumar et al. (2010) reported that the ratio of female to male population is more in Tehri Garhwal Himalayas because male population migrate other places for search Job for livelihood. The average livestock per family varies from 4.43 to 5.40 and average landholding of 0.26 ha (Manjyar) to 1.52 ha (Jaul). All surveyed families engaged in farming and their average annual income (per family per year) is very low vary from ₹ 4000 to ₹ 6500 per month. Fodder and fuelwood collection done by nearby forests and mainly women and children play role for collection and spend 3 to 4 hours. It was revealed that the grazing of animals and collection of fodder and fuelwood were the major causes of environmental depletion in the region. In a study conducted in Garhwal Himalayas (Bijalwan et al., 2011), it was found that the economy of these households was predominantly livestock-based (37–56% families), with each household having an average of 2–4 milch animal. Hence these areas have a higher fodder requirement (305.02 to 1015.17 kg/day/village to 659.53–2015.52 kg/day/village in summer and winters respectively). Since the landholding size was mostly marginal (0.03–5.6 ha per family), most of the dependence was on forests. In terms of fuelwood, the daily consumption was found to be 84.41–538.48 kg/day/village (summer) to 156.75–701.01 kg/day/village (winter).

In another sub Himalayan region of Chilapatta reserve forest in West Bengal, an investigation by Biswakarma et al. (2018) revealed that 91% of the respondents dealing with ethnobotanical plant extraction were male. The inhabitants of the forest fringe villages comprise of indigenous tribes of Mech, Rava, Oraon, Chikbaraik, and Cherwa along with some Nepali, Bengali, and Bihari communities. Young age class (35–52 years old) formed the majority with 49% of respondents falling in the age class. 91% of the respondents were literate and the majority of respondents were farmers (68%) with medium family size (52%) and marginal landholding (78%). The families were mostly belonging to a lower-income group (94%) and had lower expenses (82%). Sharma et al. (2012) conducted a study on the Van Gujjar tribe inhabiting a sub-Himalayan tract in the North-Western Himalayas of Uttarakhand State. A total of 176 households were interviewed (giving equal weight to all economic classes and family every size) using well pre-structured questionnaires. The education level was reported very low (12.9%) and the average income per household were recorded as ₹ 36000 per year. The major source of income was dairy production (80.6%) followed by labor employment (13.9%), NTFPs (4.2%) and agricultural production (1.4%). In a demographic study of Munda, Oraon, and Lohara tribes in Jharkhand, India (Islam et al., 2015) reported that majority of the respondents in these tribal villages were middle-aged (53.05%) with low education (up to primary level). The prevalent family type was nuclear (63.42%) with large family size (59.76%). Most of the respondents had marginal landholding (46.95) with an average 6–10 livestock size and income from forests was up to ₹ 8000/annum and gross annual income of ₹ 30,000/annum.

The type of extraction taking place in an area has been found to be dependent on the dominant community and their lifestyle. The majority of the people using medicinal plants for health care and livelihood security belonged to scheduled tribe in the high altitude cold desert in Leh (Kala, 2006). Pandey and Mishra (2011) found that the level of social participation among majority of the livestock farmers were low having grousing impact in rural economy of hilly region, Uttarakhand. Prakash and Sharma (2008) reported that the majority of the people participating in the forest fire control of Himachal Pradesh were having marginal and small size of landholding. The herd size or status of livestock possession among the majority of the respondents was found to be medium (Prakash and Sharma, 2008; Pal, 2009; Bijalwan et al., 2011). Kala (2006) reported that the majority of the respondents had owned one mixed house for their dwelling in the rural Leh. Sharma et al. (2015) found that agricultural and non-agricultural labor remained the main occupation of the majority of the tribal and non-tribal respondents in Arunachal Pradesh. Baba et al. (2015) found that a considerable percentage (44.74%) of the respondents belong to medium-income category, followed by low income (31.58%), high income (20.17%) and very low income (3.51%). The mean score (2.82) established the preponderance of families having low gross annual income ranging between ₹ 30001 to ₹ 60000/annum in the study area. Atta (2018) illustrated that a considerable percentage (55.34%) of the respondents belong to medium-income category, followed by low income (23.30%), high income (18.45%), and very low income (2.91%). The average income of ₹ 89094.28 established the preponderance of families having medium annual income ranging between 60001 to ₹ 90000/annum in the surveyed population. Sheikh (2015) reported that the majority (63%) of the respondents belong to medium wealth status group followed by low (20%) and high (17%). The average score of household wealth status of the respondents was recorded to be 21.95. Baba et al. (2015) reported that a considerable proportion of the respondents belonged to medium material possession category. Baba et al. (2015) also indicated that maximum households (40.35%) have urban closeness of 5–10 km followed by 10–15 km (35.09%), < 5 km (14.91%) and >15 km (9.65%). The average urban closeness of the sample households was found to 9.78 km in the locality.

8.4   FOREST RESOURCE DEPENDENCE OF FOREST FRINGE COMMUNITIES ACROSS INDIAN HIMALAYAN

India has a huge population living close to the forest with their livelihoods critically linked to the forest ecosystem. Millions of people of forest dwellers of the globe depend on forest resources for livelihood however; dependency varies from place to place (Akhter et al., 2009). The Indian Himalayan Region (IHR) is unique in the mountain ecosystems. Despite the abundance of natural resources, most of its consumption and preference of woody species and so on people are marginalized and still live on a subsistence level (Singh, 2006). In the past few decades, the Himalaya has unprecedented changes due to land use because of rapid population growth and human activities (Pandit et al., 2007). Forests are extensively used for grazing, fuelwood collection, and numerous other subsistence needs by rural people (Rahmani, 2003). Due to poverty and employment problems, the local people made their dependency on forest resources for fuelwood, fodder, non-timber forest products (NTFPs), etc. Because of this extra pressure on most of the Himalayan forests which caused less productive (Saxena et al., 2005). Thus to understand consumption pattern for assessing human-environment interactions and its effective conservation policies of this region is essential.

8.4.1 FUEL WOOD EXTRACTION

While studying availability and consumption pattern of fuelwood, fodder, and small timber in rural Kashmir, Islam (2008) revealed that the most utilized forest resource for fuelwood, fodder, and small timber availability by the people in descending order were farm forestry, social forestry, homestead forestry, forest, and pasture. Fuelwood was mostly consumed for cooking, followed by heating, cottage industries, community functions, and other purposes. Maximum quantity of fodder was consumed for cattle rearing, followed by sheepry, goatry, and horse husbandry. Similarly, a major portion of the small timber was consumed for packing cases, followed by agricultural implements, furniture, housing, cart, and carriages building, cattle shed/storehouse, fencing, scaffolding, ladder, and cremation. The level of overall forest resource utilization pattern for fuelwood, fodder, and small timber for various purposes by the people was high. Singh and Sundriyal (2009) concluded that the fuelwood consumption in Central Himalayan village is 418.86 MT and the annual fuel availability is 211.03 MT having a deficit of 207.83 MT whereas the total available fodder is 281.76 MT but the total consumption is 402.72 MT having a deficit of 207.83 MT. Sati and Song (2012) observed that forest biomass consumption varies from 13 kg/day/household (fuelwood) and 12 kg/day/household (fodder) in the lower elevation (1150 m) in comparison to 28 kg/day/household (fuelwood) and 34 kg/day/household (fodder) in the higher elevation (1900 m) of the Uttarakhand Himalaya. Balachander and Ganesan (1993) studied that Mudumalai is biologically rich wealth both plant and animal life. Increasing human population and current patterns extraction of NTFPs, fodder, and fuelwood so high that it threatens the integrity of the ecosystem. It was revealed that policies relating to grazing and basic energy requirements were needed to be re-examined, and mechanisms must be developed to increase rural incomes.

The pattern of fuelwood extraction has been different for different areas and different climatic zones. Islam et al. (2011) studied the status of fuelwood extraction and consumption in Sagipora village of Sopore tehsil and reported that the total extraction of fuelwood from different sources in the surveyed population was found to be 696.28 t/year. Out of the total extraction, 517.25 tons comes from silvi-horticulture plantation, 97.48 tons from homestead farming, 62.67 t from forest, fuelwood depots, and rest 18.88 t from community forestry. As a whole, about 74.29% of the fuelwood requirement is met from Silvi-horticulture plantation and for rest (25.71%) of the fuelwood people rely on other sources such as homestead farming, forest, fuelwood depots, and community forestry. The per capita extraction of fuelwood in the study area was recorded to be 0.28 t/year. The total consumption of the fuelwood in the study area was estimated to be 1044.41 t/year. Out of which, 438.65 t consumed in cooking, 426.74 t in heating, 136.00 t in cottage industries, 32.36 t in community function and rest 10.66 t for other purposes such as *hamam*, household rituals and religious functions. The per capita consumption of fuelwood in the study area was recorded to be 0.42 t/year. Cooking and heating are the major areas where 82.86% of the total fuelwood is consumed, whereas the combined share of cottage industries, community function, and other purposes in total fuelwood consumption is calculated to be 17.14% only. Mishra (2008) reported household behavior related to fuelwood collection and use in rural Orissa. Consumption of forest-based fuelwood was dependent on the income and wage of the household and availability of alternative to fuelwood. Pandey (2002) has analyzed to reveal the myth and reality of the fuelwood studies in India. It is evident that consumption of traditional fuel (fuelwood, crop residue and dung cake) dominates in domestic energy use in rural India, which accounts for 90% of total fuelwood consumption. Fuelwood alone accounts for 60% of total fuel consumption in rural areas, whereas in urban areas pattern of consumption is changing fast because increased use of commercial fuel (LPG, kerosene, and electricity).

Procurement of the fuel and fodder are important activities to the hilly people for which women and children spend long hours in a day. Fuelwood was generally used for cooking of food materials, preparation of food for cattle and pig, and to keep the houses warm during winter season. The consumption of fuelwood was found to fluctuate throughout seasons. Consumption of fuelwood was found higher during winter. The per capita per day consumption of fuelwood was much higher (2.97 ± 0.05 kg in winter and 2 ± 0.36 kg in summer) than people used in Jammu region (1.67 kg in winter and 1.49 kg in summer (Gupta et al., 2009). In high altitude Rudrapryog villages of Garhwal Himalayas where per capita per day were recorded 1.93 ± 0.01 kg to 2.09 ± 0.13 kg in winter and 1.14 ± 0.21 to 1.23 ± 0.16 kg in summer (Khanduri et al., 2002). Chopra and Kumar (2003) observed that forest ecosystems provide a range of products and services for human use because biodiversity inherent in them, and among the commodities, timber is extracted main by market forces as well as NTFPS extracted under a variety of arrangements. A decrease in bio-diversity of the stock may be accompanied, under a certain set of circumstances, with a rising trend in extraction, and at a rising rate. However, an increased biodiversity may imply a decreasing trend in the extraction in the future provided that present extraction does not rise at a rate faster than the rate of increase in the biodiversity. It is clear those trade-offs between timber extraction and the existence of bio-diverse forests providing a variety of goods and services.

Mushtaq et al. (2012) studied that state forests were the major source of fuelwood followed by own farm in Shiwalik Himalayas. Negi et al. (1996) reported that more than 80% of fuelwood requirement is met from forests in Himachal Pradesh. Banyal et al. (2013) conducted the study in Astingo village of Bandipora revealed that 37.48% of the land area is under agriculture followed by horticulture (20.16%), homestead farming (14.86%), agri-silviculture (AS) (12.92%), forestry (08.08%) and silvi-horticulture (06.50%), respectively. Heavy reliance on fuelwood, charcoal, and dung cakes while as electricity and LPG ranked lowest in terms of meeting energy requirements. The total requirement of fuelwood at the study site was estimated 634 t/annum against its availability 405 t/annum. Of total fuelwood consumption almost half (50.20%) used for cooking and the rest for other purposes.

8.4.2 FODDER EXTRACTION

Even though India is the highest milk producing country but still the per capita milk production is very low due to the low quality of feed provided to the cattle and high population. Hence foraging, stall feeding and feeding of agricultural byproducts (paddy straw, oats, and wheat straw) and fodder collected from forests are the predominant methods of providing nutrition and dietary requirements to the animals (Pandey and Mishra, 2011). The study revealed that there was a large gap between requirement and availability of feed at the national level. Nearly 89.44% of fodder in North-Western Himalaya, India was met out from local sources (agroforestry, horticulture, kitchen garden, cultivated grassland, forest grasses both trees + others) remaining 10.56% which were procured from external sources like purchased grass, feed, and oil. Overall, 4.15 t fodder per household is required each year to meet the requirements of livestock. In general, the highest species lopped annually, except *Olea ferruginea, Quercus floribunda, Q. leucotrichophora,* and *Salix fragilis,* which were lopped in an interval of three years. Tomar and Sharma (2002) reported that feeding of tree leaves (dry/green), paddy hay, sun-cured dry paddy, field herbage (Lowe grass) is the common traditional feeding practice of livestock owners during autumn, winter, and early spring season (November to March) in the Kashmir valley. They reported that the valley has surplus fodder production, which exceeded by 3.2 100,000 metric tons overestimated requirements (4.92 million metric tons). Fodder tree leaves alone contributed 76.8% to state fodder resource, followed by paddy hay (11.5%). Cultivated fodders (oat, maize, natural pastures, aquatic vegetations, straws, and strovers) altogether contributing 11.7% to state resources. They also recorded some features of traditional and popular feeding practices in rural Kashmir (Budgam and Srinagar) as migration of livestock on grazing pasture in summer and autumn months and fodder scarcity during the winter months and feeding paddy hay supplemented with aquatic vegetation/tree leaves to livestock with occasional incorporation of concentrates by progressive farmers. Islam et al. (2015b) reported that the average fodder requirement per household was estimated to be 47.77 kg day^{-1} with an annual requirement of 14227.34 tons of fodder in the region. The livestock owners procure about 7255.79 tons of fodder annum^{-1} inclusive of about 4277.39 tons annum^{-1} paddy straw and 2978.40 tons annum^{-1} green grasses, weeds, and other agricultural residues from agricultural fields and homesteads for their animals. Forest grazing is the main source of fodder availability contributing 6971.55 tons of fodder annum^{-1} (49.00%) of the total fodder requirement. The low economic condition, unavailability of pastures or fodder production unit and ignorance towards green fodder production results in higher intensity of grazing in the forests of this area. To relieve forests from the pressure of over-grazing, alternate source of fodder production should be developed, and the existing land resources should be efficiently exploited to get additional grass fodder, tree foliage, herbage, etc., in the area. The contribution of forest biomass in mitigation of fodder requirements of rural areas was emphasized by several workers (Bijalwan et al., 2011; Panta et al., 2011; Sati and Song, 2012; Ajake and Enang, 2012).

8.4.3 TIMBER EXTRACTION

Himalayan rural population is predominantly forest-based in terms of deriving their livelihood support and daily requirements (Shaheen et al., 2011). Wood consumption in the higher and lower altitude villages of district Bagh was found to be 3.76 and 2.19 kg/capita/day respectively with an average of 2.97 kg/capita/day. The demand of timber in India was 92 million m³, i.e., 0.098 m³ or 3.5 cft/capita/year (Rai and Chakrabarti, 2001). Islam et al. (2015b) reported that the average timber requirement per household was worked out to be 0.346 m³/annum, accounting for a total timber demand of 282.49 m³ annum⁻¹. Around 136.36 m³/annum, 69.09 m³/annum, 41.33 m³/annum 35.71 m³/annum of timber for tribal population was supplied from the sources like forests, agroforestry, community forestry sand homestead forestry. The timber extracted is mostly utilized in housing (124.66 m³/annum) followed by agricultural implements (82.71 m³/annum), rural furniture (35.25 m³/annum), carts, and carriages (17.60 m³/annum), fencing (10.23 m³/annum), cattle shed/storehouse (9.10 m³/annum) and others such as scaffolding/ladder/cremation, etc., (2.94 m³/annum) by the tribal people in the area. A number of workers have reported the enormous pressure on forest biomass for timber security among rural communities (Sapkota and Oden, 2008; Sarmah and Arunachalam, 2011; Sati and Song, 2012).

8.4.4 NON-TIMBER FOREST PRODUCTS (NTFPS) EXTRACTION

Collection, processing, and sale of various NTFPs are the main activities around which forest-based livelihood of fringe communities revolve. Among these NTFPs include Lac, edible mushrooms, bamboos, dyes, tannins, wax, aromatic oils, etc. The resources of forest are generally used for purposes of construction, food, fuelwood, fodder, medicine, agricultural implements, and many others are significant for subsistence to the villages. The findings (Islam et al., 2017) revealed that average income earned from NTFPs was ₹ 4791.16/household/annum which is differentiated as cottage industry (36.26%), fruit (13.49%), fuelwood (11.40%), toothbrush (8.55%), fodder (8.15%), mahua (*Madhuca latifolia*) flower (7.22%), oilseeds (6.74%), vegetables (4.31%) and ethnomedicine (0.88%). Among the NTFPs classes cottage industry accrued maximum (₹ 1881.09) income while ethnomedicine procured lowest (₹ 42.07) in the tribal households. Fuelwood represented the highest (67) number of species followed by fodder (37), fruit (10), vegetable (7), oilseed (5), toothbrush (4), ethno-medicine (3), cottage industry (2) and mahua flower (1). Peak seasonality of NTFPs exploitation is limited for only 3–4 months. Average gross annual income was ₹ 27894.20/household/ annum composed of agriculture (36.24%), NTFPs (17.18%), wage labor (9.75%), livestock (8.86%), business/shopkeeping (8.72%), timber (7.83%), service (6.78%), and others (4.63%). Samrah and Arunachalam (2011) studied that forest of Arunachal Pradesh support rich diversity of timber as well as non-timber yielding species. NTFPs such as fuelwood, house building materials, wild edible vegetables, and medicinal plants were mostly collected from natural habitat. The total contribution of NTFPs to annual household income was maximum (23% of the total income) in the villages of Miao circle followed by Diyun circle (21% of the total income) and Nampong circle and Vijayanagar circle (19% and 18% of the total income respectively). It was recorded minimum (11% of the total income) in the villages of Bordumsa circle. The study (Opaluwa et al., 2011) revealed that most of the NTFPs collectors were females, married, and had large household size, earning between N10000-N20000 from the sale of these products. It was further revealed that herbs, fuelwood, locust bean, bush meat, palm fruit, ogbono (*Irvingia gabonenses*), and palm wine were the major NTFPs collected in the area. Non timber Forest Produce like purun, honey fish, and fruits has been reported to be the primary income source of households in the swamp forests of Sumatra of Indonesia (Wildayana and Armanto, 2018). Other minor forest produce extracted from these forests included Wild fruits, vegetables, medicinal plants, mushrooms, nuts, nests of swallow birds, materials for handicrafts like thatching grass, brooms, dyes, etc., it was also reported that the extraction of NTFPs was declining since 1970 due to loss of biodiversity and the extent of the forest area was also decreasing sharply.

Around 85% of rural Indian population used forest species for medicinal purposes and the remoteness is the primary cause of their relatively higher dependence on forest-based traditional medicine (Fransworth, 1992; Jain, 1992). Use of such forest species in different forms of traditional system of medicine in such communities have already been reported from Tamil Nadu (Muralidharan and Narasimhan, 2012), Ladakh (Ballabh and Chaurasia, 2007) and West Bengal (Sinhababu and Banerjee, 2013). The exploration, utilization, and conservation of these ethno-botanic resources are essential for the restoration and preservation of traditional and indigenous knowledge (Narzary and Basumatary, 2013). In an attempt to document the ethnome-dicinal plant resources used in the treatment of stomach ailments in the Chilapatta Reserve Forest in West Bengal by Biswakarma et al. (2018), a total of 43 plant species were reported in which 22 were trees, 11 were herbs, 6 were shrubs and 4 were climbers. Maximum of these plant species

(15) were found effective against gastroenteritis, followed by dysentery (14) and stomach pain (9). In maximum cases, leaf (20) was the plant part used, followed by fruit (13) and bark (8). In the same area of Chilapatta Reserve Forest, Raj et al. (2018) reported a total number of 140 ethnobotanical plants used to treat 58 diseases, out of which 62 species were collected from forests, 52 species were planted, and 26 species were both planted and wild. *Rauvolfia serpentina* was the most valuable species in terms of its maximal use and higher use-value. In west Purulia district of West Bengal, 57 plant species were reported as medicinal (Mondal and Rahaman, 2012). The forest fringe communities of Birbhum district of West Bengal and Dumka district of Jharkhand were reported to use 28 medicinal plant species against 10 kinds of ailments (Fransworth, 1992). Around 46 plant species are used ethno-botanically in the form of infusion, decoction, oil pastel, and latex by the hill tribes and aborigines belonging to West Rarrh region of West Bengal as reported by Ghosh et al. (2013). Such kind of dependence on forest species for medicine was also reported from Midnapur district (Chakraborty and Bhattacharjee, 2006), Bankura (Sinhababu and Banerjee, 2013) and Purulia (Jalal and Garkoti, 2013) districts of West Bengal.

8.5 FACTORS INFLUENCING LIVELIHOOD ASSOCIATED FOREST BIOMASS CARBON OUTFLOW

The correlation analysis between overall forest resources utilization pattern and forest products consumption pattern and various independent variables of the respondents have exhibited positive and significant association with size of landholding, farm income, farming composition, fanning experience and herd size (Islam, 2008). A negative and significant association was reported for education, non-farm income, and urban contact, whereas age, family composition, parental occupation, and distance from the forest have shown insignificant relationship. Biswakarma et al. (2018) in a study at Chilapatta Reserve Forest in sub Himalayan region of West Bengal (sample size = 100) reported that casual variables like gender, caste, education, landholding, house type, material possession had a significantly positive correlation with the predicted variable of ethnobotanical plant extraction and use whereas age, family occupation and family size was found to have a negative correlation. Men (assigned higher score for evaluation) were found to be more assertive in taking the responsibilities in outside activities due to societal norms. It can also be attributed to the reason the men had more medicinal plant knowledge than woman because of the fact that they were favored in the transfer of knowledge (Teklehaymanot et al., 2007). However, in some cases men and women showed no difference in ethnobotanical plant knowledge (Fassil, 2003; Geng et al., 2016). Education also paves the way for a logical and analytical mind. Highly Literate people in many studies (Gedif and Hahn, 2003; Giday et al., 2009) were found to show less knowledge of ethnobotanical plant usage as they were more exposed to modern medicine. However, in the case of areas devoid of modernization, such difference doesn't exist. The age of a person was reported significantly affecting traditional knowledge (Bortolotto et al., 2015; Geng et al., 2016) as a higher number of respondents (both men and women) above 40 years demonstrated greater ethnobotanical plant use in comparison to respondents falling in the age group of 18–40 years. Sood et al. (2008) reported that the majority of the agroforestry farmers were middle-aged, and there was no association between on-farm tree cultivation and age of the head of the household. Further, the education of head of household did not reveal any influence on on-farm tree cultivation. Pal (2011) has shown a varied association of age of Lac growers in Kanker district of Chhattisgarh. Ajake and Enang (2012) found in rural communities of cross-river state, Nigeria that the exploitation and management of forest resources in the rainforest are significantly affected by demographic and economic factors. Regression analysis indicated a low positive relationship between demographic factors such as active age of the population household and number of literate population with the quantity of the forest products harvested. It is also reported that the quantity of forest products gathering also influenced due to increased household size and low level of literate. In another study by Adhikari et al. (2004), education was found to be negatively correlated to the extraction as education provided better employment opportunities and reduced forest dependency; indicating that a higher level of education of the family members makes fuelwood, fodder, and NTFP extraction unprofitable due to higher opportunity cost of time in collection as opposed to time spent in a job or service, etc. Singha et al. (2006) found that there was a positively significant relationship between the education level and participation level of the respondents in maintenance practices of forest resources.

Animal husbandry skills also influenced households' dependence on forest. In general, the people who practiced cattle rearing grazed them freely on pastures in the forests (Musyoki, 2012). This was attributed to the reason that the relatively low cost of forest resources, e.g., fodder, NTFPs, etc., played a major role in the choice of livelihood strategies and to survive shocks like crop failure and job loss. Livestock are considered as the capital that people fall back on to raise money to survive crisis situations.

In studying the forest resource use patterns in relation to socio-economic status, Chandra et al. (2008) observed that more than 75% of fodder and fuelwood were extracted from the forests in Garhwal Himalaya. The study revealed a positive relationship between income and livestock population (0.995) which reveals the strong role of animal husbandry in the rural economy. The equally positive relationship between incomes and fodder consumption (0.930) can be attributed to the extraction of large quantities of fodder. The correlation between income and fuelwood consumption was found to be negative (–0.882), the likely reason being poor economic conditions, leading to dependency on the forest for fuelwood as a free source of energy. Bwalya (2011) and Pascaline et al. (2011) reported that the income and consumptive demands of households for forest resources increase the dependency on forest-based resources. Panta et al. (2011) revealed that 81% of all households constantly used forest for fuelwood and fodder collection respectively, while 42% of households used forest or forest fringe for grazing in Chitwan, Nepal. The mean income was significantly different between income groups (rich, medium, and poor). The extraction of fuelwood, fodder, and other forest products was significantly different between the income groups with F-statistic = 16.480, 19.930, 29.956 at P = 0.05, respectively. Similarly, landholding size and education were also significantly different between the income groups. These findings suggested that the income status of households was the major indicator of forest dependency while poor and medium groups were highly dependent on the forests for fuelwood, fodder, and other products. Adhikari et al. (2004) in a study in the mid-hills of Nepal, found that the forest products, i.e., fodder, fuelwood was income-sensitive, and it was evident from their study that the households with larger endowments extract more forest produce than the lower-income groups. The given findings were in contrast with the study by Hussain et al. (2016) which reported that for Vangujjars of Uttarakhand, all the income groups showed the same level of fuelwood extraction as there was no available alternative source of energy and hence their sole dependency on forests irrespective of income groups. It was also found that the lower caste people, also known as "Shudras" (untouchable caste group) extracted lesser forest produce than the higher caste groups. This was because of the possession of fewer cattle animals in a lower caste family than an upper-caste one. Sapkota and Oden (2008) found that a household's wealth status exerts a strong influence on appropriating fuelwood from the forest in Terai community of Nepal. Above all, the income status of households was found to be a key determinant of household's fuelwood collection from the forest. Poor households were highly dependent on the forests for fuelwood (average annual extraction =

4561.3 kg/household) in order to sustain their day-to-day livelihood. The high dependence of poor coupled with their large population size in the region (>27%) will possibly cause forest degradation in the future. In a comparative study of fuelwood consumption between Vangujjars and adjacent villagers near Corbett Tiger Reserve (Hussain et al., 2016) observed that the family size was the determining factor in the utilization of fuelwood as lower fuelwood consumption/household was found in small families (17.80 kg/day/family) while the highest consumption/household was found in very large families (28.39 kg/day/family). Whereas; the fuelwood consumption rate per capita was found to decrease when the family size increased, i.e., 428.83 kg/capita/year in very large family as compared to 2165.66 kg/capita/year in small families. The same results in per capita fuelwood consumption were reported by Kituyi et al. (2001) and Hosier et al. (1985) in Kenya, Marufu et al. (1999) in Zimbabwe, Mahapatra, and Mitchell (1999) in India, Kumar and Sharma (2009) in Garhwal Himalayas, India, and Kersten et al. (1998) in Nigeria. These values were however higher than the ones reported for Himalayan villages, e.g., 442 kg/capita/year (Negi and Todaria, 1993); 1.23 kg/capita/day (Mahat et al., 1987) in Nepal; 1.26–1.95 kg/capita/day (Mishra et al., 1988) and 0.76–1.21 kg/capita/day (Saksena et al., 1995).

An inverse relationship has been found between the dependence on forests for fuelwood and fodder, and farm size. Sharma et al. (1989) estimated the fuelwood, timber, and fodder requirements of three farm household size groups in the hill areas of Himachal Pradesh, and determines the extent of dependence of these products on forest, own land and market sources. The results indicated a high dependence on forests, which varied on purpose due to agro-climatic conditions and income status. Mujawamariya and Karimov (2014) reported that livestock ownership, possession of skills, insecurity, and price obtained from the previous season impact on decision making to collect gum Arabic. Furthermore, household's age, experience in collecting gum Arabic and topography increase the quantities collected while gender negatively impacts amounts of collection. A study (Shrestha and Bawa, 2014) revealed that the contribution of Chinese caterpillar fungus income to total household income decreases as the household income increases making its contribution highest for the poorest households. There was a significant correlation between Chinese caterpillar fungus dependency and percentage of family members involved in harvesting, number of food-sufficient months, and total income without Chinese caterpillar fungus income. Income from Chinese caterpillar fungus is helping the poorest to educate children, purchase food, and pay debts. Opaluwa et al. (2011) reported that the outcome of the logistic regression revealed that gender and distance were found to significantly reduce the odds in favor of collecting NTFPs while family size was found to significantly increase the odds in favor of collecting NTFPs. Regression analysis revealed that education, landholding, gross annual income, proximity to the forest, forest visit and forestry resources possession significantly influenced forest dependence for employment support and the R^2 (0.786) indicated that 78.60% of the variation in the forest-based employment was explained by the household drivers (Islam et al., 2016). Results of the econometric analysis (Inoni, 2009) indicated that size of household size, education, and cost of products, forest distance, and income affected the amount of forest products exploited in a manner consistent with economic theory, and were statistically significant; though the impact of the variables on the various forest products was mixed. The adjusted R^2 values were 0.40, 0.60, and 0.78 respectively for rattan cane, firewood, and wild fruits. Gupta et al. (2009) reported that average family size, livestock holding, landholding (ha), annual income and fuelwood consumption (per family per annum) was 5.25, 6.97, 1.52, ₹ 45817/= and 6.03 to 6.387 quintals respectively in Jammu region. Family size and livestock have negative association with fuelwood consumption whereas land holding, number of trees and annual income have positive association with fuelwood consumption. These results are also supported with the help of multiple linear regressions. Larinde and Olasupo, (2011) showed that there was a positive and strong relationship between livelihood from fuelwood sales and the gender, family size, level of education and amount paid to labor; with coefficient of determination ($R^2 = 0.67556$). A negative relationship was noted between income generated from fuelwood sales and the amount paid to the government. Results showed that during the study, income generated from the sale of fuelwood increased with increasing households, education level, and amount paid to the laborers with little revenue generation into the State treasury. Results of the Tobit model indicate a positive significant relationship between total hours of collection ($b = 0.901$) with income share of the NTFPs. A negative relationship with income variables such as farm income ($b = -0.001$) and wage income ($b = -0.003$) was found to be statistically significant at the 99% confidence level. While services and allied activities ($b = -0.001$) were negative significant at 95% confidence level. These variables influence the share of NTFPs in income (Tejaswi, 2008). Islam et al. (2015a) reported that among the socioeconomic characteristics of the indigenous households, education, social participation, family composition, main occupation, housing status, farm power, farm implements, livestock possession, wealth status and gross annual income had exhibited positively significant correlation with the livelihood dependency on forest resources and age and size of landholding have non-significant association with the livelihood dependency on forest resources. All the socioeconomic characteristics of the indigenous people put together had contributed to 70.05% ($R^2 = 0.705$) variation on the livelihood dependency on forest resources. Further, among these variables, only family composition exercised a positive and significant contribution to livelihood dependency on forest resources.

Nagesha and Gangadharappa (2006) observed the existence of a positive and significant association between social participation and adoption behavior of agroforestry practices. Prakash and Sharma (2008) reported the extent of people's social participation as reflected by their membership of various socio-cultural organizations was recorded to below, and the social participation had a strong positively significant association with participation in control of forest fire. Thamban et al. (2008) found that the social participation of the farmers had a non-significant effect on the extent of participation in the field implementation of micro-irrigation technology in Kasaragod district of Kerala state. Thamban et al. (2008) observed that the family composition was found to have a significant and positive relationship with the extent of participation in the agroforestry technology in Kasaragod district of Kerala state. There was no association between family composition and on-farm tree cultivation (Chaudhary and Panjabi, 2005; Sood et al., 2008). The average total landholding/household was 0.67 ha to 0.70 ha among the forest-dwelling *Siddi* tribal community in Uttara Kannada district of Karnataka (Prakash and Sharma, 2008; Pal, 2009). The size of landholding had a positive and significant association with the adoption of agroforestry systems in northeastern districts of Karnataka (Nagesha and Gangadharappa, 2006). The size of landholding had showed no link to the people's dependence and participation in forest management, particularly fire management (Prakash and Sharma, 2008). Adhikari et al. (2004) investigated the determinants of fuelwood collection and consumption by rural households in Nepal. The results showed that fuelwood consumption per household per year was 6090 kg. The study also revealed that fuelwood consumption was highest in areas nearer to the forests.

In a study by Ofoegbu et al. (2017) in the Vhembe district of South Africa (sample size of 366 households of forest-based communities), it was found that the most common socio-economic factor for forest dependency was easy accessibility of forest resources. In the same study, it was also mentioned that the top four reasons for forest resources dependency were easy accessibility of forest resources, relative low cost of using forest resources, unemployment, and to survive shocks. Also, Bortolotto et al. (2015) found that such knowledge was more prevalent in communities more distant from the cities and towns than the ones in urban vicinity. Sharma et al. (2011) conducted a study on forest resources utilization pattern of people using their socio-economic conditions in Dudhatoli, Garhwal, Himalayas at an altitudinal range of 1750–2200 m asl. Biophysical parameters like the distance from the forests played a major role in extraction as it was found that anthropogenic pressure reported highest, on forests which was closed to the villages and the most affected pressure observed on *Quercus* spp. Tree richness and canopy cover was also lower in the forests nearer to the villages. It was also noticed that most of the respondents were females, which was attributed to the migration of men from the village for the search of better employment options. Agriculture was the main source of income, followed by labor under MGNREGA scheme and extraction of NTFPs. Average livestock holding per household ranged between 8.7 and 19.0 forests was the major source of leaf fodder and bedding material for livestock.

8.6 LIVELIHOOD SUSTAINABILITY AND FOREST DEGRADATION

Unsustainable exploitation of forest resources leads often forest degradation. Many works described the negative impact of overexploitation of resources and leads loss of diversity of native species in various regions (Armesto et al., 2001; Herrmann, 2006; Casas et al., 1994), and also to accelerated processes of erosion (Paruelo et al., 2006). Forests provide not only fuelwood but also fodder as well (stall-feed as well as grazing). Agriculture induced land-use change, grazing, and fuelwood collection are the leading causes of forest degradation (change from dense to open forests) in developing countries (Prabhakar et al., 2006; Singh, 2006; Baland et al., 2007), due to this dependence on forests for fodder, several studies have concluded that fodder extraction coupled with overstocking of livestock on pastures (Kumar and Shahabuddin, 2005) is the key reason for forest degradation (Khanduri et al., 2002; Prabhakar et al., 2006; Singh, 2006; Baland et al., 2007). Biomass extraction from forests also reduced carbon stocks, habitat of wildlife, and many other services from forests. In the middle Himalayas of India, fodder extraction estimates from forests vary from 25–66% of total fodder supply (Singh and Naik, 1987; Bajracharya, 1999; Singh, 1999; Tripathi, 1999). In recent periods, the rapid growth of animal and human populations, including unrestricted livestock grazing, has declined the density of vegetative cover and the severe impoverishment of the forests of India (MoEF, 2006). Primarily in the rural India and also in the urban sectors, most of the population depends on forests directly to meet the bulk of small timber requirement (Rai and Chakrabarti, 2001). An estimated 80% of the developing world uses NTFPs to meet some of their health and nutritional needs (Olaniyi et al., 2013).

8.7 MITIGATION STRATEGIES FOR LIVELIHOOD INDUCED CARBON OUTFLOW

Mitigation efforts to reduce the sources of or to enhance the sinks of greenhouse gas (GHG) will take time and requires international cooperation (Legesse et al., 2013). Adaptation, in contrast, can reduce climate-related risks in human-managed systems on regional and local scales and often with a short lead-time. Adaptation to climate change refers to adjustments to practices, processes, and systems to minimize current and future adverse effects of climate change and take advantage of available opportunities to maximize benefits (Smithers and Smit, 2009; Boomiraj et al., 2010; Ericksen et al., 2011). Conservation initiatives such as REDD+ are more likely to succeed if they build on the interest of forest and indigenous communities (Dkamela et al., 2009). Agroforestry is a major climate change adaptation strategy for forest communities (Verchot et al., 2007; Somorin, 2010). Agroforestry is a relevant strategy for forest carbon conservation and adaptation to climate change (Smith and Scherr, 2003; Ravindranath, 2007). Agroforestry systems are important for carbon sequestration, and at the same time they provide biophysical, economic, and social support for vulnerable communities to adapt to the negative consequences of climate change (Verchot et al., 2007). The biggest efforts have been towards tree planting and husbandry, attending capacity-building sessions on natural resources management and adoption of appropriate technologies and farming methods (Macharia et al., 2010).

A comprehensive study (Islam et al., 2017b) to evolve a region-specific forestry-based livelihood diversification strategy for livelihood security and ecological stability, the chapter has studied socioeconomic-cum-demographic aspects, i.e., population, caste, sex ratio, income, manpower, occupation, etc.), and natural resource scenario (land-use pattern, forest, soils, etc.), of Bundu block in Ranchi district of Jharkhand. The study has provided eco-friendly livelihood options like tasar sericulture, Lac culture, fruit culture, and mechanized Sal leaf plate making for tribal people livelihood diversification and socioeconomic development.

The income and employment opportunities expected from the interventions per annum are: Tasar sericulture (₹ 54.16 lakh, 21100 person-days), Lac culture (₹ 36.88 lakh, 8790 person-days), fruit culture (₹ 63.58

lakh, 19180 person-days), and mechanized Sal leaf plate making (₹ 40.26 lakh, 36900 person-days). The proposed strategy is likely to yield an income of ₹ 194.88 lakh/annum and employment potential of 85970 person-days/ annum besides securing the basic needs by mobilizing the existing natural resources. The livelihood diversification indicated that *tasar* sericulture; *Lac* culture and fruit culture on cultivable wastelands together will procure 1315 t/annum biomass of fuelwood, 691 t/annum of green herbage biomass and 303 m³/annum of timber which can relieve substantial anthropogenic pressure on the forests, alleviating forest degradation in the locality. In addition, the production of agricultural crops by intercropping, mango, and guava production through fruit culture and *ber* fruits from *Lac* culture compartments on cultivable wastelands will play a potent role in food and nutritional security for the tribes.

Development of wasteland reclamation strategy to eliminate the forest dependency for timber security in Bundu block of Ranchi district in Jharkhand, India revealed that forests contributed maximum timber (136.36 m³/annum) followed by traditional agroforestry (69.09 m³/annum), community forestry (41.33 m³/annum) and homestead forestry (35.71 m³/ annum). Timber extracted is mostly consumed in housing (124.66 m³/annum) followed by agricultural implements (82.71 m³/annum), furniture (35.25 m³/ annum), carts/carriages (17.60 m³/annum), fencing (10.23 m³/annum), cattle shed/storehouse (9.10 m³/annum) and others (2.94 m³/annum). Forests were exposed to timber pressure of 136.36 m³/annum (48.27%) posing ample deforestation and degradation. The strategy consisted of timber and bamboo plantations is designed which would secure 1065.60 m³/annum of timber, 0.455 lakh/annum of bamboo culms, 568.26 t/annum of bamboo leaf and agricultural products (Islam et al., 2017a). The strategy would yield income of ₹ 34210.78/household/annum and employment of 67.15 person-days/ household/annum). The wastelands reclamation strategy indicated that the timber (*Gmelina arborea* and *Tectona grandis*) and bamboo (*Dendrocalamus strictus*) plantations together will procure 1065.60 m³/annum of timber, 0.455 lakh/annum of bamboo culms and 568.26 t/annum of bamboo leaf besides production of field crops which can relieve substantial anthropogenic pressure on the forests alleviating forest degradation.

Low productivity of forests coupled with ever-increasing demand for timber due to huge and increasing forest fringe population contributes to the forest degradation (Aggarwal et al., 2009). Hence, the development and diversification of non-traditional and economically viable timber, bamboo plantations, and other forestry interventions on wastelands can alleviate the timber and other forest resource scarcity besides contribution to forest

184 Diversity and Dynamics in Forest Ecosystems

resources conservation and restoration of ecosystem services. The study (Islam and Quli, 2016) to examine the extent of fuelwood dependence in the forests and mitigate the pressure by evolving an eco-friendly strategy in Bundu block of Ranchi district in Jharkhand, India revealed that the total extraction of fuelwood from different sources was 598.60 t/annum @ of 0.68 t/capita/annum, of which, 308.16 t/annum was secured from forests, 133.31 t/annum from agriculture field, 90.45 t/annum from community land and 66.68 t/annum from homesteads. The fuelwood use breakup recorded 486.24 t/annum for cooking, 45.79 t/annum by cottage industries, 41.07 t/annum for heating, 18.80 t/annum for community function and 6.70 t/annum for others. The forests were exposed to fuelwood pressure of 308.16 t/annum (51.48%) posing ample deforestation and degradation. A strategy consisted of energy interventions *viz.*, biogas production (85351.60 m³/annum), agroforestry (36.84 t/annum) and energy plantation (92.10 t/annum) is proposed, the implementation of which can mitigate the fuelwood induced forest degradation besides fuelwood security of 846.14 t/annum against the present fuelwood procurement of 598.60 t/annum.

The study (Khatun and Roy, 2012) conducted in the state of West Bengal has identified has suggested the need to develop a number of strategies, especially for the poor people, to facilitate successful livelihood diversification. This includes the development of rural infrastructure in terms of road, market, electrification, telecommunication, storage facilities, etc., and also institutional innovations to reduce entry costs and barriers to poor livelihood groups.

Lac cultivation has been found to be a risk coping strategy for vulnerable cropping system in Jharkhand (Pal et al., 2009). Sericulture is advocated as an integral part of agriculture and a source of rural livelihood and poverty alleviation in Indian condition (Bhatia et al., 2011). Watershed development is considered as a landmark in the improvement of agricultural production and productivity to protect the livelihoods of people (Tilekar et al., 2009). Joint forest management (JFM) is conceived as an efficient strategy for sustainable development of the people (Bahuguna and Hilaluddin, 2011; Pandey and Mishra, 2011). Sustainable extraction of forest resources by the forest dwellers has been emphasized as a prominent strategy for their income and employment generation (Singh and Quli, 2011; Pandit, 2011). Agroforestry plantation is recognized as the most competent land-use option for ensured sustainable development, increased productivity of land, eased environmental stress and enhanced livelihoods (Roy and Tiwari, 2012; Dagar, 2012). Kalaichelvi and Swaminathan (2009) concluded that cultivation of medicinal and aromatic plants is a better alternative land use to diversify the livelihoods for forest and forest-fringe dwellers in India.

8.8 CONCLUSION

India is a country with diverse requirements and lifestyles, where livelihood based forest extraction is a common practice in every state of the country, especially the hilly states of the Himalayan region. Himalayan region has been endowed with rich natural resources, but it is also a seat of high population density and hence higher extraction. Due to the significant role of Himalayan forests in the regulation of the climate of the whole country, it is imperative to develop strategies to mitigate the livelihood-based extraction. Such strategies need to be based on the community-specific factors that delineate the extraction of forest resources and should keep the socio-economic factors into consideration. Such strategies would help in diversification of income and employment generation of forest-dependent communities in the area.

KEYWORDS

- **average cattle unit**
- **greenhouse gas**
- **Indian Himalayan Region**
- **joint forest management**
- **livelihood security**
- **non-timber forest products**

REFERENCES

Adhikari, B., Di Falco, S., & Lovett, J. C., (2004). Household characteristics and forest dependency: Evidence from common property forest management in Nepal. *Ecological Economics, 48,* 245–257.

Aggarwal, A., Paul, V., & Das, S., (2009). Forest resources: Degradation, livelihoods, and climate change. In: Datt, D., & Nischal, S., (eds.), *Looking Back to Change Track* (pp. 91–108, 219). New Delhi: TERI.

Ajake, A. O., & Enang, E. E., (2012). Demographic and socio-economic attributes affecting forest ecosystem exploitation and management in the rural communities of cross river state, Nigeria. *American Journal of Primatology, 2*(1), 174–184.

Akhter, S., Shawkat, M. S. I., Parvez, M. R., & Alamgir, M., (2009). Impact of forest and non-forest villagers on Ukhia and Inani forest Range under Cox's Bazar (South) forest division, Bangladesh. *Proceedings of Pakistan Academy Sciences, 46*(1), 13–22.

Armesto, J. J., Smith-Ramirez, C., & Rozzi, R., (2001). Conservation strategies for biodiversity and indigenous people in Chilean forests ecosystems. *Journal of Royal Society New Zealand, 31*(4), 865.

Atta, U., Islam, M. A., & Shah, M., (2018). Socio-economic profile of *Shina* community subsisting on NTFPs in Gurez valley of Kashmir. *International Journal of Advanced Research in Science and Engineering, 7*(4), 1701–1709.

Baba, M. Y., Islam, M. A., & Qaisar, K. N., (2015). Assessing the household fuelwood extraction and consumption situation in rural Kashmir, India. *International Journal of Forestry and Crop Improvement, 6*(1), 55–63.

Babulo, B., Muys, B., Nega, F., Tollens, E., Nyssen, J., Deckers, J., & Mathijs, E., (2008). Household livelihood strategies and forest dependence in the highlands of Tigray, northern Ethiopia. *Agricultural Systems, 98,* 147–155.

Bahuguna, V. K., & Hilaluddin, (2011). Contribution of joint forest management in conservation of forests, climate change and poverty alleviation. *The Indian Forester, 137*(2), 154–163.

Bajracharya, B., (1999). *Sustainable Soil Management with Reference to Livestock Production Systems.* ICIMOD, Katmandu, Nepal.

Balachander, D., & Ganesan, M., (1993). Extraction of non-timber forest products, including fodder and fuelwood in Mudumalai. *Economic Botany, 47*(3), 268–274.

Baland, J. M., Bardhan, P., Das, S., Mookherjee, D., & Sarkar, R., (2007). Managing the environmental consequences of growth: Forest degradation in the Indian Mid-Himalayas. *Indian Policy Forum, 3,* 215–266.

Ballabh, B., & Chaurasia, O. P., (2007). Traditional medicinal plants of cold desert Ladakh-used in treatment of cold, cough and fever. *Journal of Ethnopharmacy, 112,* 341–349.

Banyal, R., Islam, M. A., Masoodi, T. H., & Gangoo, S. A., (2013). Energy status and consumption pattern in rural temperate zone of western Himalayas: A case study. *Indian Forester, 139*(8), 683–687.

Bhatia, S., & Chaudhary, R. P., (2011). Wild edible plants used by the people of Manang District, Central Nepal. *Ecology of Food and Nutrition, 48,* 1–20.

Bijalwan, A., Sharma, C. M., & Kediyal, V. K., (2011). Socioeconomic status and livelihood support through traditional agroforestry systems in hill and mountain agro-ecosystem of Garhwal Himalaya, India. *The Indian Forester, 138*(12), 1423–1430.

Biswakarma, S., Pala, N. A., Shukla, G., Vineeta, K., & Chakravarty, P. S., (2018). Influence of socio-economic factors on attitude of ethno-botanical users among forest fringe communities in Sub-Himalayan Region of West Bengal, India. *Indian Journal of Hill Farming,* 27–35.

Bonan, G. B., (2008). Forests and climate change: Forcing, feedbacks, and the climate benefits of forests. *Science, 320*(5882), 1444–1449.

Boomiraj, K., Wani, S. P., Garg, K. K., Aggarwal, P. K., & Palanisami, K., (2010). Climate change adaptation strategies for agroecosystem: A review. *Journal of Agrometeorology, 12,* 145–160.

Bortolotto, I. M., De Mello, A. M. C., Neto, G. G., Oldeland, J., & Damasceno-Junior, G. A., (2015). Knowledge and use of wild edible plants in rural communities along Paraguay River, Pantanal, Brazil. *Journal of Ethnobiology and Ethnomedicine, 11,* 46–59.

Bradford, J., Weishampel, P., & Smith, M., (2009). Detrital carbon pools in temperate forests: Magnitude and potential for landscape-scale assessment. *Canadian Journal of Forest Research, 39,* 802–813.

Bwalya, S. M., (2011). Household dependence on forest income in rural Zambia. *Zambia Social Science Journal, 2,* 67–86.

Byron, N., & Arnold, J. E. M., (1999). What futures for the people of the tropical forests? *World Development, 27,* 789–805.

Cardona, W. C., (2005). Forest regulation flexibility, livelihoods, and community forest management in the northern Bolivian Amazon. In: Mery, G. A., (ed.), *Forests in the Global Balance: Changing Paradigms* (pp. 97–111). IUFRO World Series 17. Vienna: IUFRO.

Casas, A., Viveros, J. L., & Caballeros, J., (1994). *Mexican Etnobotany.* Regina publishers Mexico city, Los Angeles, S. A.

Chakraborty, M. K., & Bhattacharjee, A., (2006). Some common ethnomedicinal uses for various diseases in Purulia district, West Bengal, *Indian Journal of Traditional Knowledge, 5,* 554–558.

Chandra, R., Soni, P., & Yadav, V., (2008). Fuelwood, fodder and livestock in Himalayan watershed in Mussoorie hills, Uttarakhand, India. *The Indian Forester, 135*(10), 894–905.

Chaudhary, M. C., & Panjabi, N. K., (2005). Adoption behavior of tribal and non-tribal farmers regarding improved social forestry practices. *Rural India, 67*(6), 140–141.

Chaudhury, J. K., Biswas, S. R., Islam, S. M., Rahman, O., & Uddin, S. N., (2004). *Biodiversity of Shatchari Reserved Forest, Habiganj.* IUCN Bangladesh Country Office, Dhaka, Bangladesh.

Chia, E. L., Somorin, O. A., Sonwa, D. J., & Tiani, A. M., (2013). Local vulnerability, forest communities and forest carbon conservation: Case of southern Cameroon. *International Journal of Biodiversity and Conservation, 5,* 498–507.

Chopra, K., & Kumar, P., (2003). *Forest Biodiversity and Timber Extraction: An Analysis of the Interaction of Market and Non-Market Mechanisms.* EEE working papers series, No. 8.

Dagar, J. C., (2012). Utilization of degraded lands/habitats and poor-quality water for livelihood security and mitigating climate change. *Indian Journal of Agroforestry, 14*(1), 1–16.

Dkamela, G. P., Mbambu, F. K., Austin, K., Minnemeyer, S., & Stolle, F., (2009). *Voices from the Congo Basin: Incorporating the Perspectives of Local Stakeholders for Improved REDD Design, Working Paper.* World Resource Institute, Washington DC.

Dovie, D. B. K., (2003). Rural economy and livelihoods from the non-timber forest products trade. Compromising sustainability in southern Africa? *International Journal of Sustainable Development and World Ecology, 10,* 247–262.

Ericksen, S., Aldunce, P., Bahinipati, C. S., Martins, R. D., Molefe, J. I., Nhemachena, C., O'brien, K., et al., (2011). When not every response to climate change is a good one: Identifying principles for sustainable adaptation. *Climate and Development, 3,* 7–20.

FAO, (2010). Forests and energy: Key issues. Rome Italy: Food and Agriculture. *Organization of the United Nations* 20-8, p. 69.

FAO, (2015). *FAO, Forests and Climate Change: Working with Countries to Mitigate and Adapt to Climate Change Through Sustainable Forest Management.* Rome: FAO.

Fassil, H., (2003). Ethiopia: A qualitative understanding of local traditional knowledge and medicinal plant use. *IK Notes, 52,* 1–4.

Fransworth, N. R., (1992). Ethnopharmacology and drug development. In: Chadwick, D. J., & March, U., (eds.) *Ethnobotany and Search for New Drugs* (pp. 42–25). Ciba Foundation Symposium, 183 Wiley, Chichester.

Gangadharappa, N. R., Sajeev, M. V., Ganesamoorthi, S., Nagesha, G., Ibrahim, S., Ranganatha, A. D., & Reddy, M. V. S., (2005). Exploratory study on economic and marketing aspects of Agroforestry and their implications in Dharwad and Belgaum districts, Karnataka. *My Forest, 41*(2), 107–119.

Gedif, T., & Hahn, H., (2003). The use of medicinal plants in self-care in rural central. *Ethiopian Journal of Ethno-Pharmacology, 87,* 155–161.

Geng, Y., Zhang, Y., Ranjitkar, S., Huai, H., & Wang, Y., (2016). Traditional knowledge and its transmission of wild edibles used by the Naxi in Baidi Village, northwest Yunnan province. *Journal of Ethnobiology and Ethnomedicine, 12,* 10–30.

Ghosh, S. K., Guria, N., Sarkar, A., & Ghosh, A., (2013). Traditional herbal remedies for various ailments within the rural communities in the district of Bankura and Purulia, West Bengal, India. *International Journal of Pharmacy and Pharmaceutical Sciences, 5,* 195–198.

Giday, M., Asfaw, Z., Woldu, Z., & Teklehaymanot, T., (2009). Medicinal plant knowledge of the Bench ethnic group of Ethiopia: An ethnobotanical investigation. *Journal of Ethnobiology and Ethnomedicine, 5,* 34–43.

Gupta, T., Gupta, R. K., & Raina, K. K., (2009). Socioeconomic factors associated with fuel consumption pattern in rural habitation of Jammu region, Jammu and Kashmir. *Indian Journal of Forestry, 32*(3), 387–390.

Hajost, S., & Zerbock, O., (2013). *Lessons Learned from Community Forestry and Their Relevance for REDD+*. Washington, DC: USAID-supported Forest Carbon, Markets and Communities Program.

Herrmann, T. M., (2006). Indigenous knowledge and management of Araucaria Araucana forest in the Chilean Andes: Implications for native forest conservation. *Biodiversity Conservation, 15*(2), 647–662.

Hosier, R., (1985). Household energy consumption in Kenya. *Ambio, 14,* 255–267.

Hussain, A., Negi, A. K., Singh, R. K., Aziem, S., Iqbal, K., & Pala, N. A., (2016). Comparative study of fuelwood consumption by semi-nomadic pastoral community and adjacent villagers around Corbett tiger reserve, India. *Indian Forester, 142*(6), 574–581.

Inder, D., (2001). Problems and prospects of forage production and utilization of Indian Himalaya. *ENVIS Bulletin on Himalayan Ecology, 9*(2), 7–14.

Inoni, O. E., 2009.Effects of forest resources exploitation on the economic well-being of rural households in Delta State, Nigeria. *Agricultura Tropica Et Subtropical, 42*(1), 25–34.

Islam, M. A., & Quli, S. M. S., (2016). Non-timber forest products (NTFPs) supporting food security in Tribal Jharkhand. *Jharkhand Journal of Development and Management Studies, 14*(1), 6855–6864.

Islam, M. A., & Quli, S. M. S., (2017a). The role of non-timber forest products (NTFPs) in tribal economy of Jharkhand, India. *International Journal of Current Microbiology and Applied Sciences, 6*(10), 2184–2195.

Islam, M. A., & Quli, S. M. S., (2017b). Forestry-based livelihood diversification strategy for socio-economic development of tribes in Jharkhand, *Agricultural Economics Research Review, 30*(1), 151–162.

Islam, M. A., (2008). Availability and consumption pattern of fuelwood, fodder and small timber in rural Kashmir. *Environment and Ecology, 26*(4A), 1835–1840.

Islam, M. A., Banyal, R., Masoodi, N. A., Masoodi, T. H., Gangoo, S. A., & Sharma, L. K., (2011). Status of fuelwood extraction and consumption in rural north Kashmir: A case study. *The Indian Forester, 137*(11), 1265–1268.

Islam, M. A., Quli, S. M. S., & Baba, M. Y., (2016). Household drivers of forest dependence for employment support among tribes of Jharkhand, India. *Economic Affairs, 61*(2), 339–347.

Islam, M. A., Quli, S. M. S., & Mushtaq, T., (2017). Wasteland reclamation strategy for household timber security of tribes in Jharkhand, India, *Journal of Applied and Natural Science, 9*(4), 2264–2271.

Islam, M. A., Quli, S. M. S., Rai, R., Ali, A., & Gangoo, S. A., (2015b). Forest biomass flow for fuelwood, fodder and timber security among tribal communities of Jharkhand. *Journal of Environmental Biology, 36*(1), 221–228.

Islam, M. A., Rai, R., Quli, S. M. S., & Tramboo, M. S., (2015a). Socio-economic and demographic descriptions of tribal people subsisting in forest resources of Jharkhand, India. *Asian Journal of BioScience, 10*(1), 75–82.

Jain, S. K., (1992). Ethnopharmacology and drug development. In: Chadwick, D. J., & March, U., (eds.), *Ethnobotany and Search for New Drugs* (Vol. 153, pp. 26). Ciba Foundation Symposium, 183 Wiley, Chichester.

Jalal, J. S., & Garkoti, S. C., (2013). Medicinal plants used in the cure of stomach disorders in Kumaon Himalaya, Uttarakhand, India, *Academic Journal of Medicinal Plants, 1*(7), 116–121.

Kala, C. P., (2006). Medicinal plants of the high-altitude cold desert in India: Diversity, distribution and traditional uses. *International Journal of Biodiversity Science and Management, 2*, 43–56.

Kalaichelvi, K., & Swaminathan, A. A., (2009). Alternative land use through cultivation of medicinal and aromatic plants: A review. *Agricultural Review, 30*(3), 176–183.

Kersten, I., Baumbach, G., Oluwole, A. F., Obioh, I. B., & Ogunsola, O. J., (1998). Urban and rural fuelwood situation in the tropical rain-forest area of south-west Nigeria. *Energy, 23*(8), 87–98.

Khanduri, V. P., Sharma, C. M., Ghildiyal, S. K., & Puspwan, K. S., (2002). Forest composition in relation to socio-economic status of people at three altitudinal villages of a part of Garhwal Himalayas. *The Indian Forester, 128*(12), 1335–1345.

Khatun, D., & Roy, B. C., (2012). Rural livelihood diversification in West Bengal: Determinants and constraints. *Agricultural Economics Research Review, 25*(1), 115–124.

Kituyi, E., Marufu, L., Huber, B., Wandiga, S. O., Jumba, I. O., Andreae, M. O., & Helas, G., (2001). Biofuel consumption rates and patterns in Kenya. *Biomass and Bioenergy, 220*(2), 83–99.

Kumar, A., Avasthe, R. K., Shukla, G., & Pradhan, Y., (2012). Ethnobotanical edible plant biodiversity of Lepcha tribes. *Indian Forester, 138*(9), 798–803.

Kumar, M., & Sharma, C. M., (2009). Fuelwood consumption pattern at different altitudes in rural areas of Garhwal Himalaya. *Biomass and Bioenergy, 33*(14), 13–18.

Kumar, P., Rawat, L., & Basera, H., (2010). Socio-economic studies of Henwal Watershed, Tehri Garhwal, Uttarakhand. *Indian Journal of Forestry, 33*(2), 149–154.

Kumar, R., & Shahabuddin, G., (2005). Effects of biomass extraction on vegetation structure, diversity and composition of forest in Sariska Tiger Reserve, India. *Environmental Conservation, 32*, 248–259.

Larinde, S. L., & Olasupo, O., (2011). Socio-economic importance of fuelwood production in Gambari forest reserve area, Nigeria. *Agriculture and Social Research, 11*(1), 201–210.

Legesse, B., Ayele, Y., & Bewket, W., (2013). Small holder's perception and adaptation to climate variability and climate change in Doba district, West Hararghe, Ethiopia. *Asian Journal of Empirical Research, 3*, 251–265.

Macharia, P. N., Thuranira, E., Nganga, L. W., Lugadiru, J., & Wakori, S., (2010). Perceptions and adaptation to climate change and variability by immigrant farmers in semi-arid regions of Kenya. African Crop Science Journal, 20, 287–296.

Mahapatra, A. K., & Mitchell, C. P., (1999). Biofuel consumption deforestation and farm level tree growing in rural India. *Biomass and Bioenergy, 17*(2), 91–303.

Mahat, T. B. S., Grigffin, D. M., & Shepherd, K. P., (1987). Human impact on some forest of the middle hills of Nepal: Part 4. A detailed study in Southeast Sindhu Palanchock and Northeast Kabhere, Palanchock. *Mountain Research and Development, 7*(1), 14–34.

Mamo, G., Sjaastad, E., & Vedeld, P., (2007). Economic dependence on forest resources: A case from Dendi District, Ethiopia. *Forest Policy and Economics, 9,* 916–927.

Marufu, L., Ludwig, J., Andreae, M. O., Lelieveld, J., & Helas, G., (1999). Spatial and temporal variation in domestic biofuel consumption rates and pattern in Zimbabwe: Implication for atmospheric trace gas emission. *Biomass and Bioenergy, 16*(3), 11–32.

Mishra, A., (2008). *Determinants of Fuel Wood Use in Rural Orissa: Implications for Energy Transition South Asian Network for Development and Environmental Economics (SANDEE).* Working Paper No. 37-08.

Mishra, N. M., Mahendra, A. K., & Ansari, M. Y., (1988). A pilot survey of fuel consumption in rural areas. *Indian Forester, 114*(2), 57–62.

Misra, A., (2011). *Determinants of Fuelwood use in Rural Orissa: Implications for Energy Transition South Asian Network for Development and Environmental Economics (SANDEE).* Working Paper No. 37–08.

MoEF (Ministry of Environment and Forests), (2006). *Report of the National Forest Commission* (p. 421). New Delhi: Ministry of Environment and Forests, Government of India.

MoEF, (2009). *State of Forest Report-2009: Forest Survey of India.* Ministry of Environment and Forests, Government of India, Dehra Dun, India.

Mondal, S., & Rahaman, C. H., (2012). Medicinal plants used by tribal people of Birbhum district of West Bengal and Dumka district of Jharkhand in India. *Indian Journal of Traditional Knowledge, 11,* 674–679.

Mujawamariya, G., & Karimov, A. A., (2014). Importance of socio-economic factors in the collection of NTFPs: The case of gum Arabic in Kenya. *Forest Policy and Economics,* 42, doi: 10.1016/j.forpol.2014.02.005.

Muralidharan, R., & Narasimhan, D., (2012). Ethnomedicinal plants used against gastrointestinal problem in Gingee hills of Villupuram district, Tamil Nadu. *Journal of Applied Pharmaceutical Sciences, 2*(10), 123–125.

Mushtaq, T., Sood, K. K., & Raina, N. S., (2012). Species preferences for fuelwood in Shiwalik Himalayas: Implications for agroforestry plantations. *Indian Journal of Hill Farming, 25*(2), 18–21.

Musyoki, A., (2012). *The Emerging Policy for Green Economy and Social Development in Limpopo, South Africa.* Geneva: United Nations Research Institute for Social Development.

Nagesha, G., & Gangadharappa, N. B., (2006). Adoption of agroforestry systems in northeastern districts of Karnataka. *My Forest, 42*(4), 337–347.

Namdeo, R. K., & Pant, N. C., (1994). Role of minor forest products in tribal economy. *Journal of Tropical Forestry, 10*(1), 36–44.

Narzary, H., Brahma, S., & Basumatary, S., (2013). Wild edible vegetables consumed by Bodo tribe of Kokrajhar district of Assam, North-East India. *Archives of Applied Science Research, 5,* 182–190.

Nautiyal, S., Rao, K. S., Maikhuri, R. K., Negi, K. S., & Kala, C. P., (2002). Status of medicinal plants on way to Vashuki Tal in Mandakini Valley, Garhwal, Uttaranchal. *Journal of Non-Timber Forest Product, 9,* 124–131.

Negi, A. K., & Todaria, N. P., (1993). Studies on impact of local folk on forest of Garhwal Himalaya. *Energy from Biomass, 4,* 447–454.

Negi, Y. S., Sharma, L. R., & Singh, J., (1996). Factors affecting fuelwood consumption: A micro-level study. *Indian Forester, 112*(8), 737–740.

Ofoegbu, C., Paxie, W., Chirwa, J. F., & Folarannmi, D. B., (2017). Socio-economic factors influencing household dependence on forests and its implication for forest-based climate change interventions. *Southern Forests: A Journal of Forest Science,* 1–8.

Olaniyi, O. A., Akintonde, J. O., & Adetumbi, S. I., (2013). Contribution of non-timber forest products to household food security among rural women in Iseyin local government area of Oyo State, Nigeria. *Research on Humanities and Social Sciences, 3*(7), 41–50.

Opaluwa, H. I., Onuche, U., & Sale, F. A., (2011). Factors affecting the collection and utilization of non-timber forest products in rural communities of North Central Nigeria. *Journal of Agriculture and Food Technology, 1*(5), 47–49.

Pal, G., (2009). Resource use efficiency and level of technology adoption in Lac cultivation among trained and untrained Lac growers in Jharkhand. *International Journal of Agricultural Science, 5*(2), 615–618.

Pal, G., (2011). Socio-economic characteristics of Lac growers in Kanker district of Chhattisgarh. *The Indian Forester, 137*(11), 1294–1297.

Pandey, D., (2002). *Fuel Wood Studies in India-Myth and Reality* (pp. 72–74). Centre for International Forestry Research, Indonesia.

Pandey, R., & Gupta, A. K., (2013). Resource availability versus resource extraction in forests: Analysis of forest fodder system in forest density classes in lower Himalayas, India. *Small Scale Forestry.* doi: 10.1007/s11842-013-9253-3.

Pandey, R., & Mishra, A., (2011). Livestock fodder requirements and household characteristics in rural economy of hilly region, Uttarakhand. *Himalayan Ecology, 34*(4), 35–40.

Pandit, M. K., Sodhi, N. S., Koh, L. P., Bhaskar, A., & Brook, B. W., (2007). Unreported yet massive deforestation driving loss of endemic biodiversity in Indian Himalaya. *Biodiversity Conservation, 16,* 153–163.

Pandit, P. K., (2011). An assessment of non-timber forest products in Jhargram division. *The Indian Forester, 137*(11), 1250–1257.

Panta, M., Kim, K., & Lee, C., (2011). Household's characteristics, forest resources dependency and forest availability in central Terai of Nepal. *Journal of Korean Forest Society, 98*(5), 548–557.

Paruelo, J. M., Golluscio, R. A., Jobbagy, E. G., Canevari, M., & Aguiar, M. R., (2006). Firewood use in Bulgamogi County, Uganda: Species selection, harvesting and consumption patterns. *Biomass Bioenergy, 25*(6), 581.

Pascaline, C. L., Savadogo, P., Tigabu, M., & Oden, P. C., (2011). Factors influencing people's participation in the forest management program in Burkina Faso, West Africa. *Forest Policy and Economics, 13,* 292–302.

Prabhakar, R., Somanathan, E., & Mehta, B. S., (2006). How degraded are Himalayan forests? *Current Science, 91*(1), 61–67.

Prakash, O., & Sharma, R., (2008). Determining people's participation in forest fire control: A study of Himachal Pradesh. *Indian Journal of Forestry, 31*(1), 1–6.

Rahmani, A., (2003). Conservation outside protected areas. In: Saberwal, V., & Rangarajan M., (eds.) *Battles Over Nature: Science and Politics of Conservation* (pp. 117–138). Permanent Black New Delhi.

Rai, S. N., & Chakrabarti, S. K., (2001). Demand and Supply of fuelwood and Timber in India. *The Indian Forester, 127*(3), 23–29.

Raj, A. J., Biswakarma, S., Pala, N. A., Gopal, S. V., Kumar, M., Chakravarty, S., & Bussmann, R. W., (2018). Indigenous uses of ethnomedicinal plants among forest-dependent communities of Northern Bengal, India. *Journal of Ethnobiology and Ethnomedicine, 14*(8), 1–28.

Rajput, S. S., Shukla, N. K., & Gupta, V. K., (1985). Specific gravity of Indian timbers. *Journal of Timber Development Association of India, 3,* 12–41.

Randerson, J. T., Chapin, I. F. S., & Harden, J. W., (2002). Net ecosystem production: A comprehensive measure of net carbon accumulation by ecosystems. *Ecological Applications, 12*(4), 937–947.

Ravindranath, N. H., (2007). Adaptation and mitigation synergy in the forest sector. *Adaptation and Mitigation Strategies of Global Change, 12,* 843–853.

Ray, G. L., & Mondol, S., (2004). *Research Methods in Social Sciences and Extension Education* (pp. 66–76). Kalyani Publishers, New Delhi.

Rennaud, J. P., Ruitenbeek, J., & Tennigkeit, T., (2013). Challenges of community-forestry based carbon projects: Process, participation, performance. *Journal of Field Action, 7,* 21–34.

Roy, M. M., & Tiwari, J. C., (2012). Agroforestry for climate-resilient agriculture and livelihood in arid region of India. *Indian Journal of Agroforestry, 14*(1), 49–59.

Saha, A., & Guru, B., (2003). *Poverty in Remote Rural Areas in India: A Review of Evidence and Issues* (p. 69). GIDR Working Paper No. 139, Ahmedabad: Gujarat Institute of Development Research.

Saksena, S., Prasad, R., & Joshi, V., (1995). Time allocation and fuel usage in three villages of the Garhwal Himalaya India. *Mountain Research Development, 15,* 57–67.

Sapkota, I. P., & Odén, P. C., (2008). Household characteristics and dependency on community forests in Terai of Nepal. *International Journal of Social Forestry, 1*(2), 123–144.

Sarmah, R., & Arunachalam, A., (2011). Contribution of non-timber forest products (NTFPs) to livelihood economy of people living in forest fringes in Changlang district of Arunachal Pradesh, India. *Indian Journal of Fundamental and Applied Life Sciences, 1*(2), 157–169.

Sati, V. P., & Song, C., (2012). Estimation of forest biomass flow in the montane mainland of the Uttarakhand Himalaya. *International Journal for Soil Erosion, 2*(1), 1–7.

Schimel, D. S., (1995). Terrestrial ecosystems and the carbon cycle. *Global Change Biology, 1,* 77–91.

Schulze, E. D., Wirth, C., & Heimann, M., (2000). Climate change: Managing forests after Kyoto. *Science, 289*(5487), 2058–2059.

Shaheen, H., Qureshi, R. A., Zahid-Ullah, & Ahmad, T., (2011). Anthropogenic. *Pakistan Journal of Botany, 43*(1), 695–703.

Sharma, C. M., Butola, D. S., Gairola, S., Ghildiyal, S. K., & Suyal, S., (2011). Forest utilization pattern in relation to socio-economic status of People in Dudhatoli area of Garhwal Himalaya, *Forests, Trees and Livelihoods, 20,* 249–264.

Sharma, C. M., Gairola, S., Ghildiyal, S. K., & Suyal, S., (2009). Forest resource use pattern in relation to socioeconomic status in four temperate villages of Garhwal Himalaya, India. *Mountain Research and Development, 29*(4), 308–319.

Sharma, D., Tiwari, B. K., Chaturvedi, S. S., & Diengdoh, E., (2015). Status, utilization and economic valuation of non-timber forest products of Arunachal Pradesh. *India Journal of Forest and Environmental Science, 31*(1), 24–37.

Sharma, J., Gairola, S., Gaur, R. D., & Painuli, R. M., (2012). Forest utilization patterns and socio-economic status of the Van Gujjar tribe in sub-Himalayan tracts of Uttarakhand, India. *Forestry Studies in China, 14*(1), 36–46.

Sheikh, M. A., Kumar, M., Bussman, R. W., & Todaria, N. P., (2011). Forest carbon stocks and fluxes in physiographic zones of India. *Carbon Balance and Management, 6*(15). www.cbmjournal.com/content/6/1/15 (accessed on 4 December 2020).

Sheikh, S. S. S. G., (2015). Contribution of wicker handicraft to rural livelihood in district Pulwama of Kashmir. *MSc Forestry Thesis (Unpublished)*. Shere-Kashmir University of Agricultural Sciences and Technology of Kashmir, Srinagar, Jammu and Kashmir.

Shit, P. K., & Pati, C. K., (2012). Non-timber forest products (NTFPs) for livelihood security of tribal communities: A case study in Paschim Medinipur district, West Bengal. *Journal of Human Ecology, 40*(2), 149–156.

Shrestha, U. B., & Bawa, K. S., (2014). Economic contribution of Chinese caterpillar fungus to the livelihoods of mountain communities in Nepal. *Biological Conservation, 177*, 194–202.

Singh, A. P., Kumar, M., Nagar, B., Pala, N. A., & Bussman, W., (2019). Ethnomedicinal use of plant resources in Kirtinagar block of Tehri Garhwal in Western Himalayas. *Ethnobotany Research and Applications, 18*, 1–11.

Singh, J. S., (2006). Sustainable development of the Indian Himalayan region: Linking ecological and economic concerns. *Current Science, 90*(6), 784–788.

Singh, L., Kasture, J., Singh, U. S., & Shaw, S. S., (2011). Ethno-botanical practices of tribals in Achanakmar Amarkantak biosphere reserve. *Indian Forester, 137*(6), 767–776.

Singh, M. K., & Mascrenhans, O. A. J., (1981). Ecological analysis of a forest-based tribal village in Singhbhum (Bihar) for a follow up land resource management action. *Management and Labor Studies, 7*, 1–22.

Singh, N., & Sundriyal, R. C., (2009). Fuelwood and fodder consumption and deficit pattern in Central Himalayan village. *Natural Science, 7*(4), 85–88.

Singh, P. K., & Quli, S. M. S., (2011). Economic valuation of non-timber forest products contribution in tribal livelihood in West Singhbhum district of Jharkhand. *The Indian Forester, 137*(11), 1258–1264.

Singh, R., (1999). Smallholder dairy farming initiatives: Success and failure of milk cooperatives in the HKH. *Paper Presented at the International Symposium on Livestock in Mountain/Highland Production Systems: Research and Development Challenges into the Next Millennium*. Pokhara, Nepal.

Singh, V., & Naik, D. G., (1987). Fodder resources of central Himalaya. In: Pangtey, Y. P. S., & Joshi, S. C., (eds.), *Western Himalaya (Environment)* (Vol. I, p. 223). Shri. Almora Publication: Almora.

Singh, V., (1995). Technology for forage production in hills of Kumaon. In: Harzra, C. R., & Bimal, M., (eds.), *New Vistas in Forage Production* (pp. 197–202). AICRPF (IGFRI), Publication Information Directorate, New Delhi.

Singha, A. K., Talukdar, R. K., & Singha, J. K., (2006). Maintenance behavior of forest resources by the people of forest villagers in Assam. *Indian Journal of Forestry, 29*(1), 47–54.

Sinhababu, A., & Banerjee, A., (2013). Ethno-botanical study of medicinal plants used by tribal of Bankura district, West Bengal, India. *Journal of Medicinal Plants Studies, 1*, 98–104.

Smith, J., & Scherr, S. J., (2003). Capturing the value of forest carbon for local livelihoods. *World Development, 31*, 2143–2160.

Smithers, J., & Smit, B., (2009). Human adaptation to climatic variability and change. In: Schipper, L. E., & Burton, I., (eds.), *Adaptation to Climate Change* (pp. 15–33). Earthscan, London.

Somorin, O. A., (2010). Climate impacts, forest-dependent rural livelihoods and adaptation strategies in Africa: A review. *African Journal of Environmental Science and Technology, 4,* 903–912.

Sood, K. K., Najiar, C., Singh, K. A., Handique, P., Singh, B., & Rethy, P., (2008). Association between socioeconomic parameters and agroforestry uptake: Evidences from eastern Himalaya. *Indian Journal of Forestry, 31*(4), 559–564.

Sraku-Lartey, M., (2014). Harnessing indigenous knowledge for sustainable forest management in Ghana, *International Journal of Food System Dynamics, 5*(4), 182–189.

Tejaswi, P. B., (2008). *Non-Timber Forest Products (NTFPs) for Food and Livelihood Security: An Economic Study of Tribal Economy in Western Ghats of Karnataka, India.* MSc (Rural development) Thesis, Ghent University, Belgium.

Teklehaymanot, T., Giday, M., Medhin, G., & Mekonnen, Y., (2007). Knowledge and use of medicinal plants by people around Debre Libanos monastery of Ethiop. *Journal of Ethnopharmacology, 111,* 271–283.

Thamban, C., Vasanthakumar, J., Arulraj, S., Mathew, A. C., & Muralidharan, K., (2008). Farmer's participation in the field implementation of micro-irrigation systems. *Journal of Plantation Crops, 36*(3), 522–525.

Tilekar, S. N., Hange, D. S., Shenge, P. N., Kalhapure, S. P., & Amale, A. J., (2009). *Agricultural Economics Research Review, 22 (Conference Number),* 415–422.

Tomar, S. K., & Shama, R. L., (2002). Fodders and feeding practices of cattle and sheep in Kashmir (India). *Tropical Agricultural Research and Extension, 5*(1/2), 48–52.

Tripathi, R. S., (1999). Economics of buffalo milk production in Indian Central Himalaya. *International Journal of Animal Sciences, 14*(1), 101–108.

Tulachan, P. M., & Neupane, A., (1999). *Livestock in Mixed Farming Systems of the Hindu Kush-Himalayas: Trends and Sustainability.* ICIMOD and FAO, Kathmandu, Nepal.

Vedeld, P., Angelsen, A., Bojö, J., Sjaastad, E., & Kobugabe, B. G., (2007). Forest environmental incomes and the rural poor. *Forest Policy and Economics, 9,* 869–879.

Verchot, L., Noordwijk, M., Kandji, S., Tomich, T., Ong, C., Albrecht, A., Mackensen, J., et al., (2007). Climate change: Linking adaptation and mitigation through agroforestry. *Mitigation and Adaptation Strategies of Global Change, 12,* 901–918.

Wegener, S., & Nguyen, T. H., (2019). *Forest Extraction and Food Expenditure of Rural Households in Cambodia and Laos* (p. 23). Draft report, Institute for Environment and World Trade, Leibniz University, Hannover, Germany.

Wildayana, E., & Armanto, M. E., (2018). Utilizing non-timber extraction of swamp forests over time for rural livelihoods. *Journal of Sustainable Development, 11*(2), 52–62.

CHAPTER 9

Integrated Approach of Sustainable Agroforestry Development in Cold Arid Region of Indian Himalaya

A. R. MALIK,[1,2] D. NAMGYAL,[1] J. S. BUTOLA,[4] G. M. BHAT,[1] P. A. SOFI,[1] AJAZ UL ISLAM,[1] J. A. BABA,[3] and J. A. MUGLOO[1]

[1]Faculty of Forestry, Benhama, Ganderbal, Sher-e-Kashmir University of Agriculture Sciences and Technology of Kashmir, Jammu and Kashmir, India

[2]High Mountain Arid Agriculture Research Institute, Krishi Vigyan Kendra (SKUAST-K), Leh, Ladakh, Jammu and Kashmir, India, E-mail: malikrashid2@gmail.com

[3]KVK/ETC Malangpora, Pulwama, Sher-e-Kashmir University of Agriculture Sciences and Technology of Kashmir, Jammu and Kashmir, India

[4]Department of Forestry and Natural Resources, HNB Garhwal University (A Central University), Srinagar Garhwal 246174,Uttarakahand, India

ABSTRACT

Ladakh region of Jammu and Kashmir, India, a Cold Arid Desert, is one amongst the most elevated (2,900 m to 5,900 m asl) and coldest regions (up to $-40°C$) of the earth. Due to the limited cultivable land and short growing season (5–6 months), the indigenous population of the region mainly depends on natural resources for meeting their diverse subsistence needs, which has led to overexploitation. Traditional agroforestry system exists in the region since time immemorial. We have conducted surveys in Leh, Nyoma, and Nubra Valleys of the district Leh to study the existing traditional agroforestry systems. Results suggested for modification in the system to provide sustainable livelihood and environment security while maintaining a present standard of living. Keeping this in view, an integrated approach of agroforestry system which comprises five different models, i.e., agri-silviculture (AS), agri-horticulture (AH), agri-silvi-horticulture (ASH), silvi-pastoral (SP), and pastoral-silvi-horticulture (PSH) systems were developed and are discussed. These models have the potential to increase productivity of available farm resources without undermining ecology and environment sanctity. Further, it will open new opportunity of employment, particularly of different farming systems such as dairy, goat, and herbal. The effective implementation of these models in the region will help in social, economic, and environmental development. An appropriate strategy for promotion of agroforestry in Cold Arid Regions of India has been presented.

9.1 INTRODUCTION

The cold deserts are found in the inter mountains of North America and the interior of Asia. However, Indian Cold Deserts fall in the Himachal Pradesh (Lahaul and Spiti and Kinnaur districts), Jammu, and Kashmir (Leh and Kargil districts), Uttarakhand, Sikkim, and Arunachal Pradesh. Ladakh region is one of the most elevated and coldest regions of the earth. The region has high political and economical significance as it gives a heavenly feeling to tourists due to unique topography, scenic beauty, and great hospitality under eco-tourism system. The region is vest sandy desert, falls of golden granite dust and barren lofty mountains. It has low atmospheric oxygen, sharp temperature fluctuations and extreme aridity (Mani, 1974). The region has a very short growing season and limited cultivable land, resulted in the local population has limited crops and mostly depends on natural resources for their subsistence needs has led to overexploitation (Joshi et al., 2006). The main source of income from rearing of Pashmina goats or Changthangi goat (*Capra aegagrus hircus*) provide cashmere wool for making well known Pashmina shawl, and eco-tourism activities. The natural wealth of this place is under various biotic (over-harvesting, grazing, trampling, alien species invasion, etc.), and abiotic (natural calamities, habitat fragmentation and degradation due to human settlements, climate change, tourism activities, etc., pressures have been causing high ecological imbalance. The most parts of the region are remote, remaining mostly cut off from the rest of the country for the larger part of the year (6–8 months). According to NAEDB (1992), due to long and severe winter, human population pressure is more than carrying capacity of the vegetation.

Agroforestry has the potential to contribute to the improvement of rural livelihood offer multiple alternatives and opportunities to the farmers to enhance farm production and their income, while protecting the agricultural environment. Recently, agroforestry has also received adequate attention for its potential to carbon sequestration (Kumar et al., 2009). In Ladakh, traditional agroforestry practices is in the form of agri-silviculture (AS) system which exists since time immemorial. Increasing human population and consequently, shrinking agriculture area, the need being felt to diversify traditional agroforestry system to provide sustainable livelihood and environment security while maintaining present standard of living.

A perusal of literature indicates that the studies undertaken in the region were confined to assess biodiversity (Kachroo et al., 1977; Murty et al., 2001; Joshi et al., 2006); development of action plan for sustainable development (NAEEDB, 1992); natural resource management (Adiga, 2003) and carbon sequestration (Bhattacharyya et al., 2008; Kumar et al., 2009). To the best of our knowledge, no attempt has been made so far to develop an integrated agroforestry system sustainable development of Ladakh region. Therefore, the present study was undertaken to (i) to assess potential and prospects of Traditional agroforestry system; (ii) explore natural resources particularly floral wealth providing basic means of livelihood; (iii) develop an integrated agroforestry system for sustainable livelihood and environment security; and (iv) recommend suitable strategies for promotion of agroforestry in Cold Arid Regions of India.

9.2 MATERIALS AND METHODS

Extensive surveys were conducted in three representative areas, i.e., Nyoma (3834 m asl), Leh (4282 m asl) and Nubra (5214 m asl) of Leh district to study the Traditional agroforestry system. The areas were selected on the basis of high human and animal population, existing comprehensive model of traditional agroforestry, and social, economic, and political importance of the areas for overall development of the region. Besides physical survey of the area to study traditional agroforestry system, local inhabitants including, knowledgeable persons (#20), the farmers engaged in dairy and goat farming (#25) and representatives (#5) of lien departments (Forest, Agriculture, and Horticulture) were interviewed to gather relevant information on advantages and disadvantages of traditional agroforestry system and their suggestions for developing an integrated system for the area. Based on our experimental trials at our research station from 2009 to 2010 and long experiences in cold desert areas of the Himalaya region, an integrated agroforestry system for cold arid desert was developed.

The data on physiography, geology, soil, and land use pattern of the area were gathered from secondary sources (Wadia, 1940; Adiga, 2003; Bhattacharyya et al., 2008). The confirmation of occurrence and taxonomic identification of plant species being used by local inhabitants were done through consulting local floras (Kachroo et al., 1977; Murty et al., 2001) and time-to-time surveys of the region. Meteorological data of the area was recorded from weather station installed at our research station, Stakna-Leh, Ladakh. Physio-chemical properties of soil of Leh-Ladakh region were analyzed in our laboratory using standard scientific methods. The nutrient status of salt-affected areas was observed by analyzing the soil saturation extents, *viz.*, Na, K, Ca, Mg, Cu, Zn, and borates.

9.2.1 DESCRIPTION OF THE AREA

Ladakh region covers more than 70,000 square km geographical area of Jammu and Kashmir, India, and lies between 31° 44' 57"–32° 59' 57" N latitude and 76° 46' 29"–80° 41' 34" E longitude (Figure 9.1). It is one of the most elevated regions of the earth with an altitude range from 2,900 m asl to 5,900 m asl. It is bounded by Karakoram Mountains; East by the China border; South by Himachal Pradesh, and West by Kashmir valley. The region is sparsely populated along the riverbanks of different valleys, namely Indus, Nubra, Changthang, Zanskar, and Suru valley.

9.2.1.1 CLIMATE

Climatically, the region falls under a cold arid climate characterized by severe cold, dry winter, hot, and dry summer. There is great variation in day and night temperatures. The mean annual temperature ranges from 1°C to 8°C. The mean summer and winter temperatures are 10.2°C and –16.8°C, respectively. We observed that during 1994–2000, the extreme events of temperature remained maximum up to 35.0°C in July and minimum –30.2°C in January. The mean annual precipitation reported less than 50 mm, mostly received in the form of snowfall. The region has a heavy influx of infrared and ultraviolet radiations with low air density, i.e., reduced oxygen level. The region faces fast blowing winds 40–60 km/hr mainly in the afternoon hours. The soil moisture remains frozen during winters and low relative humidity during the summer months.

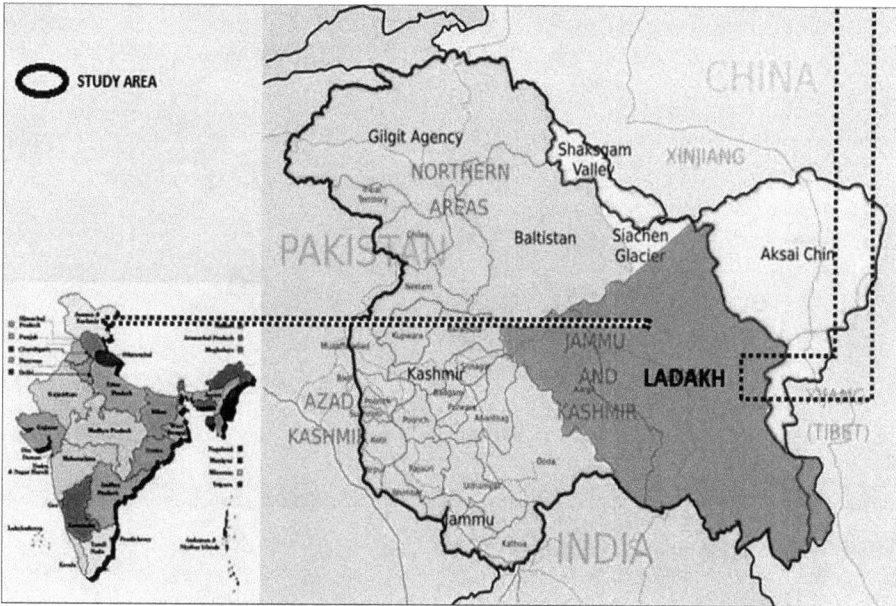

FIGURE 9.1 Map of District Ladakh.
Source: www.mapsofindia.com.

9.2.1.2 GEOLOGY AND SOIL

The region has barren topography (Figure 9.2). It represents mixed geological formation ranging from archaeans to recent crystalline rocks. Granite and gneisses of archean age occur mainly in the Zanskar ranges, Gilgit, Baltistan, and Ladakh. Slates, phyllites, schists, quartzites found in tracts of west Ladakh, while as, limestone's and shales are found extensively in the Karakorum Range. The undecomposed fragments of different rocks in this region show comparatively high calcium contents. Due to low precipitation and continuous weathering, the supply of calcium minerals is constantly increased in the soil (Wadia, 1940). In the Indus plain, soils of the region are gravelly and sandy loams on the alluvial fans to sandy and slit clay loams. Due to uneven distribution of plantation or scattered vegetation, loose sandy loam texture with, high proportion of stones and granules including their low water holding capacity and high bulk density results low soil fertility. It contains lower water retention capacity, 0.01% and 0.06% organic carbon, 0.3% and 2.1% calcium carbonate at the depth of 0.3 m and 1.5 m, respectively (Bhattacharyya et al., 2008).

FIGURE 9.2   Over view of cold desert Ladakh.

Our observations showed low organic carbon (1.32%), phosphorus (30.0 kg/ha), potash (800 kg/ha) and salt concentration (0–37 nm hos cm) and high pH (7.9) in the local soil. The nutrient status of salt-affected areas in the Ladakh region is shown in Table 9.1. It was observed that among these nutrients, boron contents was very high above the critical limit. The reclamation of boron toxicity is very difficult; however, the application of ploy phosphoric acid and subsequent flood washing may solve this problem, but it is very difficult for the region.

9.3   RESULTS AND DISCUSSION

9.3.1   USEFUL PLANT RESOURCES

The Ladakh region at the first sight seems barren and devoid of vegetation but virtually has very rich wealth of native and endemic flora. Ladakha region has three large and extremely bio-diversity rich areas, i.e., the Hemis High Altitude National Park, Karakoram Wildlife Sanctuary, and Changthang Cold Desert Sanctuary. The deposits of Aeolian nature of sand dunes are seen in Nubra valley, near Diskit and east of Nyoma. The flatlands, riverbanks, and inlands near agriculture and habitations in the region are mainly occupied by scrubs. There are *Hippophae rhamnoides* subsp. *Turkestanica, Ephedra gerardiana, Myricaria germanica, Rhododendron* spp., *Lycium ruthinicum* and *Tamarix gallica* dominated scrubs. At lower altitudes (up to 3000 m), particularly in Nubra valley, all these scrubs are present as individual species. However; in Nyoma valley (up to 4200 m), only mixed scrubs are found, i.e., *Hippophae-Ephedra, Hippophae-Myricaria, and Lycium-Tamarix. Juniperus* spp. (*J. recurva* and *J. macropoda*) are only evergreen tree species which occurs on dry, flattened as well as on the slopes. The grasslands and pastures are restricted along rivers, mostly in the eastern parts of Ladakh. The *Arabis tibetica, Atriplex crassifolis, A. hortensis, Carex nivalis, Corydalis crassifolia, Draba lasiphylla, Lychnis macrorhiza, Nepeta tibetica, Oxyria dignyta, Oxytropis leipponica, Plantego minima, Polygonum sibiricum, Potentilla multifida, Sedum eversii* and *S. tibeticum* are important grass species of the region.

TABLE 9.1 Nutrient Status of Soil of Ladakh Region

Nutrients in me/100 g		Soil Depth (cm)	
		0–10	**10–20**
Cations	Na$^+$	368.12	223.18
	K$^+$	6.85	3.98
	Ca$^+$	0.03	0.03
	Mg$^+$	0.01	0.01
	Cu$^+$	0.80	0.30
	Zn$^+$	0.55	0.45
Anion	CO$_3$$^{2-}$	19.52	9.77
	HCO$_3$$^-$	39.82	16.80
	SO$_4$$^{2-}$	59.45	18.15
	Cl$^-$	230.15	150.12
	BO$_3$$^-$	25.30	29.10
pH		9.89	11.2
Ec Mmho/cm		43.76	19.10

The diversity of medicinal herbs of the region include *Aquilegia fragrans, Origanum vulgare, Aster flaccidus, Berberis ulcina, Bergenia stracheyi, Delphinium* spp., *Dianthus angulatus, Dracocephalum heterophyllum, Epilobium angustifolium, Gentiana carinata, Lancea tibetica, Lloydea serotina, Lychnis nutans, Malva verticellata, Plantago depressa, Perovskia abrotanoides, Potentilla curviseta, Primula macrophylla, Rhodiola heterodonta, Sophora moorcroftiana, Tanacetum tibeticum, Rheum speciforme, Rosa webbiana, Tanacetum tibeticum* and *Taraxacum officinale.* Joshi et al. (2006) have observed the rare and threatened medicinal plants species in the region including: *Aconitum violaceum* (Critically Endangered-CR), *Arnebia euchroma* (Endangered-E), *Artemisia maritima* (E), *Dactylorhiza hatagirea* (CR), *Hippophae rhamnoides* (Low risk, not threatened; LR-NT), *Hyoscyamus niger* (LR-NT), *Juniperus recurva* (Rare-R), *Lancea tibetica* (R), *Meconopsis aculeata* (CR), *Physochlaina praealta* (Vulnerable-Vu), *Rheum speciforme* (Vu), *Saussurea bracteata* (R), *Saussurea gnaphaloides* (R) and *Saussurea obvallata* (Vu).

9.3.2   TRADITIONAL AGROFORESTRY SYSTEM

The land utilization pattern of Leh, Nubra, and Nyoma showed that only 2.64% area is arable and 9.66% area is under vegetation cover (Table 9.2). The lower percentage of arable land is restricted only to flat valleys and in lower slopes. The snowfall is important for species survival because absence of water and therefore only a few species are cultivated by farmers. AS system is exist in traditional agroforestry system where agricultural crops combinated with boundary plantations of Willow (*Salix* spp.) and Popular (*Populus* spp.) species (Figure 9.3). The Nubra valley has more than 5,75,000 plants of Willow and Popular are the main source of fuel and fodder. According to an estimation, every year, these species are contributing 400 t of leaf litter to the ground and thus, being a great source of organic carbon and responsible for sequestration of more than 75,000 tons of carbon (Kumar et al., 2009). Apart from these species, sea buckthorn (*Hippophae rhamnoides*), a multipurpose values thorny shrub used by the villagers for food, fuel, fodder, medicine, and fencing their fields (Figure 9.4). Being a nitrogen fixing species, it is planted in the Igoo-PHE canal for soil rehabilitation measures and support agricultural crops. The integration of agricultural crops with fruit trees as apple (*Malus pumila* or *M. sylvestris*), apricot (*Prunus armeniaca*), peach (*Prunus persica*), mulberry (*Morus alba*) and walnut (*Juglans regia*) are rarely seen. The raising of some fruit tree species in the kitchen garden is also a well-established tradition of the region.

TABLE 9.2   Land Use and Vegetation Pattern in Ladakh Region

Land Use/Vegetation Class	Leh		Nobra		Nyoma	
	Area (Km²)	Percentage	Area (Km²)	Percentage	Area (Km²)	Percentage
Arable land	39.18	2.04	8.95	0.53	1.39	0.07
Barren rocks	932.39	48.52	719.38	42.83	911.93	47.07
Barren rocks under shadow	548.71	28.56	561.15	33.41	494.90	25.54
Ephedra dominated scrub	0.00	0.00	2.91	0.17	0.00	0.00
Built-up area	20.05	1.04	0.00	0.00	0.00	0.00
Grasslands/pastures	2.89	0.15	0.00	0.00	40.93	2.11
Hippophae dominated scrubs	7.45	0.39	42.06	2.50	0.00	0.00
Myricaria dominated scrubs	0.00	0.00	5.11	0.30	2.63	0.14
Marshy/waterlogged area	0.94	0.05	0.31	0.02	4.93	0.25
Mixed scrub	2.30	0.12	0.70	0.04	14.68	0.76
River/water body	8.50	0.44	18.06	1.08	8.18	0.42
Riverine sand	9.06	0.47	150.43	8.96	117.55	6.07
Salt affected area	0.94	0.05	0.00	0.00	18.28	0.94
Sandy area	183.27	9.54	31.91	1.90	321.81	16.61
Snow	141.25	7.35	127.59	7.60	0.00	0.00
Willow/popular plantations	24.59	1.38	10.87	0.65	0.28	0.01
Total	**1921.53**	**100.00**	**1679.44**	**100.00**	**1937.50**	**100.00**

Source: Adiga (2003).

The traditional crops species of the region include: barley (*Hordeum vulgare*), grim (*Hordeum aegiceras*), wheat (*Triticum aestivum*), buckwheat (*Fagopyrum tataricum* and *F. esculentum*), millets (*Panicum miliaceum*), and oat (*Avena sativa*). Besides, a chunk of the cropped area is occupied by pea (*Pisum sativum*), potato *(Solanum tuberosum),* and mustard (*Brassica* spp.)*. Allium cepa, A. stracheyi, Coriandrum sativum,* and *Carum carvii* are cultivated mostly in kitchen gardens and great demand for spice and medicinal purposes.

FIGURE 9.3   Traditional agroforestry system in cold arid desert, Ladakh.

FIGURE 9.4   Live fence of *Hippophae rhamnoides* plantation.

Besides cultivated crops, some wild plants also used as vegetable include *Amaranthus spinosus, Capsella thomsonii, Allium thomsonii, Lactuca dolicho-phylla, Chenopodium foliolosum, Lepidium latifolium, Orobanche hansii,* and *Polygonum aviculare.*

9.3.3 MAJOR SET BACK IN TRADITIONAL AGROFORESTRY SYSTEM

- Choice of tree components of farmers is limited: i.e., willow, and poplar.
- Research-extension linkages are poor therefore, local people are not aware in propagation and plantation techniques of newly species of tree and agricultural crops.
- Incentives and subsidies are lacking for agroforestry development.
- Integrated sustainable multi-disciplinary approach is lacking.
- Research strategies are ineffective or insufficient.
- Shifting of local people from agriculture to other enterprises and ecotourism activities.
- Moisture and temperature limitations: water availability is restricted due to very low precipitation in the region.
- Climatic uncertainties.

9.3.4 INTEGRATED AGROFORESTRY SYSTEM

Agroforestry models that can be adopted by the farmers of cold arid desert of the Ladakh region are discussed in subsections (Table 9.3).

9.3.4.1 AGRI-SILVICULTURE (AS)

The system refers to the integration of agricultural crops and forest trees, including shrubs leading to concurrent production of food, fiber, fodder, fuel, and small timber. In this system, the agricultural component acts as primary importance and tree component as secondary. The system includes a combination of Wheat/Barley/Alfa-Alfa/Malilotus can be grown with Willow/Poplar/Robinia. *Medicago lupulina* is also a promising fodder for the region. Nearby forests are the major source of fodder includes; *Artemisia dracunculus, Astragalus adesmifolius, A. confertus, A. oxydon, Bromus inermis, B. oxydon, Calamogrostis emodensis, Eragrostis pilosa, Festuca kashmeriana, Heracleum pinnatum, Lactuca tatarica, Lindelofia anchusoides, L. stylosa, Oxytropis cahsemiriana, O. tatarica, Poa bulbosa, P. falconeri, P. stapfiana, Stipa sibirica* and *Ulmus wallichiana* (Joshi et al., 2006). Including that legumes are associated with agricultural crops act as valuable sources of fodder and soil enrichment.

TABLE 9.3 Integrated Agroforestry System for Cold Arid Desert of Ladakh Region

SL. No.	Suitable Models	Combinations
1.	Agri-silviculture (AS)	Wheat (*Triticum aestivum*)/Barley (*Hordeum vulgare*)/Alfa-Alfa Vern Ol (*Medicago sativa* and *M. lupulina*)/Malilotus (*Melilotus alba, M. indica* and *M. officinalis*) + Willow (*Salix* spp.)/Poplar (*Populus* spp.)/Robinia (*Robinia pseudoacacia*)
2.	Agri-horticulture (AH)	Wheat/Barley/Alfa-Alfa/Malilotus/Vegetable crops + Apple (*Malus pumila* or *M. sylvestris*)/Apricot (*Prunus armeniaca*)/Cherry (*Prunus abium* or *P. cerasus*)/Plum (*Prunus domestica* or *P. salicina*)/Pear (*Pyrus communis*)/Pomegranate (*Punica granatum*)
3.	Agri-silvi-horticulture (ASH)	Wheat/Barley/Alfa-Alfa/Malilotus/Vegetable crops + Apple/Apricot/Cherry/Plum/Pear/Pomegranate/Willow/Poplar/Robinia
4.	Silvi-pastoral (SP)	Willow/Robinia + Wheat/Barley/Alfa-alfa/Malilotus
5.	Pastoral-silvi-horticulture (PSH)	Sea buckthorn (*Hippophae rhamnoides*) + Willow + Forage crops + Vegetable crops
6.	Energy plantation	Plantation of fast-growing native species (e.g., *Hippophae rhamnoides, Salix* spp.) on fertile as well as marginal and wasteland for fuel or other purposes.
7.	Development of agroforestry based enterprises	Goat farming, Dairy farming, and Herbal farming systems.

9.3.4.2 AGRI-HORTICULTURE (AH)

In this system, fruit trees like Apple/Apricot/Cherry/Plum/Pear/Pomegranate can be grown in combination with agricultural crops, i.e., Wheat/Barley/Alfa-Alfa/Malilotus/Vegetable crops. Recently farmers have diversified their agricultural crops including vegetable crops, i.e., *Brassica rappa, B. oleracea* (both cauliflower and cabbage), *B. nigra, B. caulorapa, Chenopodium album, Cucurbita maxima, Cucumis melo, Lycopersicon esculentum, Solanum melongena,* and *S. tuberosum.* They use these crops both for self-consumption and for sale (government and defense personnel). Strawberry (*Fragaria vesca*) has been introduced in combination with agricultural crops.

9.3.4.3 AGRI-SILVI-HORTICULTURE (ASH)

The system refers to the combined production system with agricultural crops, i.e., Wheat/Barley/Alfa-Alfa/Malilotus/Vegetable crops as well as forest and fruit trees, i.e., Apple/Apricot/Cherry/Plum/Pear/Pomegranate/ Willow/Poplar/Robinia.

9.3.4.4   SILVI-PASTORAL (SP)

This system refers to combined production of fodder trees, i.e., Willow/ Robinia and Grasses or Forage crops, i.e., Wheat/Barley/Alfa-Alfa/Malilotus.

9.3.4.5   PASTORAL-SILVI-HORTICULTURE (PSH)

This system refers to the cultivation and management of grasslands supporting forest and fruit trees simultaneously. The perennials in this system provide fodder or function as live fences around grazing lands, e.g., a combination of sea buckthorn, willow, forage crops, vegetable crops, and fruit trees.

9.3.4.6   ENERGY PLANTATION

The requirement of fuelwood during winters (during temperature up to $-30°C$) met through cutting of dry as well as green plants or collection of fallen twigs in nearby forests, this unsustainable harvesting practice posed huge pressure on wild stock. The energy plantation of fast-growing native species on fertile as well as marginal and wasteland for fuel or other purposes will divert the dependency of local people from exclusive wild source. Almost all the woody species are used as a source of fuelwood in the valley. Apart from various species of Willow and Popular, the demand of fuelwood is met from wild plant species, *viz., Atriplex hortensis, Ephedra gerardiana, Hippophae rhamnoides, Sophora moorcroftiana, Tanacetum tibeticum, Rosa webbiana, Berberis ulicina, Myricaria germanica,* and *Tamarix gallica.*

9.3.5   DEVELOPMENT OF AGROFORESTRY BASED ENTERPRISES

9.3.5.1   DAIRY FARMING

Due to various reasons, including lack of grazing areas as well as availability of grazing areas for a short period of time and lack of green fodder, dairy farming is very restricted in the region. Moreover, adverse climatic conditions confine diversification in rearing of milch animals. From Changthang and Zanaskar valleys, owing to scarcity of fodder during winter, animal death have been reported. Due to unique topography, scenic beauty, and environmental conditions, the inflow of outsiders is very high in the region. This generates high demand of milk and milk products. Therefore, the need of hour is dairy farming which could be emerged as the backbone of economy of local people. Several research workers have shown that dairy farming as profitability system over arable farming (Tomar et al., 1982; Singh et al., 1986). The integration of livestock with the above mentioned agroforestry system will be helpful in developing dairy farming at an optimum pace.

9.3.5.2   GOAT FARMING

Goats are considered as a poor man's cow. Goat farming plays a main role in the economy. Pashmina native to the Himalayas is well acclimatized in the high altitudes cold climate of the Lakakhi Chanthangi or Baltistan (Kashmir region) and neighboring areas of Tibet. It is used pack animal and source of finest cashmere wool. The high demand for cashmere wool, the Pashmina goats produce and also commercially raised in the Gobi Desert in Inner and Outer Mongolia (pastures are more fertile) where similar weather as the Himalayas region. The Pashmina goat is most profitable because used for cashmere wool and also meat during winter months. The Pashmina shawl fetches high prices ($ 108–$ 2173 per shawl) and its industry is fast emerging well developed and very popular in the world. Implement of modern tools and techniques are needed to make this industry economically sound for more profitable and sustainable. Currently, unavailability of green grazing areas with long period of stall-feeding due to prolonged snowfall period reduced mass scale rearing of Pashmina goat. By introducing suitable agroforestry system, this industry can flourish very well that will provide great economic stability to the local inhabitants.

9.3.5.3   HERBAL FARMING

Diversification in the present cropping system seems to be the need of the day to cope up with the ever-increasing demand for a variety of products and assured income. The region has great medicinal plants diversity (Joshi et al., 2006). Cultivation of medicinal plants with existing agroforestry systems is a viable option. Wherever markets are established, medicinal plants are remunerative alternative intercrops to the traditionally grown annual crops (Zou and Sanford, 1990). Economically viable and highly demanded medicinal herbs like *Aconitum heterophyllum, Arnebia euchroma, Artemisia maritime, Dactylorhiza hatagirea, Hyoscyamus niger, Inula racemosa, Mentha* spp., *Picrorhiza kurrooa, Rheum spp., Saussurea costus, S. gossypiphora, S. obvallata* and *S. simpsoniana* and shrub, *Ephedra gerardiana* are suitable for cultivation in the area. Among the species, Ephedra, the source of Ephedrine hydrochloride, Ephedrine alkaloid, and Physachlain; Hyoscyamus, crude source of Atropine sulfate and Carum, the important only source of cumin used in Indian System of Medicine and flavoring agent for wines and soaps have already been grown in the region. Our country imports these active constituents in large quantities. Most of these species are native to Himalaya and can adopt local climatic conditions, thus to be intercropped with local trees. However, the selection of species will depend on the size and intensity of its canopy shade, tree spacing, and management, especially pruning of branches and nature of the medicinal plants, i.e., their light/shade, moisture, and nutrient requirements. For most of the above species, their cultivation with Willow and Popular is suitable. Jha and Gupta (1991) have observed that only 10 out of 64 herbaceous medicinal plants tried in intercropping with two-year-old poplar (*Populus deltoides*) spaced 5 m apart provide poor performance. In the tropical region of India, some successfully intercropped medicinal plants with fuelwood trees (e.g., *Acacia auriculiformis, Albizia lebbeck, Eucalyptus tereticornis, Gmelina arborea*, and *Leucaena leucocephala*), include *Safed Musli* (*Chlorophytum borivilianum*), rauvolfia (*Rauvolfia serpentina*), turmeric (*Curcuma longa*), wild turmeric (*C. aromatica*), *Curculigo orchioides*, and ginger (*Zingiber officinale*) have been reported (Chadhar and Sharma, 1996; Mishra and Pandey, 1998; Prajapati et al., 2003).

Saussurea costus is not available naturally but has the potential to be culti-vated in the region. It is a critically endangered species, which has mentioned in *Appendix I* of CITES (Convention of International Trade in Endangered Species of Wild Fauna and Flora). Its illegal trade is strictly prohibited under Foreign Trade Development Act-1992. Presently, it is being cultivated in the Lahaul and Spiti district of Himachal Pradesh. Owing to the fluctuating market prices and difficulty in getting its trade license, the cultivation is restricted into a few villages. Demand for unprocessed material and finished herbal products is very high even in local areas due to the presence of tourists and defense personnel. Besides, there is a well-developed 'Amchis System of Medicine' in the region, which includes a number of herbs for curing various diseases of over 60% tribal population. In this system, the combination of two or more than two species is used for the treatment of a single ailment. Low availability of wild medicinal herbs 'Amchis' sometimes have to travel far to search these medicinal plants (Joshi et al., 2006). Cultivation of medicinal reduces the over-exploitation pressure on natural medicinal plant populations. Medicinal plants (*Achillea millefolium, Aconitum violaceum, Arnebia euchroma, Bunium persicum, Capparis spinosa, Carum carvi, Dactylorhiza hatagirea, H. Rhamnoides, Hyoscyamus niger, Juniperus* spp., *Lancea tibetica, Meconopsis aculeata, Medicago sativa, Mentha longifolia, Ocimum basilicum, Origanum vulgare, Physochlaina praealta, Rheum* spp., *Rhodiola imbricata, Rosa webbiana, Saussurea costus, S. gnaphaloides, S. obvallata, S. bracteata, Scrophularia scabiosaefolia* and *Inula rhizocephala*) are used as major ingredients of 'Amchis system of medicine.' Besides, the Field Research Laboratory, Leh, the world's highest research laboratory of Defense Research and Development Organization (DRDO), has also formulated certain high value medicated herbal products in the past few years (Ballabh et al., 2007). Medicinal properties of these products help to provide better adaptational and functional health to army personnel and tribal communities residing at high altitudes and outsiders, i.e., tourists, surveyors, researchers, and common people.

It is recognized in India that many medicinal trees can be integrated into croplands by planting them on field bunds or as scattered trees (Pushpangadan and Nayar, 1994). In the study region, plantation of *H. rhamnoides* and *Juniperus* spp. in the boundaries of cultivated fields will be fruitful. The soil fertility is improved by planting sustainable native agroforestry tree species (Yadav et al., 2009). *H. rhamnoides* is sparsely present in traditional agroforestry system can be augmented through plantation at appropriate distance and place. *J. macropoda* is high value medicinal and aromatic plant and source of Juniper oil and local people used its fruit and leaves for aromatic sticks (Dhoop). Due to various biotic disturbances (population pressure, grazing, fire) and lack of scientific knowledge of propagation the natural regeneration is negligible. There is a dire need to develop propagation, multiplication, and plantation techniques of this species to harness its potential for socio-economic and ecological development of the region.

Hops (*Humulus lupulus*), a climber used for making alcoholic drink, is cultivated in a few areas of the region. During 1995–2000, due to high market demand, it was extensively cultivated in the region, but due to some conflicts, its cultivation is restricted to a very few number of villages.

9.4 STRATEGIES FOR PROMOTION OF AGROFORESTRY IN COLD ARID REGIONS OF INDIA

Agroforestry is an integral rural development approach. The following points can be considered as an appropriate strategy for agroforestry development in cold arid regions:

- Popularization of integrated agroforestry system for its diversified sustainable uses.
- Establishment of demonstration blocks/garden/parks of integrated location-specific agroforestry system to create awareness.
- Agroforestry knowledge center to establish linkages between different stakeholders, i.e., farmers, NGOs, rural cooperatives, government officials, and other interested persons and supply of quality planting material to the farmers.
- Capacity building and skill development of the farmers and other stakeholders regarding integrated agroforestry development.
- Involvement of government or corporate sector in regulating marketing of wood and wood-based products or minimum support price at par with agricultural crops.
- Loan based facilities for wood-based industries.
- Since agroforestry as a science is just two to three decades old, therefore, to promote research and extension activities, financial support from different funding agencies should be provided. It is the ultimate need to address the important issues related to agroforestry in the future.

9.5 CONCLUSION

In the present study, we assessed potential of traditional agroforestry system in the Ladakh region. Keeping sustainable livelihood and environmental security in view, we offered an integrated agroforestry system for the region. We envisage that effective implementation of the aforementioned agroforestry system in farmer's fields will be fundamental in achieving sustainable development in the region. The following points can be taken as remedial measures in the process: (i) By integrating the agricultural crops including medicinal plants with horticulture and forest trees, the demand of fuelwood, fodder, small timber, crude drug, etc., can be fulfilled; this will ultimately generate employment opportunities through development of Dairy farming,

Goat farming and marketing of medicinal plants (ii) Increasing productivity of available farm resources will be helpful in reducing pressure on natural resources (iii) Increasing vegetation cover will be helpful in maintaining ecological balance and making the climate more salubrious to the inhabitants of the region (iv) Through developing and maintaining green cover the local inhabitants will be benefited from carbon credit system in the near future.

ACKNOWLEDGMENTS

We are grateful to the Assistant Director, Research, and Extension, Agriculture Research Station/KVK, SKUAST-K, Leh, for providing necessary facilities and logistic support to conduct the present study. We are also thankful to the local inhabitants of Ladakh for providing relevant information on traditional agroforestry system in the region.

KEYWORDS

- **agri-silviculture**
- **agroforestry**
- **cold arid desert**
- **integrated approach**
- **pastoral-silvi-horticulture**
- **sustainability**

REFERENCES

Adiga, S., (2003). *Natural Resource Management in Ladakh Region: A Remote Sensing Based Study* (p. 192).

Ballabh, B., Chaurasia, O. P., & Ahmed, Z., (2007). Herbal products from high altitude plants of Ladakh Himalaya. *Current Science., 92*(12), 1664, 1665.

Bhattacharyya, T., Pal, D. K., Chandran, P., Ray, S. K., Mandal, C., & Telpanda, B., (2008). Soil carbon storage capacity as a tool to prioritize areas for carbon sequestration. *Current Science, 95,* 482–494.

Chadhar, S. K., & Sharma, M. C., (1996). Survival and yield of four medicinal plant species grown under tree plantations of Bhataland. *Vaniki Sandesh., 20*(4), 3–5.

Jha, K., & Gupta, C., (1991). Intercropping of medicinal plants with poplar and their phenology. *Indian Forester, 7*, 535–544.

Joshi, P. K., Rawat, G. S., Padilya, H., & Roy, P. S., (2006). Biodiversity characterization in Nubra valley, Ladakh with special reference to plant resource conservation and bioprospecting. *Biodiversity and Conservation, 15*, 4253–4270.

Kachroo, P., Sapru, B. L., & Dhar, U., (1977). *Flora of Ladakh: An Ecological and Taxonomic Appraisal.* Bishen Singh Mahendra Pal Singh, Dehradun, India.

Kumar, P. G., Murkute, A. A., Gupta, S., & Singh, S. B., (2009). Carbon sequestration with special reference to agroforestry in cold desert of Ladakh. *Current Science, 97*(7), 1063–1068.

Mani, M. S., (1974). *Fundamental of High-Altitude Biology* (p. 196). Oxford and IBH Publishing Co., New Delhi.

Mishra, R. K., & Pandey, V. K., (1998). Intercropping of turmeric under different tree species and their planting pattern in agroforestry systems. *Range Management Agroforestry, 19*, 199–202.

Murty, S. K., (2001). *Flora of Cold Deserts of Western Himalaya* (Vol. 1). (Monocotyledons). Botanical Survey of India, Dehradun, India.

NAEEDB, (1992). *Action Plan on Cold Desert: An Integrated Approach for Sustainable Development* (p. 500). National Afforestation and Ecological Development Board, Dr. Y.S. Parmar-UHF, Nauni Solan, H.P.

Prajapati, N. D., Purohit, S. S., Sharma, A. K., & Kumar, T., (2003). *A Handbook of Medicinal Plants* (p. 553). Agribios (India).

Pushpangadan, P., & Nayar, T. S., (1994). Conservation of medicinal and aromatic tree species through agroforestry. In: Thampan, P. K., (ed.), *Trees and Tree Farming* (pp. 265–284). Peekay tree crops development foundation, Cochin, India.

Singh, N., Malik, B. S., Kadian, V., & Mehra, O. P., (1986). Comparative studies of various farming systems. *Haryana Vet. J., 1*(25), 30–33.

Tomar, S. P., Sairam, R. K., Hasika, A. S., & Ganguly, T. K., (1982). Comparative efficiency of dairy and mixed farming system. *Forage Research, 1*, 93–98.

Wadia, D. N., (1940). In: *Geology of India* (pp. 223–225). Tata Mc Grew-Hill Publishing Co, New Delhi.

Yadav, S. K., Juwarkar, A. A., Kumar, G. P., Thawale, P. R., Singh, S. K., & Chakrabarti, T., (2009). Bioaccumulation and phytotranslocation of arsenis, chromium and zinc by *Jatropa curcas* L.: Impact of dairy sludge and biofertilizer. *Bioresource Technology, 100*(20), 4616–22.

Zou, X., & Sanford, R. L., (1990). Agroforestry systems in China: A survey and classification. *Agroforestry System, 11*, 85–94.

CHAPTER 10

Traditional Practices in Forest Conservation: Experience from Indian Himalaya

NAZIR A. PALA,[1] MUNEESA BANDAY,[1] M. M. RATHER,[1]
MEGNA RASHID,[1] PEERZADA ISHTIYAK,[1] and A. K. NEGI[2]

[1]*Faculty of Forestry, Benhama, Ganderbal, Sher-e-Kashmir University of Agricultural Sciences and Technology of Kashmir, Jammu and Kashmir, India, E-mail: nazirpaul@gmail.com (N. A. Pala)*

[2]*Department of Forestry and Natural Resources, HNB Garhwal University (A Central University) Srinagar, Uttarakhand–246174, India*

ABSTRACT

The involvement of local people and their knowledge/experience of resource values and management options have an important role to play for forest conservation. Simultaneously to conserve forest resources, the role of cultural and spiritual values becomes of prime importance. These community conserved sites of Garhwal Himalaya are very important from the conservation viewpoint of plant species diversity and dependence of the local inhabitants for their livelihood activities like collection of fuelwood, fodder, small timber, and NTFP. The Uttarakhand state, also called 'Dev Bhumi' or abode of Gods, is unique in this regard. The landscape in this state is dotted with many holy places of worship. These places are often of small to medium-sized with natural vegetation as a sacred grove of the deity. The present book chapter has given the status of some community-conserved forests sites from Garhwal Himalaya for its contribution in plant diversity conservation, regeneration of the tree species, ecosystem services/biodiversity value, and role of belief systems in conservation practices. Article 8 (j) of the Convention on Biological Diversity calls for respecting, preserving, and maintaining knowledge, innovations, and practices of indigenous and local communities embodying traditional lifestyles relevant for the conservation and sustainable use of Biological diversity according to the national legislation. In this context, sacred conservation practices assume particular importance. The present book chapter highlights status of some community-conserved forests having religious significance apart from getting conservation from communities from Garhwal Himalaya for its contribution in plant diversity conservation, regeneration of the tree species, ecosystem services/biodiversity value, and role of belief systems in conservation practices.

10.1   INTRODUCTION

Cultural practices like religious ones have evolved to protect a group of trees or a patch of vegetation protected by the local people in the form of sacred groves (Isreal et al., 1997). These conserved sites are of immense importance for both tangible and intangible uses like high diversity of medicinal, rare, and endemic plants and as these are mostly undisturbed and hence provide, as refugia for relict flora of a region (Whittaker, 1975; Jeeva et al., 2007; Pala et al., 2012, 2013). A tag of "mini biosphere reserves" has been given to acknowledge the conservation potential of these groves (Gadgil and Vartak, 1975). Conservation approaches like socio-cultural, religious, and other traditional practices outside of protected areas has made India rich with biodiversity and an important conservation centers for conserving ecosystem diversity of various life forms (Dash, 2005). The protected areas range from a spectrum of intensive use and management approaches to biodiversity loss and are strictly protected. Community-conserved areas (CCAs), mostly managed by inhabitant communities have been found effective for biodiversity conservation in the past few decades (Bray et al., 2003; Kothari, 2006; Pala et al., 2015).

Worldwide ecologists are talking about the fragmentation of forest resources by human habitats, and conversion of forestland into non-forest activity has been reported as one of the major cause (Prentice and Parish, 1990). Prioritizations of internationally collaborative research, native studies in terms of habitat fragmentation and factors leading to it shall be taken into account for in-depth study. The need is urgent for preservation because direct benefits arising from the conservation of those components viz., ecosystems services, biological resources, and social benefits (Heywood, 1995) are of high significance for global conservation and sustainable development efforts. Biodiversity conservation, long-term productivity, sustainability, and global climate change require consideration of broad geographic scales (landscapes to regions) and long time frames (decades to centuries). Some forest assessments (He et al., 1998; Spies et al., 2002) have applied simulation models to forest stands in a geographic information system (GIS) to examine regional landscape change and likewise distribution of forest resources and uses across multiple-ownership regions, as well as changes in landscape patterns and forest conditions over time. Shall be studied for policymaking and implementation.

Gokhale et al. (2011) described Haryali Devi sacred grove and its biodiversity status. Gokhale and Pala (2011) have also reported the status of ecosystem services from more than 100 sacred natural sites of Uttarakhand. Pala et al. (2012–2015) have worked on several sacred groves of Uttarakhand and described their ecological status, carbon stock potential, ecosystem services and medicinal plant conservation potential. Remote sensing tools and GISs based forest studies are gaining more grounds for implementation of conservation (Bridgewater, 1993; Kupfer, 1995). These types of studies using different spatial and temporal scales has not only become a national obligation under the convention of biological diversity (CBD), but also has emerged as an important discipline of biodiversity science in this new millennium (UNEP, 1992). Biodiversity as a part of our daily live and livelihood constitutes the resource base upon which our future generations depend (Pushpangadan et al., 1997). Maintenance and periodic assessment of diverse ecosystems and a whole range of biological diversity therein are, therefore, crucial for the long-term survival of humans (Berkes et al., 1998; Ayensu et al., 1999). The present book chapter has given the status of some community-conserved forests sites from Garhwal Himalaya for its contribution in plant diversity conservation, regeneration of the tree species, ecosystem services/biodiversity value, and role of belief systems in conservation practices.

10.2   MATERIALS AND METHODS

10.2.1   RECONNAISSANCE SURVEY AND SITE SELECTION

Reconnaissance field survey was carried out in various parts of Garhwal Himalaya for the present study. Based on the reconnaissance survey for detailed study, six groves/forests were selected. The main selection criteria of forest were the conservation by local communities because of their sacred-ness and associated belief on them. Other factors like size, vegetation, and altitude were also taken into consideration. All selected sites were from the temperate forests. The selected sites were of reserve forest, communal forest,

Van Panchyat, or a combination of these. The present study was conducted in six selected forests, i.e., Chanderbadni, Jameshwar, Ulkagari, Sem Mukhem, Ansuiya Devi, and Maroor located in four districts viz Rudraprayag, Chamoli, Pauri, and Tehri of Garhwal Himalaya having faith-based conservation system apart from having legal conservation system (Figure 10.1).

10.2.2 METHODOLOGY

After the selection of sites for the study, an intensive field survey was carried out along with the local communities associated with the selected sites to recognize the actual area around these temples considered sacred and protected. Ground truthing was done along the selected boundary by taking geo-coordinates at various places. A detailed field survey was conducted within these sacred/protected forests for ecological studies. For ecological studies, quadrats of a size 10 m × 10 m were used for tree layer, 5 m × 5 m for shrubs, and 1 m × 1 m for herbs. The GBH (girth at breast height, 1.37 m) measured with tape to calculate the basal area. Plant species present in the forest were listed, and vegetation data was quantitatively analyzed. The indices like important value index (IVI) and Shannon diversity index (1963) were worked out. For the study of regeneration pattern, 50 quadrats each of a size 10 m × 10 m were laid down at each site. In each 10 × 10 m quadrate, individuals having >31.5 cm CBH (circumference at breast height, i.e., 1.37 m above the ground) were considered as trees and counted individually and species-wise. Individuals having <10.4 cm circumference were considered as seedlings, and individuals intermediate position with respect to these circumferences were considered as saplings (Knight, 1963). The number of seedlings and saplings of each tree species was used to determine the regeneration status of that species. The regeneration status of the sampled species was assessed based on (Uma, 2001; Pala et al., 2012) as (a) Good regeneration, if seedlings> saplings >adults; (b) Fair regeneration, if seedlings >or ≤ saplings ≤ adults; (c) Poor regeneration, if the species survives only at sapling stage, but no seedlings (saplings may be>, < or = adults), (d) No regeneration, if a species is present only in adult form, (e) New regeneration, if the species has no adults but only seedlings or saplings.

The discussion held with the local inhabitants revealed that the majority of the plant species present were having different use-values, but not utilized by the local people. In the present chapter, the majority of the utilization and ecosystem services of these plants used were from available secondary literature and from our previous published chapters (Gokhale and Pala, 2011;

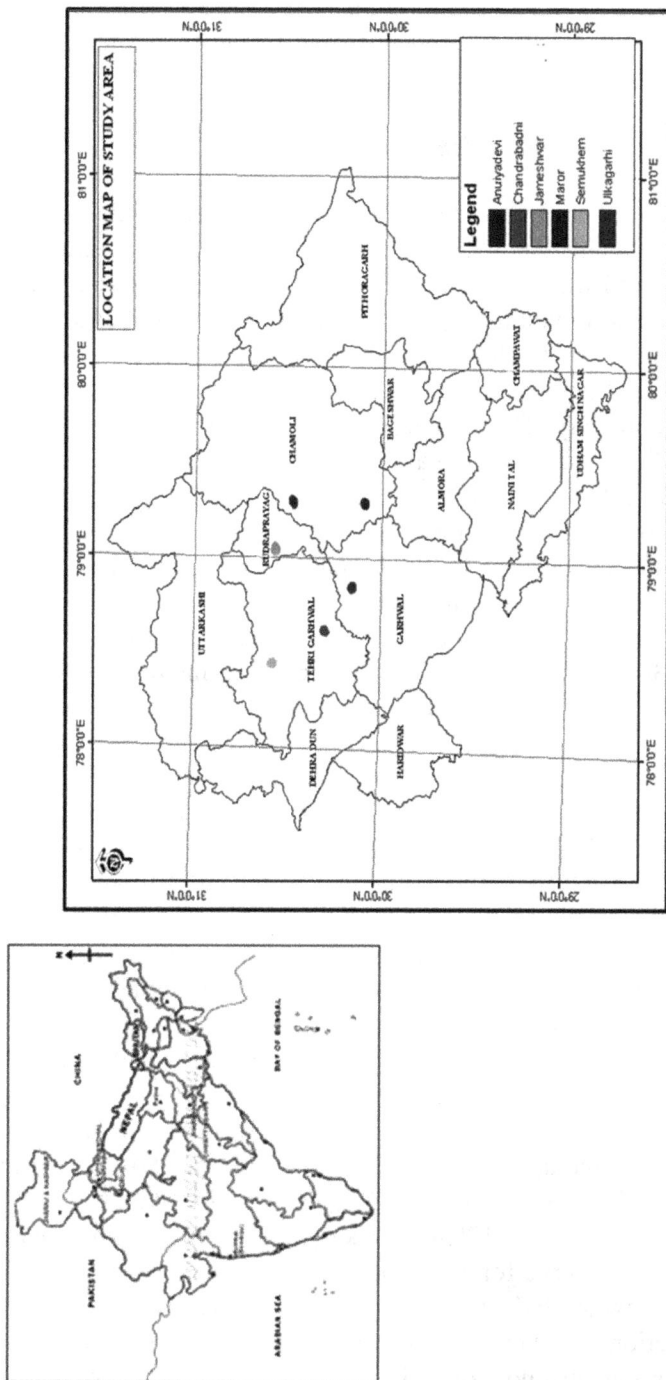

FIGURE 10.1 Map showing distribution of study sites in Garhwal Himalaya.

Pala et al., 2012). The conservation management strategies have also been taken from the previous published chapter (Gokhale and Pala, 2016).

10.3 RESULTS AND DISCUSSION

The forest area around Chanderbadni temple having religious significance as per the inhabitants has land-use classes like forest, agriculture, and non-forest (barren and wasteland). Forest was named *Quercus leucotrichophora* forest based on dominant tree species. Non-Forest land covering an area of 5 ha includes mostly rocky, hilly terrain and some patches cleared by villagers near the village boundary, which is devoid of vegetation and has been mixed with village boundary. The area around Jameshwar temple having religious significance falls under altitudinal range 1812 m to 2096 m above msl with N 30° 34′ 23″–30° 33′ 58″ to E 78° 02′ 6.6″–78° 2′ 2.9″. Some part of this area is reserve forest while most of it is a community-conserved area (Van Panchyat) dedicated to deity. This forest had *Quercus floribunda* the dominant tree species. The area around the Sem Mukhem temple considered religious comes under altitudinal range of 2205 m to 2628 m above msl with N 30° 34′ 41″–30° 34′ 25″ to E 78° 56′ 29″–78° 26′ 25″. This forest was dominated by *Quercus floribunda*. Non-forest/forest blank class of this area is mella (fair) ground and some area around the boundary has been converted into agricultural land whereas some patches inside forest are devoid of vegeta-tion. The area of Ulkagari falls under the altitudinal range of 1839 m to 2065 m above msl with N 30° 9′ 58″–30° 9′ 33″ to E 78° 51′ 13″–78° 50′ 55″. The whole area dedicated to deity falls under single land-use class (Forest). The dominant tree species of the area was *Quercus leucotrichophora followed by Rhododendron arboreum, Lyonia ovalifolia, Pinus roxburghii, Benthamidia capitata,* and *Sympoloccus rasmisissima*. The area around Ansuiya temple considered sacred comes under altitudinal range from 1600 m to 2300 m above msl with N 30° 27′ 39″–31° 28′ 42″ E 79° 16′ 25″–79° 31′ 58″. Forest was named *Daphniphyllum himalense* forest based on the dominant tree species. Non-Forest class spread around the temple has some settlements in the form of darmshalla, huts, and few houses. The forest of Maroor area falls under altitudinal range of 1900 m to 2470 m above msl with N 30° 4′ 42″–30° 5′ 59″ to E 79° 16′ 36″–79 17′ 36″. This forest is actually Van Panchyat area, and some part is reserve forest, but having several temples inside the forest, people have now started offering this forest to these deities for protection from destruction and hence held sacred now. The dominant tree species was *Quercus leucotrichophora*. Some blank forest areas includes grazing patches, damage due to forest fire and some part cleared for practicing agriculture crops during summer season. Community Conserved Areas (CCAs), or biodiversity rich areas having local conservation management, has gained considerable importance by effectively conserving species (Bray et al., 2003; Kothari, 2006; Pala et al., 2013, 2015). The linking of conservation goals to specific territories (mapping for conservation) is a practice that finds expression in an expanding map of protected areas (Zimmerer et al., 2004). Sacred conserved practices have been increasingly threatened and fragmented due to encroachments and changes in different land uses (Kushalappa and Kushalappa, 1996).

The increased fragmentation of the groves could undermine the utility of these groves in serving as refugia for the rare, endangered, and threatened (RET) species (Tambat et al., 2004). The land-use system in many sacred groves is now threatened (Chandrasekhar and Sankar, 1998) and must be studied simultaneously to understand and minimize the ecological impact of humans on forest (Williams, 2002). The condition of sacred groves all over the world appears to be deteriorating, and researchers often point out the erosion of values and consequently weaker restrictions and taboos as the main reason for the worsening conditions of the groves (Gadgil and Vartak, 1976). Thus it can be argued that the weakening of restrictions is often influenced or even driven by other factors like such as market demand for wood-based products. Chanderakanth et al. (2004) identified commercial agriculture, changing demographics, and weak property-right system as some other reasons causing the disintegration of groves.

A total of 254 plant species representing 179 genera belonging to 77 families were recorded from the six studied sacred/protected groves. Of these, 49 were trees representing 38 genera under 26 families. Eighty species were shrubs belonging to 58 genera from 28 families. The number of herbaceous species was 125, representing 92 genera under 44 families (Table 10.1). Asteraceae with 25 species was the dominant family. A maximum number of species (119) was found in Ansuiya Devi, whereas the lowest number of species (80) was found in Chanderbadni. In Chanderbadni sacred forest *Quercus leucotrichophora* with IVI (116.63) was dominant among tree species followed by *Banthimidia capitata* (25.72). Among shrubs *Berberis aristata* with IVI value of (30.23) was the dominant species followed by *Sinarundinaria falcate* (21.23). Out of 32 herbaceous species, *Andropogon munroi* with IVI (38.84) was found dominant herb species followed by *Cynodon dactylon* (37.42). In Sem Mukhem sacred forest *Quercus floribunda* with IVI (95.63) was dominant among tree species followed by *Quercus leucotrichophora* (32.37) and *Lyonia ovalifolia* (31.49). *Berberis aristata*

TABLE 10.1   List of Flora in Different Study Sites under Community Conserved Forests of Garhwal Himalaya

Name of Tree Species	Family	UGSF	SMSF	MRSF	JMSF	CBSF	ADSF
Abies spectibalis (D. Don) Mirl.	Pinaceae	–	–	–	–	–	+
Acer caesium Wallich ex Brandis	Aceraceae	–	+	+	+	–	+
Acer oblongum Wallich. Ex DC.	Aceraceae	–	–	–	–	–	+
Aesculus indica (Wall. Ex Camb.) Hook.f	Hippocastanaceae	+	–	–	+	–	+
Albizia julibrissin Durazzini,	Mimosaceae	–	–	–	+	–	–
Alnus nepalensis D. Don	Betulaceae	–	–	+	+	–	+
Benthamidia capitata (Wallich ex Roxb.) Hara	Cornaceae	+	+	+	–	+	–
Buxus wallichiana Baill,	Buxaceae	–	–	–	+	–	+
Betula alnoides Buch-Ham. Ex D. Don	Betulaceae	–	–	+	+	+	–
Carpinus faginea Lindl.	Betulaceae	–	–	–	–	–	+
Carpinus viminea Lidle.	Betulaceae	–	–	+	+	–	+
Cedrus deodara (Roxb. Ex D. Don) G. Don,	Pinaceae	+	–	+	–	+	–
Celtis australis L.	Ulmaceae	+	–	–	–	+	–
Cinnamomum tamala (Buch-Ham.) Nees and Ebermaeir	Lauraceae	+	–	–	–	+	–
Cupressus torulosa D. Don in Lambert	Cupressaceae	+	–	+	+	+	+
Daphniphyllum himalense Wall. Ex Steud.	Thymelaeaceae	–	–	–	+	–	+
Engelhardtia spicata Leschenault ex Blume	Juglandaceae	–	–	–	–	+	–
Eurya acuminata DC	Theaceae	–	+	+	–	–	+
Ficus auriculata Lour.	Moraceae	–	–	–	–	–	+
Ficus neriifolia Smith	Moraceae	–	+	+	–	–	–
Fraxinus micrantha Lingelsheim	Oleaceae	+	+	+	+	–	+
Ilex dipyrena Wallich,	Aquifoliaceae	–	+	–	–	–	+
Juglans regia L.	Juglandaceae	+	+	+	+	+	+
Lindera pulcherrima (Nees) Benth.ex Hook.f.	Lauraceae	–	–	+	+	+	–
Lyonia ovalifolia (Wallich) Drude,	Ericaceae	+	+	+	–	+	+

TABLE 10.1   (Continued)

Name of Tree Species	Family	UGSF	SMSF	MRSF	JMSF	CBSF	ADSF
Myrica esculenta Buch-Ham. Ex D. Don,	Myricaceae	+	+	+	−	+	+
Neolitsea cuipala (Buch-Ham, ex. D. Don) Kostermans	Lauraceae	−	−	−	+	−	−
Persea duthiei (King ex Kook.f.)	Lauraceae	+	+	+	+	+	+
Pinus roxburghii Sargent	Pinaceae	+	+	+	−	+	+
Pinus wallichiana A.B. Jackson	Pinaceae	−	+	−	−	−	−
Populus ciliata Wallich ex Royle	Salicaceae	+	+	−	−	−	−
Prunus cerasoides D. Don.	Rosaceae	+	+	+	+	+	−
Pyrus communis L.	Rosaceae	+	−	+	−	+	−
Pyrus pashia Buch-Ham, ex. D. Don	Rosaceae	−	+	+	+	+	+
Quercus floribunda Lindley ex Rehder	Fagaceae	−	+	+	+	−	+
Quercus glauca Thunb.	Fagaceae	−	−	−	−	−	+
Quercus leucotrichophora A. Camus	Fagaceae	+	+	+	+	+	+
Quercus semecarpifolia J.E. Smith.	Fagaceae	−	+	−	+	−	−
Rhamnus viratus Roxb.	Rhamnaceae	−	+	+	−	−	−
Cotoneaster confuses Klotz		−	+	+	+	−	−
Rhododendron arboreum Smith, Exot. Bot.	Ericaceae	+	+	+	+	+	+
Rhus punjabensis J.L. Stewart	Anacardiaceae	−	−	−	−	+	−
Sorbus aucuparia (Spach) Hedlund,	Rosaceae	+	−	−	−	+	−
Swida macrophylla (Wallich) Sojak	Cornaceae	+	+	+	+	+	+
Symplocos paniculata (Thunb.). Miq,	Symplocaceae	+	−	−	−	+	−
Symplocos ramosissima Wallich ex G. Don	Symplocaceae	−	+	+	+	−	−
Taxus baccata L. SSP	Taxaceae	−	−	−	+	−	+
Toona ciliata Roemer	Meliaceae	−	−	−	−	+	−
Toona serrata (Royle) Roem	Meliaceae	+	−	−	−	−	+

TABLE 10.1 *(Continued)*

Name of Tree Species	Family	UGSF	SMSF	MRSF	JMSF	CBSF	ADSF
Shrubs							
Arachne cordifolia (Decne.) Hurusawa	Euphorbiaceae	–	+	–	–	+	+
Artemisia japonica Thunb.,	Asteraceae	+	–	–	+	–	–
Artemisia roxburghiana Wallich ex Berser	Asteraceae	+	+	+	–	–	+
Asparagus adscendens Buch-Ham. Ex Roxb.,	Liliaceae	+	–	–	–	–	–
Asparagus curilus Buch-Ham. Ex Roxb.	Liliaceae	–	–	–	–	–	+
Asparegus racemosus Willd.,	Liliaceae	+	–	–	–	+	–
Barleria cristata L.	Acanthaceae	–	–	–	+	+	–
Berberis aristata DC.,	Berberidaceae	+	+	+	+	+	+
Berberis asiatica Roxb. Ex DC	Berberidaceae	+	+	+	–	–	+
Berberis lycium Royle.	Berberidaceae	–	–	–	–	–	+
Berchemia edgeworthii Lawson	Rhamnaceae	–	+	+	–	+	–
Boehmeria macrophylla D. Don	Urticaceae	–	–	–	–	–	+
Boehmeria platyphylla D. Don	Urticaceae	–	–	–	–	–	+
Buddleja paniculata Wall.	Buddlejacea	–	–	–	–	–	+
Caryopteris foetida (D. Don) Thellung,	Verbenaceae	+	–	–	–	+	–
Caryopteris odorata (D. Don) B.L. Robinson,	Verbenaceae	+	–	–	–	–	–
Colebrookia oppositifolia J.E. Smith	Lamiaceae	+	+	+	+	+	+
Cotoneaster bacillaris Wallich	Rosaceae	+	+	+	+	+	+
Cyathula tomentosa (Roth) Moq.	Amaranthaceae	–	+	+	+	–	+
Daphne papyracea Wallich ex Steudei	Thymelaeaceae	+	+	+	+	+	+
Debregeasia salicifolia (D. Don) Rendle	Urticaceae	–	+	–	+	–	+
Desmodium concinnum DC	Fabaceae	+	–	–	–	–	+
Desmodium elegans DC.	Fabaceae	–	+	+	–	+	+

TABLE 10.1 *(Continued)*

Name of Tree Species	Family	UGSF	SMSF	MRSF	JMSF	CBSF	ADSF
Duetzia compacta Craib	Fabaceae	–	+	–	–	–	–
Elsholtzia flava (Benth.) Benth	Lamiaceae	+	–	–	–	–	–
Elsholtzia fruticosa (D. Don) Rehder	Lamiaceae	–	–	–	–	–	+
Eupatorium adenophorum Sprengel	Asteraceae	+	+	+	+	+	–
Flacourtia indica (Burm. F.) Merrill	Flacourtiaceae	+	–	–	–	–	–
Girardinia diversifolia (Link) Friis	Urticaceae	–	+	–	–	–	+
Goldfussia dalhousiana Nees	Acanthaceae	–	–	–	+	–	–
Hypericum elodeoides Choisy,	Hypericaceae	+	+	+	+	–	–
Hypericum uralum Buch-Ham. Ex D. Don	Hypericaceae	+	–	+	+	–	–
Indigofera cassioides Rottler ex DC	Fabaceae	–	–	+	–	–	+
Indigofera dosua Buch-Ham.ex D. Don	Fabaceae	–	+	+	–	–	–
Indigofera heterantha Wallich ex Brandis	Fabaceae	–	–	+	+	+	–
Inula cappa (Buch-Ham. Ex D. Don) DC.,	Asteraceae	+	–	+	+	+	–
Inula cuspidata (DC.) C.B. Clarke	Asteraceae	–	–	+	+	–	–
Lantana camara L.	Verbenaceae	+	–	–	–	–	–
Leptodermis lanceolata Wallich	Rubiaceae	+	+	+	+	–	+
Linderbergia grandiflora (Buch-Ham. Ex D. Don) Benth	Scrophulariaceae	+	–	–	–	–	–
Lonicera quinquelocularis Hardwicke,	Caprifoliaceae	–	–	–	–	–	+
Myrsine africana L.	Myrsinaceae	+	+	+	+	+	–
Nepeta govaniana (Benth.) Benth	Lamiaceae	+	+	+	–	+	–
Phyllanthus parvifolius Buch-Ham.ex D. Don	Euphorbiaceae	–	–	–	–	–	+
Phyllanthus virgatus G. Forster	Euphorbiaceae	–	–	–	+	–	–
Plectranthus japonicus (Burm. F.)	Lamiaceae	–	–	–	–	–	+
Pogostemon benghalense (Burm.f.) Kuntze	Lamiaceae	–	+	+	+	–	+

Diversity and Dynamics in Forest Ecosystems

TABLE 10.1 *(Continued)*

Name of Tree Species	Family	UGSF	SMSF	MRSF	JMSF	CBSF	ADSF
Prinsepia utilis Royle	Rosaceae	+	+	+	+	+	+
Pteracanthus angustifrons (C.B. Clarke) Bremek	Acanthaceae	+	+	+	–	–	–
Pyracantha crenulata (D. Don) M. Roemer	Rosaceae	–	+	+	+	+	+
Rhamnus persica Boissier	Rhamnaceae	–	–	–	+	+	–
Rhamnus purpureus Edgew	Rhamnaceae	–	–	–	+	–	–
Rhamnus virgatus Roxb.	Rhamnaceae	–	–	–	–	–	+
Rhus parviflora Roxb.	Rosaceae	+	+	+	+	+	–
Rosa brunonii Lindley	Rosaceae	–	+	+	+	–	+
Rosa macrophylla Lindley	Rosaceae	+	+	+	–	+	–
Rubia manjith Roxb. Ex Fleming	Rubiaceae	–	–	+	–	–	–
Rubus ellipticus Smith	Rosaceae	+	+	+	+	+	+
Rubus foliolosus D. Don	Rosaceae	–	+	–	–	+	–
Rubus niveus Wallich ex G. Don	Rosaceae	+	–	–	–	–	+
Rubus paniculatus Smith,	Rosaceae	+	–	–	–	–	–
Sarcococca saligna (Don) Munell.	Buxaceae	–	–	–	–	–	+
Saxifraga diversifolia Wallich ex Seringe	Saxifragaceae	–	–	+	–	–	–
Segereetia filiformis (Roth) G. Don. Syst.	Rhamnaceae	–	–	+	–	–	–
Sinarundinaria falcata (Nees) Chao and Renvoize	Poaceae	–	+	+	+	+	–
Skimmia anquetilia Taylor and Airy Shaw	Rutaceae	–	+	+	–	–	+
Smilax aspera L.	Smilacaceae	–	+	+	+	–	+
Smilax glaucophylla Klotz.	Smilacaceae	–	–	–	–	–	+
Solanum incanum L.	Solanaceae	–	–	–	–	–	+
Spermadictyon saaveoleus Roxb.,	Rubiaceae	+	–	–	+	–	–
Spiraea canescens D. Don	Rosaceae	+	–	+	+	–	–
Taxillus articulatum Var. liquidambaricolum	Loranthaceae	–	–	–	–	–	+

TABLE 10.1 *(Continued)*

Name of Tree Species	Family	UGSF	SMSF	MRSF	JMSF	CBSF	ADSF
Thamnocalamus falconeri Hook. F. ex Munro	Poaceae	–	–	–	–	–	+
Thamnocalamus spathiflora (Trinius) Murno	Poaceae	–	–	–	–	–	+
Urtica ardens Link.	Urticaceae	–	–	–	–	–	+
Urtica dioica L.	Urticaceae	+	+	+	+	–	+
Viburnum cordifolium Wallich ex DC	Caprifoliaceae	–	+	+	–	+	–
Viburnum mullaha Buch-Ham ex D. Don	Urticaceae	–	+	–	–	–	–
Woodfordia fructicosa (L.) Kurz	Lythraceae	–	+	+	–	+	+
Zanthoxylum armatum DC	Rutaceae	–	–	+	–	+	–
Herbs							
Achyranthes bidentata Blume	Amaranthaceae	–	+	–	–	–	+
Aechmanthera gossypina (Nees) Nees	Acanthaceae	+	–	–	–	–	–
Agrostis pilosula Trinius,	Poaceae	–	+	–	–	–	–
Ainsliaea aptera DC	Asteraceae	+	–	–	–	–	+
Ainsliaea latifolia (D. Don) Schultz-Bipontinus	Asteraceae	+	–	–	+	–	+
Ajuga brachystemon Maxim	Lamiaceae	–	+	–	–	–	+
Anaphalis adnata Wallich ex DC	Asteraceae	–	+	+	–	+	+
Anaphalis busua (Buch-Ham. Ex D. Don) DC.,	Asteraceae	+	–	–	+	+	+
Anaphalis contorta (D. Don) Hook.f,	Asteraceae	–	+	+	–	–	–
Anaphalis triplinervis (Sims) C.B. Clarke	Asteraceae	+	+	+	–	–	–
Andropogon munroi C. B. Clarke	Poaceae	+	+	+	+	+	+
Angelica glauca Edgew.	Apiaceae	–	–	+	–	–	–
Anthraxon prionodes (Steuel) Dandy	Poaceae	+	–	–	–	–	–
Apluda aristata L.	Poaceae	–	+	+	–	–	–
Apluda mutica L.	Poaceae	+	+	+	+	–	+
Argostemma verticillatum Wallich,	Rubiaceae	–	–	–	–	–	+

TABLE 10.1   *(Continued)*

Name of Tree Species	Family	UGSF	SMSF	MRSF	JMSF	CBSF	ADSF
Arisaema tortuosum (Wallich) Schott	Araceae	–	–	+	+	–	–
Arundinella bengalensis (Spreng.) Druce	Poaceae	–	–	–	–	–	+
Arundinella birmanica Hook. F	Asteraceae	+	–	–	+	–	–
Arundinella nepalensis Trin	Poaceae	–	–	–	–	–	+
Arundinella nervosa (Roxb.) Neex ex Hook. and Arn	Poaceae	+	–	+	+	+	–
Bergenia ciliata (Haworth) Sterbn	Saxifragaceae	+	+	+	+	+	+
Bidens bipinnata L.	Asteraceae	–	+	+	–	–	–
Bidens pilosa L.	Asteraceae	+	+	+	–	+	–
Boenninghausenia albiflora (Hook.) Reichb. Ex Meisn	Rutaceae	+	–	+	+	+	+
Boerhavia diffusa L.	Nyctaginaceae	–	–	+	–	+	–
Buddleja paniculata Wallich.	Buddlejaceae	–	+	–	–	–	–
Bupleurum falcatum L.	Apiaceae	–	–	+	–	–	–
Bupleurum hamiltonii Balakrishnan	Apiaceae	–	+	–	+	–	–
Cannabis sativa L.	Cannabiaceae	–	–	–	+	–	–
Carex caricina (D. Don) Ghildyal and Bhattacharyya	Cyperaceae	+	–	–	–	+	–
Carum carvi L.	Apiaceae	–	+	–	–	–	–
Centella asiatica (L.) Urban	Apiaceae	–	–	–	–	–	+
Chenopodium album L.	Chenopodiaceae	–	+	–	–	–	+
Chrysopogon gryllus (L.) Trinius	Poaceae	–	–	–	+	–	–
Circa alpina L.	Onagraceae	–	–	–	–	+	+
Coelogyne cristata Lindley	Orchidaceae	–	–	–	–	–	+
Commelina paludosa Blume,	Commelinaceae	–	–	–	–	–	+
Convolvulus arvensis L.	Convolvulaceae	+	–	–	+	–	–
Conyza stricta Willd.	Asteraceae	–	–	+	–	–	+
Cucurbita maxima Duchesne	Cucurbitaceae	–	–	–	–	–	–

TABLE 10.1 (Continued)

Name of Tree Species	Family	UGSF	SMSF	MRSF	JMSF	CBSF	ADSF
Curcuma aromatica Salisbury	Zingiberaceae	–	+	+	–	–	–
Cuscuta europaea L.	Cuscutaceae	–	+	–	–	–	–
Cuscuta santapaui	Cuscutaceae	+	–	–	–	–	+
Cymbopogon martini (Roxb.) W. Watson	Poaceae	+	+	+	+	–	–
Cynodon dactylon (L.) Pearsoon	Poaceae	+	+	+	+	+	–
Cynoglossum zeylanicum (Valh ex Hornem.) Thunb. Ex Lehmann	Boraginaceae	–	–	–	–	–	+
Danthonia schneiedri Pilger.	Poaceae	–	+	–	–	–	+
Datura fastuosa L.	Solanaceae	–	–	+	–	–	–
Datura stramonium L.	Solanaceae	–	+	–	–	–	+
Desmodium triflorum (L.) DC	Fabaceae	+	–	–	+	–	–
Dichrocephala integrifolia (L.f.) Kuntze	Asteraceae	–	–	–	–	–	+
Dryopteris cochleata (Buch-Ham. Ex D. Don)C. chr	Dryopteridaceae	+	–	–	+	–	+
Dryopteris cochleata (Don.) C.	Dryopteriaceae	+	–	–	–	–	–
Dryopteris justaposita Christ	Dryopteriaceae	+	–	–	+	–	+
Dryopteris molliscula (Wall.) Presl.	Dryopteridaceae	–	+	–	–	–	–
Dryopteris wallichiana (Sprengin.L.) Hayland	Dryopteridaceae	–	+	–	–	–	–
Drypteris redactopinnata Basu Et Panigr	Fabaceae	–	–	+	–	–	–
Drypteris wallichiana (Sprengin.L) Hayland	Dryopteridaceae	–	–	+	–	–	–
Drypteris xyloides (Ktze) C. Chr	Dryopteridaceae	–	–	+	–	–	–
Duchesnea indica (Andrews) Focke	Rosaceae	–	–	–	–	+	–
Dyropteris nigropaleacea (Jankins) Jankins	Dryopteridaceae	–	–	–	–	–	+
Epilobium royleanum Haussknecht	Onagraceae	–	–	–	–	–	+
Eragrostis nigra Nees ex Stud.) Meld.	Poaceae	–	–	–	–	–	+
Euphorbia pilosa L.	Euphorbiaceae	+	–	–	+	–	–

TABLE 10.1   (Continued)

Name of Tree Species	Family	UGSF	SMSF	MRSF	JMSF	CBSF	ADSF
Fragaria nubicola Lindley ex Lacaita	Rosaceae	+	+	–	+	+	–
Galium acutum Edgew	Rubiaceae	–	–	+	–	+	–
Galium aparine L.	Rubiaceae	–	+	–	–	–	+
Galium elegans Wallich,	Rubiaceae	+	+	–	+	–	+
Gebera gossypina (Royle) G. Beauv.	Asteraceae	–	–	–	–	–	+
Gentiana capitata Buch-Ham ex D. Don	Gentianaceae	+	–	–	+	–	+
Gentiana pedicellata (D. Don) Wallich	Gentianaceae	+	–	–	+	–	–
Geranium wallichianum D. Don ex Sweet	Geraniaceae	–	+	–	–	–	+
Gerbera gossypina (Royle) G. Beaue.,	Asteraceae	+	–	–	+	–	–
Gerbera maxima (D. Don) G. Beauv.,	Asteraceae	–	–	–	–	+	–
Heteropogon contortus (L.) P. Beauv. Ex Roemer and Schultes,	Poaceae	+	–	–	+	+	–
Hypericum elodeoides Choisy,	Hypericaceae	–	–	–	–	–	+
Hypericum uralum Buch-Ham. Ex D. Don	Hypericaceae	–	–	–	+	–	+
Imperata cylindrica (L.) P. Beauv	Poaceae	+	–	–	+	–	–
Isachne albens Trinius	Poaceae	+	–	–	+	–	+
Leucas lanata Benth	Lamiaceae	–	–	–	–	+	–
Linderbergia indica (L.) Vatke	Scrophulariaceae	+	–	–	+	–	–
Mentha arvensis L.	Rubiaceae	–	+	+	–	–	–
Micromaria biflora (Buch-Ham. Ex D. Don) Benth	Lamiaceae	+	–	–	–	+	–
Neanotis calycina (Wallich ex Hook. F.) W.H. Lewis	Rubiaceae	–	+	–	–	–	+
Origanum vulgare L.	Lamiaceae	–	+	–	–	+	+
Oxalis corniculata L.	Oxalidaceae	+	+	–	–	+	+
Parthenium hysterophorus	Asteraceae	–	+	–	–	–	–
Paspalum scrobiculatum L.	Poaceae	–	–	–	–	–	+

TABLE 10.1 *(Continued)*

Name of Tree Species	Family	UGSF	SMSF	MRSF	JMSF	CBSF	ADSF
Perilla frutescens (L.) Britton	Lamiaceae	+	+	-	+	-	+
Phytolacca acinosa Roxb.	Phytolaccaceae	-	-	-	-	+	-
Pimpinella diversifolia DC	Apiaceae	-	+	-	-	-	+
Polygonatum multiflorum (L.) Allion	Liliaceae	-	+	-	-	-	-
Polystichum squarrosum (D. Don) Fée	Dryopteridaceae	-	+	+	-	-	-
Polystichum stimulana	Dryopteriaceae	+	-	+	-	-	+
Potentilla gerardiana Lindley ex Lehmann	Rosaceae	-	+	+	-	+	+
Potentilla fulgens Wallich ex Hook	Rosaceae	+	+	+	+	-	+
Primula denticulata Smith	Primulaceae	-	-	+	-	+	-
Reinwardtia indica Dumortier	Linaceae	+	+	+	+	+	-
Rosularia adenotricha (Wallich ex Edgew) Jansson,	Crassulaceae	-	+	-	-	+	-
Rosularia rosulata (Edgew.) Ohba	Crassulaceae	-	+	-	-	+	-
Rumex hastatus D. Don	Polygonaceae	-	+	+	-	+	+
Salvia lanata Roxb.	Lamiaceae	+	+	+	+	+	-
Salvia nubicola Wallich ex Sweet	Lamiaceae	+	-	-	+	-	-
Scutellaria grossa Wallich ex Benth	Lamiaceae	+	-	-	+	-	-
Sedum multicaule Wallich ex Lindley.	Crassulaceae	-	+	+	-	-	+
Senecio nudicaulis Buch-Ham ex D. Don	Asteraceae	+	+	+	+	+	-
Senecio rufinervis DC.	Asteraceae	-	-	+	-	-	+
Siegesbeckia orientalis L.	Asteraceae	+	-	-	-	-	+
Smilax aspera L.	Smilacaceae	-	-	+	-	-	-
Smilax glaucophylla Klotzsch	Smilacaceae	-	-	+	-	-	-
Sonchus oleraceus L.	Asteraceae	+	-	-	-	-	-
Swertia chirayita (Roxb. Ex Fleming) Karsten	Gentianaceae	-	+	+	+	-	+
Taraxacum officinale Weber,	Asteraceae	+	-	-	-	-	-

Diversity and Dynamics in Forest Ecosystems

TABLE 10.1 *(Continued)*

Name of Tree Species	Family	UGSF	SMSF	MRSF	JMSF	CBSF	ADSF
Thalictrum foliolosum DC.	Ranunculaceae	+	–	–	–	–	–
Themeda anathera (Nees ex Steud.) Hack	Poaceae	–	–	–	–	–	+
Themeda arundinacea (Roxb.) Ridley	Poaceae	–	–	+	–	–	+
Triplostegia glandulifera Wall ex. DC	Dipsacaceae	–	+	–	–	+	–
Valeriana jatamansii Jones,	Valerianaceae	–	–	–	+	–	–
Veronica biloba L.	Scrophulariaceae	+	–	–	–	–	+
Vicatia coniifolia DC	Apiaceae	–	–	–	–	+	–
Vicia tenera Graham ex Benth	Fabaceae	–	+	–	+	–	–
Viola betonicifolia J. Snith	Violaceae	+	–	–	–	+	–
Viola biflora L.	Violaceae	–	+	+	+	+	–
Viola canescens Wallich	Violaceae	+	+	–	–	–	–

Source: Pala (2012).

with IVI (24.10) was dominant among shrubs, followed by *Pyracantha crenulata* (14.18). *Anaphalis adnata* with IVI (21.22) was dominant among herbs followed by *Andropogon munroi* (11.46). In Ansuiya Devi sacred forest, among tree species *Daphniphyllum himalense* with IVI (37.94) was dominant, followed by *Quercus leucotrichophora* (37.03). Among shrubs *Sarcococca saligna* with IVI (26.89) was dominant, followed by *Daphne papyraceae* (21.27). *Andropogon munroi* among herbs was dominant with IVI (27.14) followed by *Boenninghausenia albiflora* (12.67) and *Bergenia ciliata* (11.31).

Total basal area for tree species ranged from 32.63 m^2/ha in Chanderbadni to 93.27 m^2/ha in Sem Mukhem. The vegetation density for tree species was found lowest in Chanderbadni (6.88 trees/100 m^2), whereas the highest density (13.52 trees/100 m^2) was found in Maroor. The highest value of shrub density (18.78 m/25 m^2) was found in Maroor, whereas the highest value of density for herbaceous flora (26.7/m^2) was observed in Ulkagari sacred forest. In the case of tree species, the diversity index (H') showed almost similar range. The higher diversity index (2.93) for trees was found in the Ansuiya Devi, which also showed the highest value for species richness (SR) (4.35) with 27 tree species. The lowest value for diversity index (H') (2.35) and SR (3.29) with 21 tree species was recorded for Ulkagari. For shrubs, highest (H') diversity index (3.42) was found in Ansuiya Devi and lowest (2.49) in Ulkagari. In the herb strata, diversity index (H') was found highest (3.51) for Ansuiya Devi and lowest (2.49) for Chanderbadni. For all the three layers evenness showed a similar trend in all sites. In case of tree species, Margalef index was found positively correlated with altitude (r = 0.75, p = 0.05). Shannon diversity (H') was highly positively correlated with Marglef index (r = 0.81, p = 0.05). In case of shrubs Margalef index (SR) showed highly positive correlation (r = 0.98, p = 0.01) with density/25 m^2. Simpson's diversity (CD) was found highly negatively correlated (r = –0.79, p = 0.05) with Shannon diversity (H). Whereas in case of herbs, Simpsons diversity (CD) was found negatively correlated with both density/m^2 (r = –0.75, p = 0.05) and Shannon diversity (H') (r = –0.87, p = 0.05) respectively. Shannon diversity index (H') was found highly positively correlated (r = 0.85, p = 0.01) with Marglef index (SR).

Khumbongmayum et al. (2006) reported 173 species representing 145 genera under 70 families from four sacred groves of Manipur, including 81 tree, 24 shrubs, and 62 herb species. Sukumaran et al. (2008) reported 329 plant species belonging to 251 genera and 100 families from 40 miniature sacred groves in the Kanyakumari district of Tamil Nadu. This included 139

234 Diversity and Dynamics in Forest Ecosystems

trees, 95 shrubs, and 79 herb species. Ghildiyal et al. (2008) reported 372 species belonging to 78 families from the Tarkeshwar sacred grove in the Pauri district of Garhwal Himalaya. Basu (2009) reported 114 species from six sacred groves in Bankura district (West Bengal), including 62 trees, 14 shrubs, and 23 herb species. The tree density values of the present study are supported by results of Sinha and Maikhuri (1998) who reported density values of 1399 trees/ha and 1144 trees/ha in the core zone and interactive zone of Hariyali sacred forest in Garhwal Himalaya. Pande et al. (2001) study as per aspect and altitude on quantitative vegetation analysis in Garhwal Himalayan forests and observed tree density in between 885 to 1111 trees/ha, total basal cover (TBC) ranged of 32.63 m²/ha in Chanderbadni to 93.27 m²/ha in Sem Mukhem. The variation in the TBC in different study sites may be due to variation in a number of tree species in different sites. Vidyasagaran et al. (2005) reported the average TBC value of 25.79 m²/ha in sacred groves of the Thrissur district of Kerala. Sinha and Maikhuri (1998) also reported TBC values of 47.59 to 26.87 m²/ha in the core and interactive zone of Hariyali sacred forest of Garhwal Himalaya is comparable to the present study.

10.3.1 REGENERATION STATUS/PATTERN

In Chanderbadni forest, a total of 13 species were found in the sapling stage and 10 species in the seedling stage out of 21 trees species. *Quercus leucotrichophora* was the dominant species at both seedling and sapling stage with density (5.42 seedlings/100 m²) and (4.14 saplings/100 m²) respectively. The overall density for seedlings was (11.76 seedlings/100 m²). Highest value of IVI (82.7 and 72.26) was also observed for *Quercus leucotrichophora* at both seedling and sapling stages, respectively. As far as regeneration status of Chanderbadni forest is concerned, 14.29%, 38.57% and 9.25% of species showed fair, good, and poor regeneration whereas 38.1% species showed no regeneration. In Jameshwar, a total of 17 and 22 species were observed surviving at seedling and sapling stages, respectively, out of 23 trees. *Quercus floribunda* was reported dominant at the seedling stage with a density value of 3.02 seedlings/100 m² and IVI (75.47). *Daphniphyllum himalense* was dominant at the sapling stage with density (2.14 saplings/100 m²) and IVI (67.70). Maximum percentage (42.31%) of species showed fair regeneration, 26.92% showed good, 15.38% poor and 15.38% of species did not show regeneration. In Sem Mukhem, a total of 11 and 18 species were found surviving in seedling and sapling stage, respectively. *Quercus floribunda* was the dominant species at seedling and sapling stages with density (6.78 seedlings/100 m²), and (4.84 saplings/100 m²), respectively. Highest value of IVI (123.34) was again observed for *Quercus floribunda* at both sapling and seedling stages. Maximum (30.43%) species showed poor regeneration, 21.74% showed fair, 26.09% showed good, whereas 21.74% of species were not regenerating. Highest value of IVI (116.94) and (116.76) was found for *Quercus leucotrichophora* for both seedling and sapling stages, respectively. Maximum (34.78%) of tree species were not regenerating, whereas 26.09% species showed fair regeneration, 30.43% species regenerating good and only 8.70% of species showed poor regeneration (Table 10.2). Distribution and density of individuals was maximum in lower girth classes which decreased with the increase in size of girth class in all study sites. The reverse J shaped curve was observed in all the CBH classes. The maximum portion of the curve was constituted by 0–15 CBH class across different study sites. Factors of locality may be the reasons for new regeneration of some species and non-regenerating potential of other species. Assessing the potential colonization rates of individual species is a must for future predicting for determining how forest ecosystems respond to rapid changes in climate that have been forecasted (Solomon and Kirilenko, 1997).

TABLE 10.2 Regeneration Status of Tree Species Available in Different Sites

Name of Species	Site and Status				
	Chanderbadni	Jameshwar	Ulkagari	Maroor	Ansuiya Devi
Abies spectibilis	–	–	–	–	No
Acer caesium	–	Fair		No	No
Acer oblongum	–	–	–	–	No
Aesculus indica	–	Fair	No		No
Albizia julibrissin	–	No	Poor	–	–
Alnus nepalensis	–	No		Fair	Poor
Buxux wallichiana	–	Fair			Fair
Benthamidia capitate	Good	–	–	Good	–
Betula alnoides		No		No	–
Carpinus faginea	–	–	–	–	No
Carpinus viminea		No			Fair
Cedrus deodara	Fair	Poor	Fair	No	New
Celtis australis		Fair		–	–
Cinnamomum tamala	No	Poor	Poor	–	New

TABLE 10.2　*(Continued)*

Name of Species	Site and Status				
	Chanderbadni	**Jameshwar**	**Ulkagari**	**Maroor**	**Ansuiya Devi**
Cotoneaster confuses		No	No	Fair	–
Cupressus torulosa	No	No	Poor	No	No
Daphniphyllum himalayense		No	–	–	Good
Engelhardtia spicata	No	–	–	–	
Eurya acuminate	–	–	–	–	No
Ficus neriifolia	–	Poor		No	–
Ficus auriculata	–	–	–	–	Poor
Fraxinus micrantha	–	Fair	No	No	Poor
Ilex dipyrena	Poor	–	–	–	Fair
Juglans regia	Poor	Fair	Poor	No	No
Lindera pulcherrima	No	–		Poor	
Lyonia ovalifolia	Good	Fair	Good	Fair	Fair
Myrica esculenta	Good	–	Good	Fair	Fair
Neolitsea cuipala		Fair	–	–	–
Persea duthiei	No	Fair	Poor	Fair	Fair
Pinus roxburghii	Good	–	Good	Good	Poor
Populus ciliate	–	–	No	–	–
Prunus cerosoides	Fair	Fair	Poor	Poor	–
Pyrus communis			No	No	–
Pyrus pashia	Fair	Poor	Good	Good	Fair
Quercus floribunda	–	No	–	Good	Fair
Quercus glauca	–	–	–	–	No
Quercus leucotrichophora	Good	Good	Good	Good	Good
Quercus semecarpifolia		Good	–	–	–
Rhododendron arboreum	Good	Fair	Fair	Fair	Good
Rhus punjabenensis	No	–	–	–	–
Sorbus aucuparia	Fair	–	–	–	–
Swida macrophylla	Good	Fair	Poor	Good	Good
Symplocos paniculata	No	–	–	–	New
Symplocos ramosissima	–	Good	Fair	Good	–
Taxus baccata	–	–	–	–	Poor
Toona ciliata Roemer	No	–	–	–	–
Toona serrata	–	–	–	–	No

Source: Pala (2012); Pala et al. (2013).

10.3.2 BIODIVERSITY VALUE

Out of total plant species reported from the study sites, 205 species were having different tangible or intangible benefits relevant to society. Out of 205 plant species, having different biodiversity value, 158 are of medicinal importance. Edible parts of 38, vegetables from 14 dyes from 12, oil, and food items of 19 species are used by local communities. Twenty species are used for timber/construction/building material, 12 for agricultural implements, 8 for minor construction, 17 for resin/varnishes/paints/turpentine, and 17 for fiber/ropes/walking sticks. Sixteen species were found helpful in stopping erosion/soil binder/reclamation for wastelands, etc., whereas 17 species are useful in bee-forage and 4 in apiculture (Table 10.3) (Pala et al., 2012, 2013).

10.3.3 CONSERVATION STRATEGIES AND BELIEF SYSTEM

The belief system and conservation practices were mainly attributed by the local inhabitants to the presence of some famous temples and goddesses present in these forests. In Chanderbadni, the goddess associated with the temple is Durga Mata Sita. It is said that Shri Jagdambshwar saint worshiped Shiva at Jameshwar temple. His wife Reenuka was also with him and the stream flowing from the north of the temple is known as Reenuka. The Ulkagari forest is having a centrally located temple situated at the top, which is dedicated to Ulkeshwari Devi and another temple dedicated to Gandiyal devta is also present there. The myth behind the conservation is that the deity Ulkeshwari Devi was a speaking deity. The Ansuiya Devi forest is considered sacred because of the presence of some temples of great saints like Ansuiya Devi temple, Attramuni ashram, Amdar Devta temple, and Dodga Devta. Maha Rishi Attri is one of the seven saints on the list of great Maharishis after Kashyap, Ustreey, Jagdambeshwar, Bharadwaj, Vishwa Mitra, and Gautam. Nagraj Mela on 25 and 26 November after every three years is celebrated in memory of Lord Krishna, who is believed to have occupied Sem temple in Sem Mukhem. Temples of Gandiyal Devta, Bahrav Devta, and Bhagwati Devi exist within Maroor forest, which is actually Van Panchyat and because of the presence of these temples has given it sacred nature which is also the reason for its protection. Celebration of festivals and mellas on different occasions is representing true socio-culture heritage of region as well as of whole Uttarakhand. Communal forests have

TABLE 10.3 The Biodiversity Values/Ecosystem Services of Different Species

Name of Species	Family	Biodiversity Values/Ecosystem Service
	Trees	
Abies spectabilis (D. Don) Mirl.	Pinaceae	Medicinal, fuelwood, timber, thatching
Acer caesium Wallich ex Brandis	Aceraceae	Medicinal
Acer oblongum Wallich. ex DC.	Aceraceae	Agriculture implements, biofertilizer, apiculture, and bee-forage
Aesculus indica (Wall. ex Camb.) Hook.f	Hippocastanaceae	Medicinal, fodder, food items
Albizia julibrissin Durazzini	Mimosaceae	Ornamental reclamation of forests
Alnus nepalensis D. Don	Betulaceae	Fodder, soil conservation, small timber, agriculture implements, tannin, dye, reclamation
Buxus wallichiana Baill.	Buxaceae	Fuelwood, fodder, crafts wood, medicinal
Betula alnoides Buch.-Ham. ex D. Don	Betulaceae	Edible, medicinal, minor construction
Carpinus viminea Lidl.	Betulaceae	Wooden objects
Cedrus deodara (Roxb. ex D. Don) G. Don.	Pinaceae	Construction, medicinal
Celtis australis L.	Ulmaceae	Edible, fodder, dye, medicinal, agroforestry
Cinnamomum tamala Nees	Lauraceae	Medicinal, aromatic, tea, spice
Cupressus torulosa D. Don	Cupressaceae	Fuelwood, timber, medicinal, windbreak, nitrogen-fixing, ornamental
Daphniphyllum himalayense Wall. ex Steud.	Daphniphyllaceae	Agriculture implements, fuel
Engelhardtia spicata Leschen. ex Blume	Juglandaceae	Fuel, medicinal
Eurya acuminata DC.	Theaceae	Fuel, fodder, agriculture implements
Ficus auriculata Lour.	Moraceae	Edible, religious, vegetables
Ficus neriifolia Smith	Moraceae	Edible, fodder
Fraxinus micrantha Lingelsh.	Oleaceae	Fuel, agriculture implements, medicinal

TABLE 10.3 (*Continued*)

Name of Species	Family	Biodiversity Values/Ecosystem Service
Ilex dipyrena Wallich	Aquifoliaceae	Agriculture implements, fuel
Juglans regia L.	Juglandaceae	Furniture, construction, edible, oil, tannin, dye, medicinal, insecticide, social forestry
Lyonia ovalifolia (Wallich) Drude	Ericaceae	Fuel, medicinal
Myrica esculanta Buch.-Ham. ex D. Don,	Myricaceae	Edible, medicinal, dye, fuel, agriculture implements
Neolitsea cuipala (Buch.-Ham ex D. Don) Kosterm	Lauraceae	Fuel, fodder, medicinal, agriculture implements
Persea duthiei (King ex kook.f.)	Lauraceae	Construction, fuelwood, fodder
Pinus roxburghii Sarg.	Pinaceae	Construction, resin, varnishes, paints, turpentine, medicinal
Pinus wallichiana A.B. Jackson	Pinaceae	Construction, resin, varnishes, paints, medicinal
Populus ciliata Wallich ex Royle	Salicaceae	Paper, match, sports goods, medicinal, fodder
Prunus cerasoides D. Don	Rosaceae	Fodder, edible, medicinal, small timber, used in rituals
Pyrus communis L.	Rosaceae	Edible, apiculture, bee-forage
Pyrus pashia Buch.-Ham ex D. Don	Rosaceae	Fodder, edible, medicinal, erosion control, small timber
Quercus floribunda Rehder	Fagaceae	Fuelwood, fodder, household items, edible
Quercus glauca Thunb.	Fagaceae	Fuelwood, fodder, agriculture implements
Quercus leucotrichophora A. Camus	Fagaceae	Construction, agriculture implements, fodder, edible, medicinal, social forestry.
Quercus semecarpifolia J.E. Smith	Fagaceae	Construction, fuelwood, fodder, social forestry
Rhamnus virgatus Roxb.	Rhamnaceae	Medicinal, fuelwood
Cotoneaster confusus klotz.	Rosaceae	Fuel wood, agriculture implements
Rhododendron arboreum Smith	Ericaceae	Fuelwood, charcoal, medicinal, bee-forage, edible
Rhus punjabenensis J.L. Stewart	Anacardiaceae	Fuelwood, medicinal
Sorbus aucuparia (Spach) Hedlund	Rosaceae	Edible, medicinal

TABLE 10.3 (Continued)

Name of Species	Family	Biodiversity Values/Ecosystem Service
Swida macrophylla (Wallich) Sojak	Cornaceae	Edible, agriculture implements, bee-forage, fodder, social forestry
Symplocos paniculata (Thunb.). Miq.	Symplocaceae	Dye, fodder, necter by bees, medicinal
Symplocos ramosissima Wallich ex G. Don	Symplocaceae	Fodder, fuelwood, small timber
Taxus baccata L.	Taxaceae	Edible, medicinal
Toona ciliata Roem.	Meliaceae	Timber, dye, tannin, building material, shelter, fire breaks, reclamation, ornamental
Toona serrata (Royle) Roem	Meliaceae	Fodder, timber, plywood, shade during summer, ornamental
Shrubs		
Arachne cordifolia (Decne.) Hurusawa	Euphorbiaceae	Medicinal
Artemisia japonica Thunb.	Asteraceae	Medicinal, sacred
Artemisia roxburghiana Wallich ex Besser	Asteraceae	Medicinal
Asparagus adscendens Buch.-Ham. ex Roxb.	Liliaceae	Vegetables, medicinal
Asparagus curileus Buch.-Ham. ex Roxb.	Liliaceae	Vegetables, medicinal
Asparagus racemosus Willd.	Liliaceae	Medicinal
Barleria cristata L.	Acanthaceae	Medicinal, bee-forage, soil binder
Berberis aristata DC.	Berberidaceae	Medicinal, edible, dye
Berberis asiatica Roxb. ex DC.	Berberidaceae	Medicinal, edible, dye
Berberis lycium Royle	Berberidaceae	Edible, medicinal
Berchemia edgeworthii Lawson	Rhamnaceae	Medicinal
Boehmeria platyphylla D. Don	Urticaceae	Fodder, fiber
Buddleja paniculata Wallich	Buddlejacea	Medicinal, fuel, soil binder
Colebrookea oppositifolia J.E. Smith	Lamiaceae	Medicinal, bee-forage

TABLE 10.3 *(Continued)*

Name of Species	Family	Biodiversity Values/Ecosystem Service
Cotoneaster bacillaris Wallich	Rosaceae	Medicinal, sticks
Cyathula tomentosa (Roth) Moq.	Amaranthaceae	Medicinal, fodder
Daphne papyracea Wallich ex Steudee	Thymelaeaceae	Ropes, medicinal, religious
Debregeasia salicifolia (D. Don) Rendle	Urticaceae	Fodder, ropes, cordages, edible, medicinal
Desmodium elegans DC.	Fabaceae	Medicinal
Deutzia compacta Craib	Hydrangeaceae	Medicinal, bee-forage
Elsholtzia flava (Benth.) Benth	Lamiaceae	Medicinal
Eupatorium adenophorum Spreng.	Asteraceae	Medicinal
Flacourtia indica (Burm. f.) Merr.	Flacourtiaceae	Edible, fodder, medicinal, bee-forage
Girardinia diversifolia (Link) Friis	Urticaceae	Ropes, cordages, medicinal
Hypericum uralum Buch.-Ham. ex D. Don	Hypericaceae	Medicinal
Indigofera cassioides Rottler ex DC	Fabaceae	Vegetables, medicinal, fodder
Indigofera heterantha Wallich ex Brandis	Fabaceae	Vegetables, fodder, medicinal, baskets
Inula cappa (Buch.-Ham. ex D. Don) DC.	Asteraceae	Medicinal
Inula cuspidata (DC.) C.B. Clarke	Asteraceae	Medicinal
Lantana camara L.	Verbenaceae	Fuel, medicinal, soil binder
Leptodermis lanceolata Wallich	Rubiaceae	Medicinal, fodder
Linderbergia grandiflora (Buch-Ham. ex D. Don) Benth.	Scrophulariaceae	Medicinal
Lonicera quinquelocularis Hardw.	Caprifoliaceae	Edible, sticks, bee-forage
Myrsine africana L.	Myrsinaceae	Medicinal
Phyllanthus parvifolius Buch.-Ham.ex D. Don	Euphorbiaceae	Fodder
Phyllanthus virgatus G. Forst.	Euphorbiaceae	Medicinal

TABLE 10.3 *(Continued)*

Name of Species	Family	Biodiversity Values/Ecosystem Service
Pogostemon benghalense (Burm.f.) Kuntze	Lamiaceae	Medicinal, soil binder
Prinsepia utilis Royle	Rosaceae	Oil, edible, medicinal
Pyracantha crenulata (D. Don) M. Roem.	Rosaceae	Sticks, tool handles, edible, soil binder
Rhamnus persica Boiss.	Rhamnaceae	Edible, fodder
Rhamnus purpureus Edgew	Rhamnaceae	Agricultural implements, fodder, manure, medicinal
Rhamnus virgatus Roxb.	Rhamnaceae	Medicinal, fuel
Rhus parviflora Roxb.	Anacardiaceae	Edible, food substances, medicinal
Rosa brunonii lindl.	Rosaceae	Medicinal, bee-forage, soil binder
Rosa macrophylla Lindl.	Rosaceae	Edible, fuel, medicinal
Rubia manjith Roxb. ex Fleming	Rubiaceae	Medicinal
Rubus ellipticus Smith	Rosaceae	Edible, medicinal, soil binder
Rubus foliolosus D. Don	Rosaceae	Edible
Rubus niveus Wallich ex G. Don	Rosaceae	Edible, medicinal
Rubus paniculatus Smith	Rosaceae	Edible, medicinal
Sarcococca saligna (D. Don) Muell.-Arg	Buxaceae	Manure, walking sticks, medicinal, soil binder
Saxifraga diversifolia Wallich ex Ser.	Saxifragaceae	Medicinal
Sinarundinaria falcata (Nees) Chao and Renvoize	Poaceae	Cottage, wicker works, fodder
Skimmia anquetilia Taylor and Airy Shaw	Rutaceae	Medicinal, agricultural implements
Smilax aspera L.	Smilacaceae	Vegetables, medicinal, bee-forage
Smilax glaucophylla Klotz.	Smilacaceae	Medicinal, fodder, vegetables
Solanum nigrum L.	Solanaceae	Edible, medicinal, beverages
Spermadictyon suaveoleus Roxb.	Rubiaceae	Medicinal, fodder, bee-forage
Spiraea canescens D. Don	Rosaceae	Baskets, soil binder, fodder, apiculture

TABLE 10.3 *(Continued)*

Name of Species	Family	Biodiversity Values/Ecosystem Service
Thamnocalamus falconeri Hook. f. ex Munro	Poaceae	Mats, fodder
Thamnocalamus spathiflora (Trin.) Murno	Poaceae	Medicinal, baskets, mats
Urtica ardens Link	Urticaceae	Fiber, ropes, sacs, medicinal
Urtica dioica L.	Urticaceae	Fiber, ropes, sacs, medicinal
Viburnum nervosum D. Don	Caprifoliaceae	Medicinal, edible
Viburnum mullaha Buch.-Ham. ex D. Don	Caprifoliaceae	Fuel, edible
Woodfordia fructicosa (L.) Kurz	Lythraceae	Dye, medicinal, refreshing drinks, soil binder
Zanthoxylum armatum DC.	Rutaceae	Medicinal, wash, and tooth
Herbs		
Achyranthes bidentata Blume	Amaranthaceae	
Aechmanthera gossypina (Nees) Nees	Acanthaceae	Ropes, netting, bee-forage,
Agrostis pilosula Trin.	Poaceae	Fodder
Ainsliaea aptera DC.	Asteraceae	Medicinal
Ainsliaea latifolia (D. Don) Sch.-Bip.	Asteraceae	Medicinal
Ajuga brachystemon Maxim.	Lamiaceae	Febrifuge
Anaphalis adnata Wallich ex DC.	Asteraceae	Medicinal, fire by friction (agela)
Anaphalis busua (Buch.-Ham. ex D. Don) DC.,	Asteraceae	Medicinal
Anaphalis contorta (D. Don) Hook.f	Asteraceae	Insect repellent, medicinal
Anaphalis triplinervis (Sims,) C.B. Clarke	Asteraceae	Medicinal
Andropogon munroi C. B. Clarke	Poaceae	Fodder
Angelica glauca Edgew.	Apiaceae	Essential oil, spice, medicinal
Apluda aristata L.	Poaceae	Fodder
Apluda mutica L.	Poaceae	Fodder, thatching, brooms

TABLE 10.3 *(Continued)*

Name of Species	Family	Biodiversity Values/Ecosystem Service
Arundinella nepalensis Trin.	Poaceae	Thatching, fodder
Bergenia ciliata (Haworth) Sternb.	Saxifragaceae	Medicinal, food items
Bidens bipinnata L.	Asteraceae	Vegetables, medicinal
Bidens pilosa L.	Asteraceae	Fodder, vegetable, medicinal
Boenninghausenia albiflora (Hook.) Meissn.	Rutaceae	Medicinal
Boerhavia diffusa L.	Nyctaginaceae	Vegetable, medicinal
Buddleja paniculata Wallich	Buddlejaceae	Fuelwood, bee-forage, soil binder
Bupleurum falcatum L.	Apiaceae	Medicinal, spice
Bupleurum hamiltonii N. P. Balakr.	Apiaceae	Fodder, medicinal
Cannabis sativa L.	Cannabaceae	Medicinal, ropes, fuelwood, edible
Carum carvi L.	Apiaceae	Aromatic, vegetable, medicinal
Centella asiatica (L.) Urban	Apiaceae	Medicinal
Chenopodium album L.	Chenopodiaceae	Vegetables
Coelogyne cristata Lindl.	Orchidaceae	Medicinal
Commelina paludosa Blume	Commelinaceae	Vegetable
Convolvulus arvensis L.	Convolvulaceae	Medicinal
Cucurbita maxima Duch. Ex lam	Cucurbitaceae	Medicinal, bee-forage
Curcuma aromatica Salisb.	Zingiberaceae	Dye, mendhi (Hena)
Cuscuta europaea L.	Cuscutaceae	Medicinal
Cymbopogon martinii (Roxb.) Wat.	Poaceae	Medicinal,
Cynodon dactylon (L.) Pers.	Poaceae	Medicinal, religious, medicinal
Cynoglossum zeylanicum Thunb. ex Lehm.	Boraginaceae	Medicinal
Datura fastuosa L.	Solanaceae	Medicinal

TABLE 10.3 *(Continued)*

Name of Species	Family	Biodiversity Values/Ecosystem Service
Datura stramonium L.	Solanaceae	Medicinal
Desmodium trifloum (L.) DC.	Fabaceae	Fodder
Dryopteris cochleata (Buch.-Ham. ex D. Don) C. Chr	Dryopteridaceae	Medicinal
Duchesnea indica (Andrews) Focke	Rosaceae	Edible, medicinal
Elsholtzia fruticosa (D. Don) Rehder	Lamiaceae	Spice, medicinal
Epilobium royleanum Haussk.	Onagraceae	Medicinal
Euphorbia pilosa L.	Euphorbiaceae	Medicinal
Fragaria nubicola Lindl. ex Lacaita	Rosaceae	Edible, medicinal
Galium acutum Edgew.	Rubiaceae	Medicinal
Galium aparine L.	Rubiaceae	Medicinal
Galium elegans Wallich	Rubiaceae	Medicinal
Geranium wallichianum D. Don ex Sweet	Geraniaceae	Medicinal, dye
Gerbera gossypina (Royle) P. Beauv.	Asteraceae	Medicinal, dye, fiber
Herteropogon contortus (L.) P. Beauv. ex Roem. and Schult.	Poaceae	Fodder
Hypericum uralum Buch.-Ham. ex D. Don	Hypericaceae	Medicinal
Imperata cylindrica (L.) P. Beauv.	Poaceae	Medicinal
Leucas lanata Benth.	Lamiaceae	Medicinal, vegetable
Linderbergia indica (L.) Vatke	Scrophulariaceae	Medicinal
Mentha arvensis L.	Lamiaceae	Flavoring, sauce, medicinal
Micromaria biflora (Buch.-Ham. ex D. Don) Benth.	Lamiaceae	Medicinal
Origanum vulgare L.	Lamiaceae	Medicinal, flavor, vegetable
Oxalis corniculata L.	Oxalidaceae	Vegetable, medicinal

TABLE 10.3 *(Continued)*

Name of Species	Family	Biodiversity Values/Ecosystem Service
Parthenium hysterophorus L.	Asteraceae	Medicinal
Paspalum scrobiculatum L.	Poaceae	Medicinal
Perilla frutescens (L.) Britton	Lamiaceae	Flavoring agent, condiments, medicinal
Phytolacca acinosa Roxb.	Phytolaccaceae	Vegetable, medicinal, beverages
Pimpinella diversifolia DC.	Apiaceae	Medicinal
Polygonatum multiflorum (L.) All.	Liliaceae	Medicinal
Potentilla gerardiana Lind. ex Lehm.	Rosaceae	Medicinal
Potentilla fulgens Wallich ex Hook.	Rosaceae	Edible, medicinal
Primula denticulata Smith	Primulaceae	Medicinal
Reinwardtia indica Dumort.	Linaceae	Medicinal, apiculture, bee forage
Rumex hastatus D. Don	Polygonaceae	Sauce, medicinal
Salvia lanata Roxb.	Lamiaceae	Medicinal, bee-forage
Salvia nubicola Wallich ex Sw.	Lamiaceae	Medicinal
Sedum multicaule Wallich ex Lindl.	Crassulaceae	Medicinal
Senecio nudicaulis Buch.-Ham. ex D. Don	Asteraceae	Medicinal
Sisgesbeckia orientalis L.	Asteraceae	Medicinal
Smilax aspera L.	Smilacaceae	Vegetables, medicinal, bee-forage
Smilax glaucophylla Klotz.	Smilacaceae	Fodder, medicinal, vegetable
Sonchus oleraceus L.	Asteraceae	Medicinal, vegetable
Swertia chirayita (Roxb. ex Fleming) Karsten	Gentianaceae	Medicinal
Taraxacum officinale Weber	Asteraceae	Medicinal, vegetable
Thalictrum foliolosum DC.	Ranunculaceae	Medicinal
Themeda anathera (Nees ex Steud.) Hack.	Poaceae	Fodder

TABLE 10.3 *(Continued)*

Name of Species	Family	Biodiversity Values/Ecosystem Service
Themeda arundinacea (Roxb.) Ridley	Poaceae	Medicinal, fodder, thatching
Valeriana jatamansii Rom	Valerianaceae	Medicinal, insecticide, aroma
Vicia tenera Grah. ex Benth.	Fabaceae	Fodder
Viola betonicifolia J. Smith	Violaceae	Medicinal
Viola biflora L.	Violaceae	Medicinal
Viola canescens Wallich	Violaceae	Medicinal

Source: Pala et al. (2012, 2013).

been raised around these sacred groves to reduce the pressure of external forces. Demarcation of boundary by stones or red flags is also a practice of conservation. Based on the findings of the present study, it is concluded that the sacred grove of Garhwal Himalaya is still free from encroachment and can satisfy the environmental, scientific, cultural, livelihood, and esthetic needs of mankind under the current social situation. Out of various management options followed in SNS of Uttarakhand, three options are mainly dominant, namely Social fencing, Boundary demarcation, and customary rights. The faith of the local people in the deity is the most obvious aspect of the SNS and which gets reflected from the social fencing as the dominant management option followed by the local communities (Figure 10.2). There have been three major suggestions which dominate out of total eight suggestions for conservation management of the SNS in Uttarakhand. The most important suggestion (26%) is awareness and education for not only to the younger generation but also to the practitioners and the government departments such as the State Forest Department (Figure 10.3). The present study highlights the importance of identifying the areas under sacred conservation practices, because there is potential for future conflicts of interest within these Sacred Conserved Areas. The potential utility of mapping can be helpful in detecting changes in the future. Article 8 (j) of the Convention on Biological Diversity calls for respecting, preserving, and maintaining knowledge, innovations, and practices of indigenous and local communities embodying traditional lifestyles relevant for the conservation and sustainable use of Biological diversity according to the national legislation. In this context, sacred conservation practices assume particular importance. In India, local communities have traditionally protected species or entire ecosystems due to sacred beliefs, associated with them. If these sites are provided legal recognition, there would be incentives for local communities to continue their nature-friendly beliefs while leading to the conservation of critical species and ecosystems. Furthermore, amendments to the Wild Life (Protection) Act 1972 have opened up new opportunities to involve the communities in the conservation efforts by declaring these fragmented patches as conservation reserves (on government and private/community land, respectively). These amendments will recognize (the largely isolated) community conservation efforts and integrate these with the mainstream conservation strategies. In this context, sacred natural sites such as sacred groves can be potential areas to expand the protected area network (PAN).

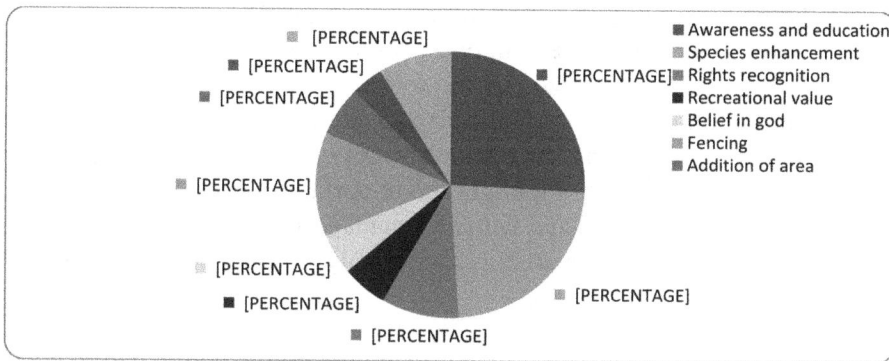

FIGURE 10.2 Various conservation measures followed for SNS in Uttarakhand.
Source: Gokhale and Pala (2016).

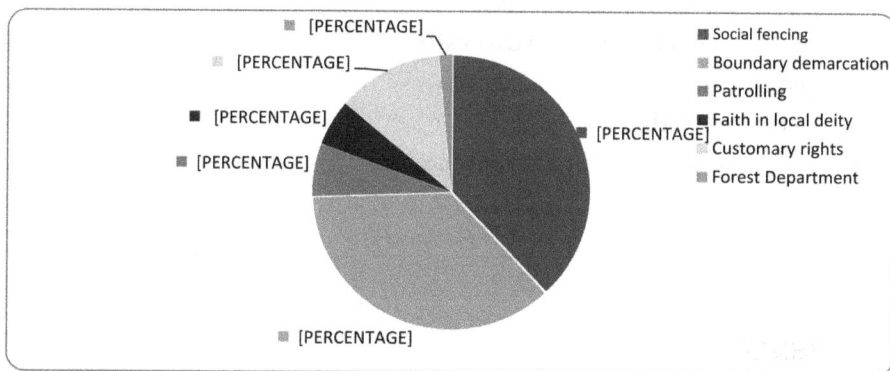

FIGURE 10.3 Management options and perception for conservation of SNS of Uttarakhand.
Source: Gokhale and Pala (2016).

10.4 CONCLUSION

The present book chapter highlights that sacred grove/community conserved areas of Garhwal Himalaya are still free from encroachment and can satisfy the environmental, scientific, cultural, livelihood, and esthetic needs of mankind under the current social situation. The importance of identifying the areas under sacred conservation practices, because there is potential for future conflicts of interest within these Sacred Conserved Areas. These are ecologically very important and are of much conservation importance outside

PAN. The present status of sacred groves in Garhwal Himalaya is a matter of concern as the belief system among communities is gradually declining, especially among youth. Devotees and tourists visit these shrines regularly and by formation of concrete paths around the temple, which is a sign of disturbance to regenerating vegetation. In the present study assessment of floral diversity was done to highlight the importance of these sacred groves in the conservation of flora and fauna for managerial aspects as these can be strong candidates to be declared as community or conservation reserve to expand the PAN.

KEYWORDS

- community conserved areas
- convention of biological diversity
- geographic information system
- important value index
- rare, endangered, and threatened
- total basal cover

REFERENCES

Ayensu, C. R. D., Collins, M., Dearing, A., Fresco, L., Gadgil, M., Gitay, H., Glaser, G., et al., (1999). International ecosystem assessment. *Science, 286*, 685–686.

Berkes, F., Kislalioglu, M., Folke, C., & Gadgil, M., (1998). Exploring the basic ecological unit: Ecosystem-like concepts in traditional societies. *Ecosystems, 1*, 409–415.

Bray, D. B., Merino-Perez, L., Negreros-Casillo, P., Segura-Warnholtz, G., Torres-Rojo, & Vester, H. F. M., (2003). Mexico's community forests as a global model for sustainable landscapes. *Conservation Biology, 17*, 672–677.

Bridgewater, P. B., (1993). Landscape ecology, geographic information systems and nature conservation. In: Haines-Young, R., Green, D. R., & Cousins, S. H., (eds.), *Landscape Ecology and GIS* (pp. 23–36). Taylor and Francis, London.

Chandrakanth, M. G., Bhat, M. G., & Accavua, M. S., (2004). Socio-economic changes and sacred groves in South India. Protecting a community-based resources management institutions. *Natural Resources Forum, 28*(2), 102–111.

Chandrashekara, U. M., & Sankar, S., (1998). Ecology and management of sacred groves in Kerala. *Forest Ecology and Management, 112*, 165–177.

Dash, S. S., (2005). Kabi sacred grove of North Sikkim. *Current Science, 89*(3), 427–428.

Gadgil, M., & Vartak, V. D., (1975). Sacred groves of India: A plea for continued conservation. *Journal of Bombay Natural History Society, 72*, 314–320.

Gadgil, M., & Vartak, V. D., (1976). Sacred groves of Western Ghats of India. *Ecological Botany, 30*, 152–160.

Gokhale, Y., Pala, N. A., Negi, A. K., Bhat, J. A., & Todaria, N. P., (2011). Sacred landscapes as repositories of biodiversity: A case study from Hariyali Devi sacred landscape, Uttarakhand. *International Journal of Conservation Sciences, 2*(1), 37–44.

He, S. H., Mladenoff, D. J., Radeloff, V. C., & Crow, T. R., (1998). Integration of GIS data and classified satellite imagery for regional forest assessment. *Ecological Applications, 8*, 1072–1083.

Heywood, V. H., (1995). *Global Biodiversity Assessment* (p. 1140). UNEP, Cambridge University Press.

Israel, E., Vijai, C., & Narasimhan, D., (1997). Sacred groves: Traditional ecological heritage. *International Journal of Ecology and Environmental Sciences, 23*, 463–470.

Jeeva, S., Kingston, C., Kiruba, S., Kannan, D., & Jasmine, T. S., (2007). Medicinal plants in the sacred forests of Southern Western Ghats. In: *National Conference on Recent Trends on Medicinal Plants Research (NCRTMPR-2007)*. Organized by Centre for Advanced Studies in Botany, University of Madras, Guindy Campus, Chennai.

Knight, D. H., (1963). A distance method for constructing forest profile diagrams and obtaining structural data. *Tropical Ecology, 4*, 89–94.

Kothari, A., (2006). Community conserved areas: Towards ecological and livelihood security. *Parks, 16*, 703–733.

Kupfer, J. A., (1995). Landscape ecology and biogeography. *Progress in Physical Geography, 19*, 18–34.

Kushalappa, C. G. & Kushalappa, K. A. (1996) Preliminary report of the project on impact of working in Western Ghats forests of Kodagu, College of Forestry, Ponnampet.

Pala, N. A., (2012). *Plant Diversity and Regeneration Status of Sacred and Protected Groves in Garhwal Himalaya* (p. 172). PhD thesis submitted to HNB Garhwal University.

Pala, N. A., Gokhale, Y., Negi, A. K., Razvi, S., & Todaria, N. P., (2012). Local deities in conservation: A conservation practice in Banju Nami Tok sacred grove in Tehri Garhwal, Uttarakhand. *Indian Forester, 138*(8), 710–713.

Pala, N. A., Negi, A. K., & Todaria, N. P., (2014). The religious, social, and cultural significance of forest landscapes in Uttarakhand Himalaya, India. *International Journal of Conservation Science, 5*(2), 215–222.

Pala, N. A., Negi, A. K., & Todaria, N. P., (2015). Ecological status and socio-cultural significance of Sem Mukhem temple landscape in Garhwal Himalaya, India. *Indian Forester, 141*(5), 496–504.

Pala, N. A., Negi, A. K., Gokhale, Y., & Todaria, N. P., (2013). Tree Regeneration status of sacred and protected landscapes in Garhwal Himalaya, India. *Journal of Sustainable Forestry, 32*, 230–246. doi: 10.1080/10549811.2013.762492.

Pala, N. A., Negi, A. K., Gokhale, Y., Razvi, S., & Todaria, N. P., (2012). Medicinal plant resources in sacred forests of Garhwal Himalaya. *Journal of Non-Timber Forest Products, 19*(4), 291–296.

Pala, N. A., Negi, A. K., Shah, S., & Todaria, N. P., (2013). Floristic composition, Ecosystem services and Biodiversity value of sacred forests in Garhwal Himalaya. *Indian Journal of Forestry, 36*(3), 353–362, 2013.

Prentice, C., & Parish, D., (1990). *Conservation of Beat Swamp Forest: A Forgotten Ecosystem* (pp. 128–141). Proceed. of the int. conf. on trop. biodiverse, in harmony with nature. Kuala Lumpur.

Shannon, C. E., & Wiener, W., (1963). *The Mathematical Theory of Communication*. University of Illinois Press, Urbana.

Spies, T. A., Reeves, G. H., Burnett, K. M., McComb, W. C., Johnson, K. N., Grant, G., Ohmann, J. L., et al., (2002). Assessing the ecological consequences of forest policies in a multi-ownership province in Oregon. In: Liu, J., & Taylor, W. W., (eds.), *Integrating Landscape Ecology into Natural Resource Management* (pp. 179–207). Cambridge University Press, Cambridge, UK.

Tambat, B., Rajanikanth, G., Ravikanth, G., Uma, S. R., Ganeshaiah, K. N., & Kushalappa, C. G., (2004). Seedling mortality in two vulnerable tree species in the sacred groves of Western Ghats, South India. *Current Science, 88*(3), 350–352.

Uma, S. R., (2001). A case of high tree diversity in a Sal (*Shorea robusta*) dominated lowland forest of Eastern Himalaya: Floristic composition, regeneration and conservation. *Current Science, 81*, 776–786.

Whittaker, R. H., (1975). *Communities and Ecosystems*. Macmillan Publishing Company, New York.

Williams, L. G., (2002). Tree species richness complementarily, disturbance and fragmentation in a Mexican tropical montane cloud forest. *Biodiversity and Conservation, 11*, 1825–1845.

Zimmerer, K., Ryan, E. G., & Margaret, V. B., (2004). Globalization and multi-spatial trends in the coverage of protected-area conservation (1980–2000). *Ambio, 33*(8), 520–529.

CHAPTER 11

Timber Volume Increments in the Woodlots of Guadalcanal Island

DAVID LOPEZ CORNELIO

School of Natural Resources and Applied Sciences,
Solomon Islands National University, P.O. Box R113, Honiara,
E-mail: david.cornelio@sinu.edu.sb

ABSTRACT

In line with a new strategy of the Solomon Islands government to restructure and further develop the forestry sector, this chapter aims to quantify the growth performance of four exotic timber species commonly planted in the country. The results are based on three woodlots in Guadalcanal Island of 5, 11.25, and 11.5 years of age. A total of 726 trees were measured, and results of mean annual increments (MAIs) were statistically compared. Tree heights were estimated by trigonometry and volumes by considering the main stem as a frustum of paraboloid. The chapter expects to contribute to the optimization of timber yields prediction, and consequently contribute on the planning and management of plantations in the country.

11.1 INTRODUCTION

The Solomon Islands is a group of islands located in the Southwest Pacific between 155°30' and 170°30'W longitude and between 5°10' and 12°45'S latitude (Whitmore, 1969). Guadalcanal Island of volcanic origin is the largest island of the Solomon Islands with an area of 5,302 km². Its highest point at Kavo Range reaches 2330 m asl. Its coasts, sometimes lined with mangroves, are crossed by short, rapid streams coming down from the mountains (www.britannica.com). Temperatures are stable throughout the year (30/31°C) from November to April, with a slight drop from May to

October. Rainfall ranges from 3,000 to 5,000 mm per year, with a slight
decrease from May to November (Meteorological Services Division, 2019).
The soils are generally deep, weathered, and leached on the lowlands, and
shallow on the hillsides with colluvial rock debris. Most soils are strongly
acid (pH 3 to 5) clays with low plant nutrient contents (Lee, 1969), in some
areas, they are rich alluvium soils (Solomon Islands Historical Encyclopedia,
2013). The forestry sector has until 1997 accounted for about 45–55% of
foreign exchange and 20–30% of government revenue. The annual rate
of extraction is 750,000 cubic meters which is three times the sustainable
extraction level. Currently, the government policy aims to restore sustainable
levels of logging rates. Fisheries and agriculture are key contributors to GDP,
but subsistence agriculture is the dominant economic activity. Agricultural
exports are primarily palm oil, palm kernels, copra, coconut, and cocoa; the
country's fish resource is substantial but the sustainable annual catch level
of 120,000 tons was not yet attained (Figure 11.1 and Table 11.1) (Solomon
Islands Initial National Communications, 2015).

FIGURE 11.1   Location of study sites: Tetere (T): 1.2 ha (396 trees in total), Aruligo (A): 10
ha (sample of 165 trees), and Golf (G) (126 trees in total).

The total commercial area of timber plantations in 2005 was 28,000 ha,
7600 ha less than in 2003 (Table 3). Planted total area in 2003 was 35,600 ha
at Alu, Kolombangara, Viru, Gizo, Choiseul Bay, Moli, Allardyce, and Santa
Cruz with *Gmelina arborea, Campnosperma brevipetiolatum, Eucalyptus*

deglupta, Terminalia calamansanai, T. brassiai, Acacia spp. and *Swetenia macrophylla.* After harvest replanting is done mainly with *G. arborea, E. deglupta, Tectona grandis,* and *S. macrophylla.* The yield projected for a 30 years rotation is of 5 m^3/ha/year (Pauku, 2009).

TABLE 11.1   Tree Species Commonly Planted in the Solomon Islands

Industrial Plantation	Percentage (%)	Village Plantation	%
Eucalyptus deglupta	28%	*Tectona grandis*	65
Gmelina arborea	19%	*Swetenia macrophylla*	14
Campnosperma brevipitiolata	14%	*Eucalyptus deglupta*	11
Swetenia macrophylla	14%	*Gmelina arborea*	9
Terminalia spp.	9%	*Others*	1
Agathis macrophylla	7%	**Total**	**100**
Tectona grandis	3%		
Acacia sp	2%		
Others	4%		
Total	**100%**		

Source: Ministry of Forests and Research (2012).

Growth is defined as the increase in dimensions of one or more individuals in a forest stand over a given period of time, while yield refers to their dimensions at the end of a certain period of time. In even-aged stands, a growth equation may predict the growth of diameter, basal area, or volume in units per annum as a function of age and other stand characteristics, whereas a yield equation may predict the diameter, stand basal area or total volume production attained at a specified age (Vanclay, 1994).

Tectona grandis (Teak) is a native tree to Southeast Asia and India; it is one of the most widely planted hardwoods in the world (2.25 million ha) (Ball et al., 1999). Its volume mean annual increment (MAI) ranges from 2 to over 15 m^3/ha/year at the half rotation age of 30–40 years (White, 1991). *Acacia mangium* (Acacia) is a secondary tree species of Indonesia and Papua New Guinea (Gunn and Midgley, 1991), widely planted through Asia and the Pacific. It grows fast, has quality wood, and is tolerant to different soils and environments (National Research Council, 1983). Its girth growth reaches 15 cm in three years, slows down after the 5th year, and levels off at around 25 cm by the age of 8 years. The height growth follows a similar trend, in the first 2–3 years increases moderately up to 10–15 m, reaching 25 m by year five and leveling off afterwards (Krisnawati et al., 2011). *Swietenia macrophylla* (Mahogany) is a fast-growing, light-demanding specie that appears late in the succession when there is an opening or canopy disturbance (Grogan et al., 2005). In Belize they grow 0.6–0.80 cm/year depending on the site, on logged areas the growth decreases to 0.17–0.74 cm/year (Bird, 1998). Its height MAI exceeds 1 cm/year, with slightly higher growth rates in trees of over 50 cm diameter at breast height. The growth rate among trees differs considerably, with the fastest-growing individuals growing at rates greater than 2 cm/year (Shono and Snook, 2006); although others found the range to be from 0.6–2.5 cm/year (Lamb, 1966; Shono and Snook, 2006; Grogan et al., 2010) to 0.16–0.38 cm/year (Snook, 2000; Negreros-Castillo and Mize, 2006). Young trees reach 20 cm of dbh in 30 years; afterwards the diameter increases 0.8 cm/year (Snook, 2003).

Eucalyptus sp. (Eucalyptus) is some of the fastest growing trees. In Madang (Papua New Guinea) the dominant trees reached 38 m in height and 39 cm in diameter at breast height (dbh) over bark in three years, which corresponds to an MAI of 80–90 m³/ha/year (Eldridge et al., 1993). MAIs of 2–4 years old stands in Costa Rica ranged between 2 and 39 m³/ha/year (Sánchez, 1994). The maximum yield recorded was of 89 m³/ha/year on 4.5 years (Ugalde, 1980). The average of diameter and height increments of four eucalyptus species in 10 years old plantations at New Zealand were of 2.1 cm/year and 1.9 m/year, respectively (Miller et al., 2000).

The research questions in this chapter were: Which of the tree species grow faster?; Is the tree location a significant factor on volumetric increase?; How do these tree species perform in the province compared with similar plantations overseas?; and What are the annual increments of basal area and volume per specie per year?.

11.2   METHODOLOGY

The results are based on three woodlots in Guadalcanal Island of 5, 11.25, and 11.5 years of age. A total of 726 trees were measured and results of mean annual volume increment were statistically compared. Tree heights were estimated by trigonometry and volumes by considering the main stem as a frustum of paraboloid and a taper factor of 0.7. A 100% inventory was performed at Tetere and Golf sites, a sampling at Aruligo (along transects until at least 50 trees were measured for each species). All the trees were coded with white paint at 1.4 m above the ground. Tree heights were estimated with a clinometer by trigonometry, the basal area at breast height with metric tapes (BA = πr^2) and tree volumes were estimated as the product of both considering a taper factor of 0.7. The x, y coordinates in meters of each tree at Tetere were surveyed along two transects of 350 m and two of 50 m; then recalculated in a single gridline before entering the values into Ilwis GIS to carry out a points interpolation. Statistical tests for data analysis were performed with JASP, SISA, and VassarStats open software. One-way ANOVA tests were performed to find significant differences in growth between species and among sites, with TUKEY HSD when K>2 and the analysis of variance yielded a significant F-ratio. Volume tabulations and regressions were performed in excel (Figure 11.2).

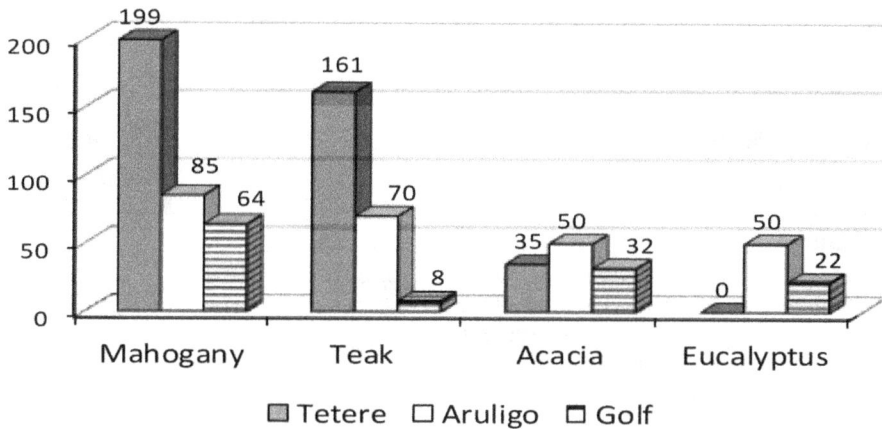

FIGURE 11.2   Total number of trees surveyed per species per site.

11.2.1   TETERE SITE

Tetere site is located in rural Guadalcanal, on the northern plains within the boundaries of the current Correctional Camp of Guadalcanal. The plantation was done with the participation of inmates, locals, students, and foreign economic assistance (Figures 11.3–11.6).

The three species at Tetere site had a mean basal area of 0.05 m², a mean volume of 0.03 m³ per tree for the three species and commercial heights of 8 m (Mahogany), 5 m (Teak) and 6 m (Acacia). The histograms for Mahogany show a more normal distribution than the other two species due to their higher number in the plantation (almost half of the total of trees); however, the merchantable heights of Acacia and Teak trees are more uniform than those of Mahogany (Figures 11.7–11.12).

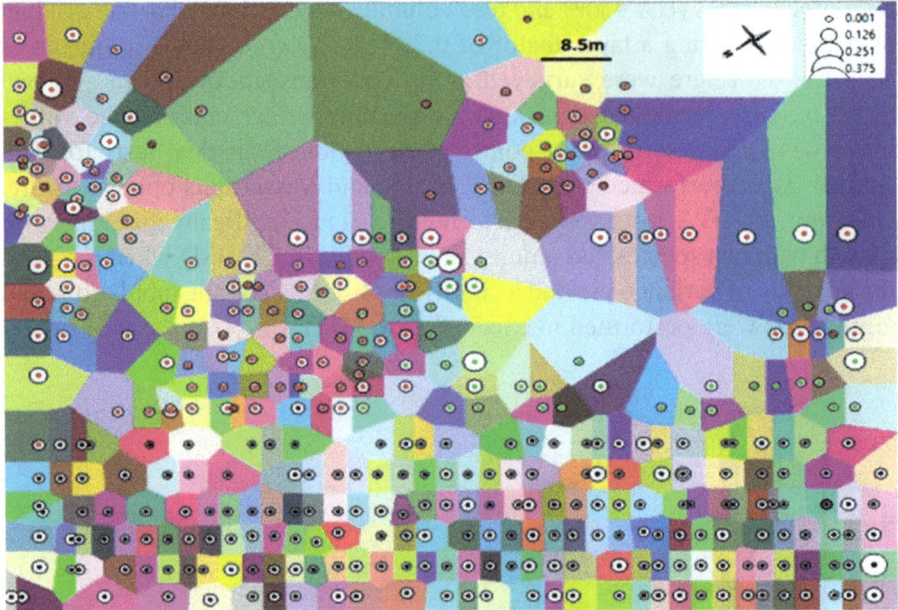

FIGURE 11.3 Map of trees location at Tetere site with a total area of 125 × 100 m. Mahogany, teak, and acacia trees are depicted in black, red, and green. Trees with bigger diameter appear within larger circles. The background map is a *Thiessen* polygons map which allocates space to the nearest point feature (tree stems); every location is nearer to this point than to all the others (Boots, 1999).

Source: Reprinted with permission from Boots, 1999. © John Wiley.

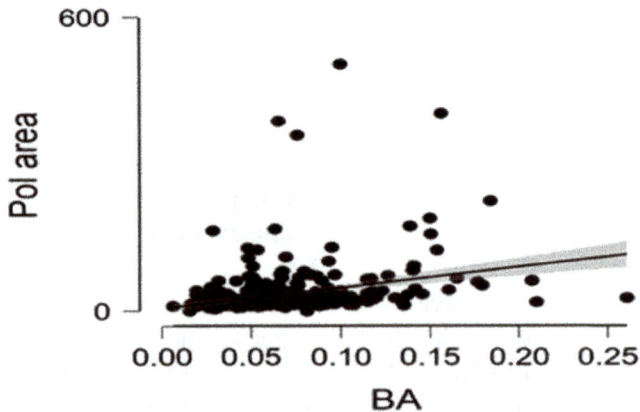

FIGURE 11.4 Correlation curve between basal areas and their respective available space for tree growth with a CI of 95%, F-value 15.467, and p-value 0.0001. Trees with more available area for growth have larger diameters at breast height.

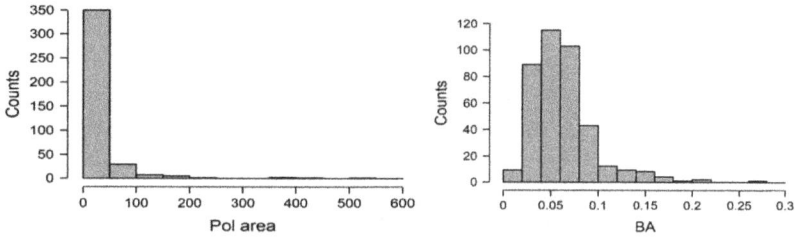

FIGURE 11.5 (a) Polygon areas are expressed in pixels of 0.1 × 0.1 m; and (b) basal areas are expressed in m²/tree.

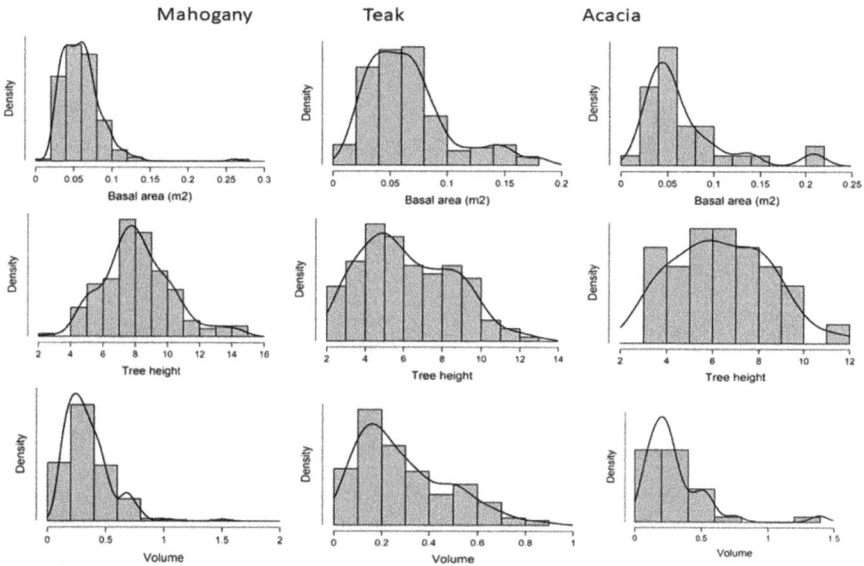

FIGURE 11.6 Histograms for basal areas, commercial heights, and volumes of Mahogany, Teak, and acacia trees at Tetere plantation site.

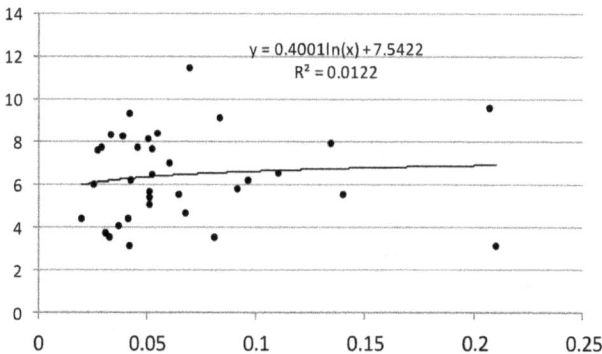

FIGURE 11.7 Basal area-commercial height regression for Acacia trees at Tetere site.

$y = 0.3507\ln(x) + 1.3137$
$R^2 = 0.6368$

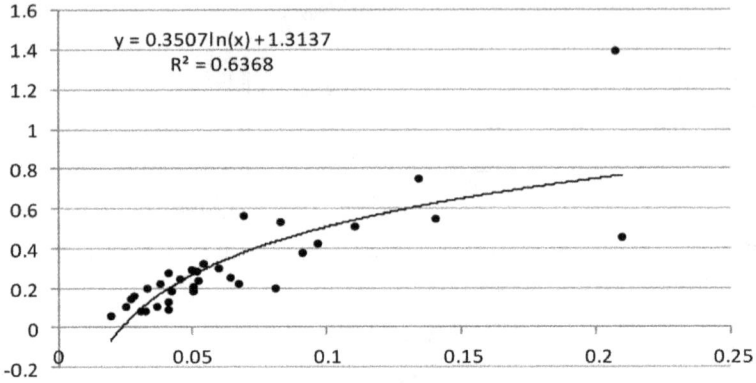

FIGURE 11.8   Basal area-volume regression for Acacia trees at Tetere site.

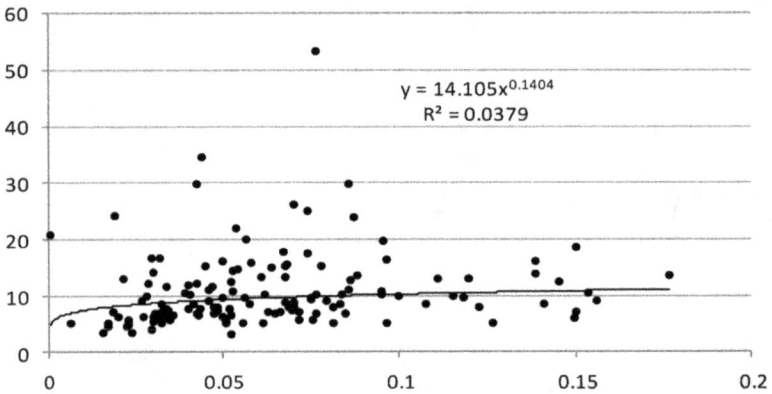

$y = 14.105x^{0.1404}$
$R^2 = 0.0379$

FIGURE 11.9   Basal area-commercial height regression for Teak trees at Tetere site.

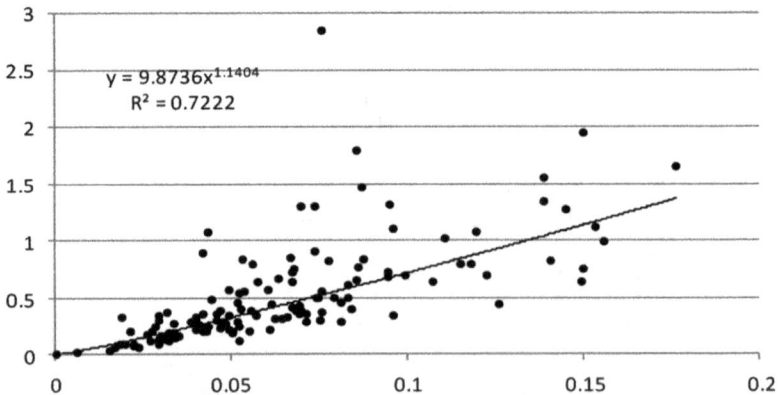

$y = 9.8736x^{1.1404}$
$R^2 = 0.7222$

FIGURE 11.10   Basal area-volume regression for Teak trees at Tetere site.

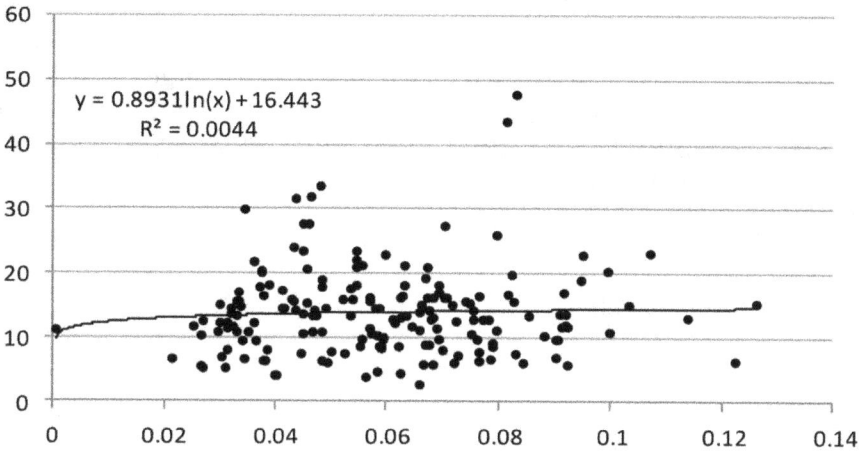

FIGURE 11.11 Basal area-commercial height regression for Mahogany trees at Tetere site.

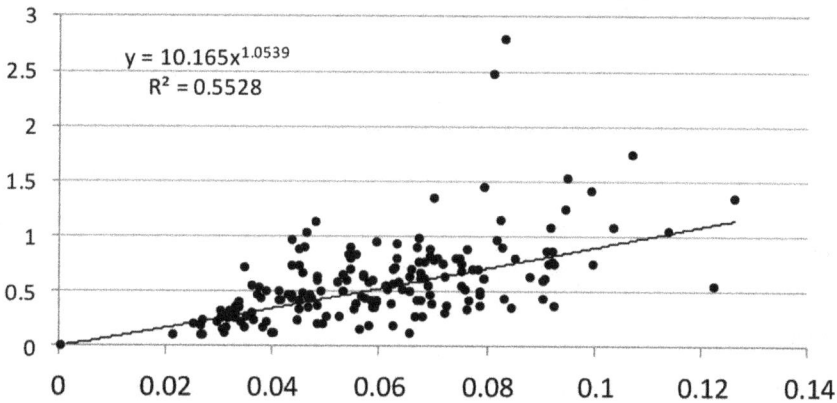

FIGURE 11.12 Basal area-volume regression for Mahogany trees at Tetere site.

At Tetere the regressions between basal areas and heights were not as pronounced as in the case of basal areas with commercial volumes for the three species; with no significant differences ($p < 0.01$) between the basal areas, commercial heights and volumes (Figures 11.13–11.15).

11.2.2 ARULIGO SITE

The plantation at Aruligo site was planned by the provincial government in order to support the relocation of displaced locals after major

floods in the island. In July 1965 heavy rains caused landslides along
the coast near Avuavu on the Weather Coast destroying some villages
and adjacent gardens; in February 1967 they caused large landslides in
central Guadalcanal, some of them more than 300 meters long; on April
21, 1977 an earthquake caused the largest land movement in memory
(Webber, 2011). The main planted species were Teak, Eucalyptus, and
Mahogany (Figure 11.16).

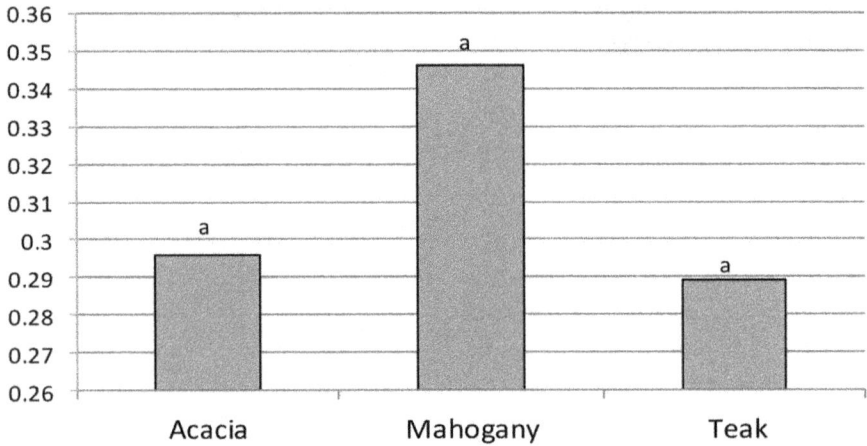

FIGURE 11.13   Mean volume (m³) per species at Tetere site. Bars with different letters
denote significant differences (p < 0.001).

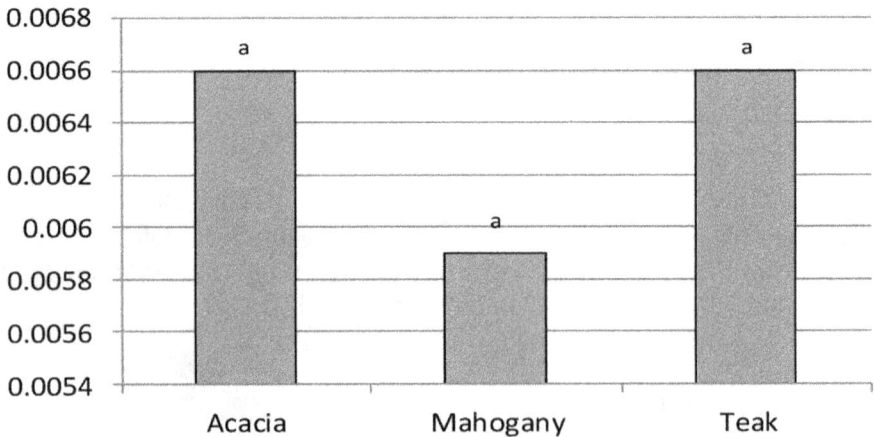

FIGURE 11.14   Mean annual increment of basal area (m²/year) at Tetere site. Bars with
different letters denote significant differences (p < 0.001).

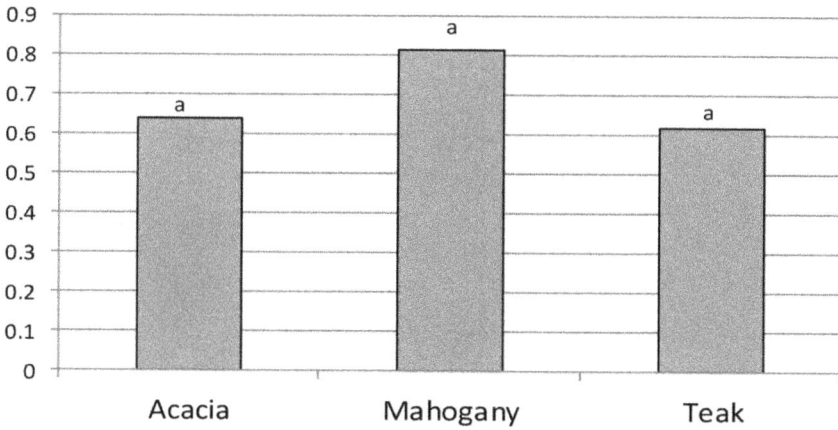

FIGURE 11.15 Mean annual increment of commercial height (m/year) at Tetere site. Bars with different letters denote significant differences (p < 0.001).

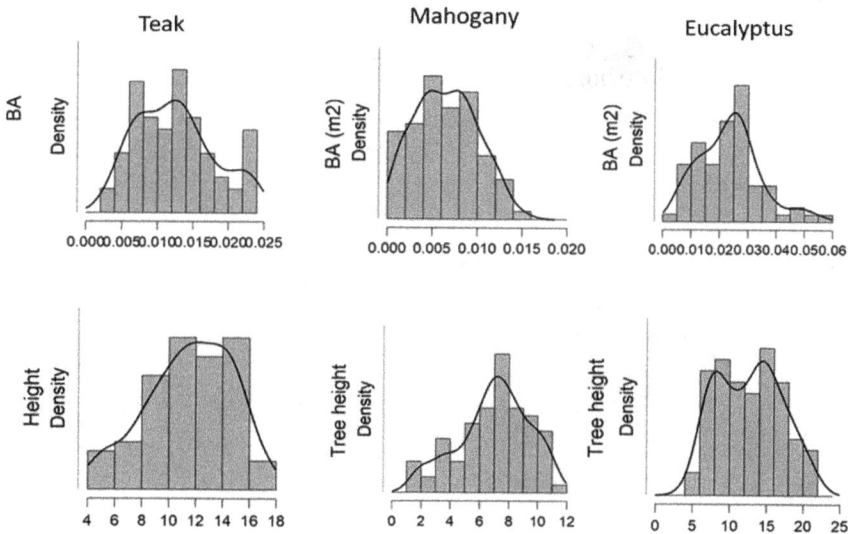

FIGURE 11.16 Histograms for basal areas, commercial heights, and volumes of Teak, Mahogany (center) and Eucalyptus trees at Aruligo plantation site.

At Aruligo the basal area MAI of Eucalyptus trees was nearly three times those of Mahogany trees and twice as much as of Teak trees. Mahogany trees didn't perform well due to the low soil water retention capacity of the convex terrain. In nature Mahogany trees are normally found along riverbanks (Figures 11.17–11.24) (Orwa et al., 2009).

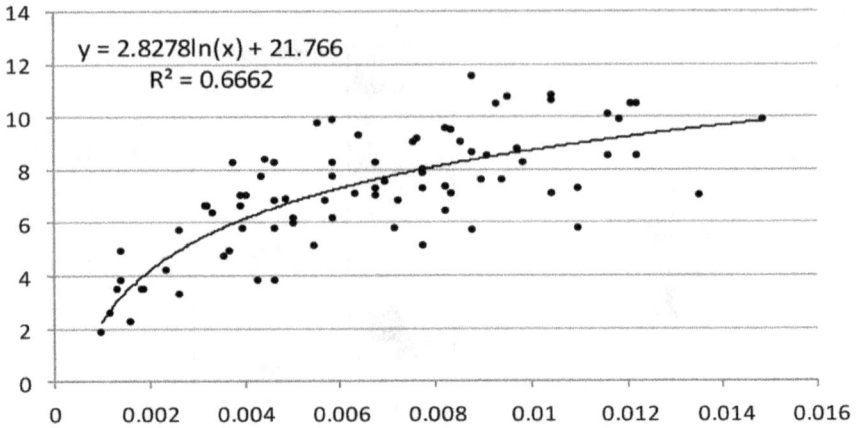

FIGURE 11.17   Basal area-commercial height regression for Mahogany trees at Aruligo site.

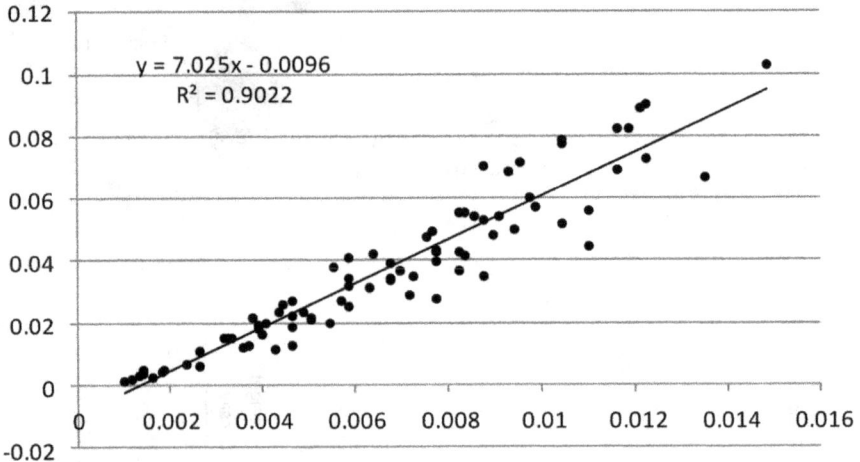

FIGURE 11.18   Basal area-volume regression for Mahogany trees at Aruligo site.

11.2.3   *GOLF SITE*

Honiara Golf Club was built in 1957 on the site that was the Fighter II airfield at Kukum during WWII. By the mid-1950s the airfield began to deteriorate with grass growing through the runway, therefore a new airfield was built at Henderson in 1958 and the Fighter II airfield that housed the Immigration and Customs Services was refurbished as a Golf Clubhouse by expanding the green areas (Figure 11.25) (Solomon Islands Historical Encyclopedia, 2013).

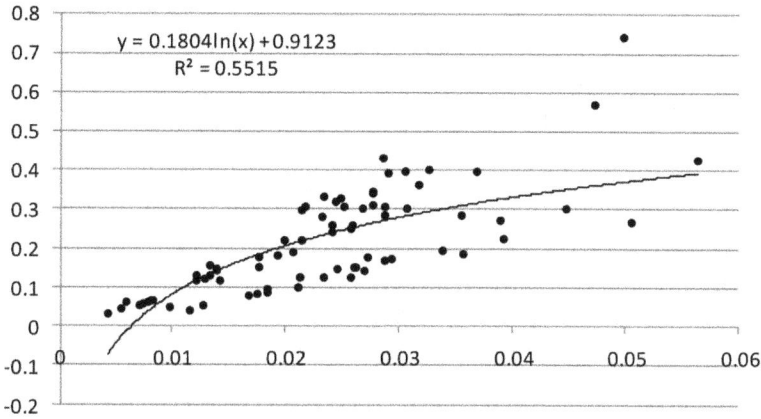

FIGURE 11.19 Basal area-total height regression for Eucalyptus trees at Aruligo site.

Aruligo Teak BA - T. Height

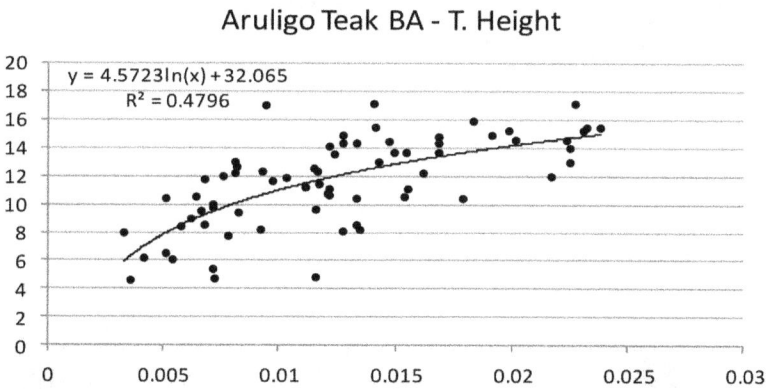

FIGURE 11.20 Basal area-volume regression for Eucalyptus trees at Aruligo site.

Aruligo Teak BA - T. Height

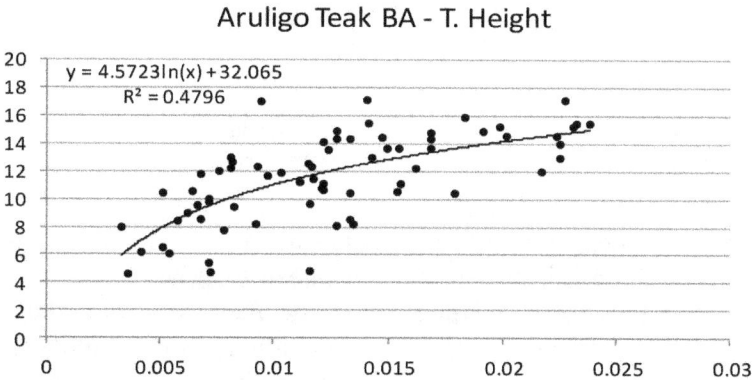

FIGURE 11.21 Basal area-total height regression for Teak trees at Aruligo site.

$$y = 138.44x^2 + 7.9153x - 0.0156$$
$$R^2 = 0.9156$$

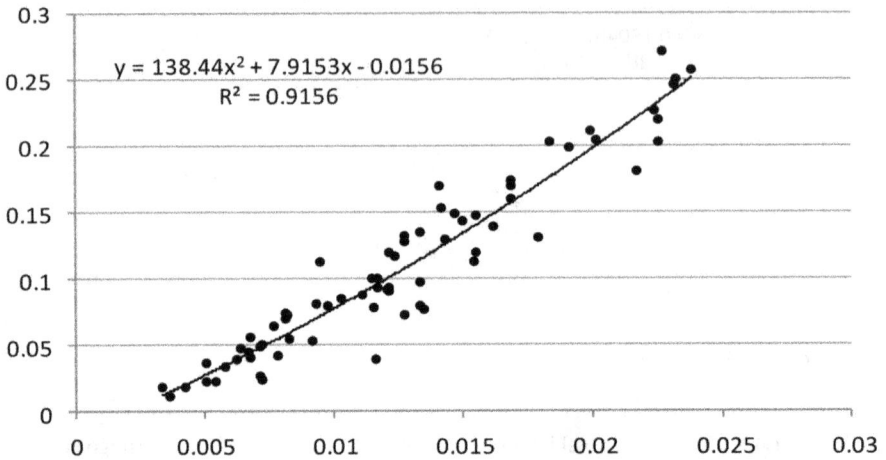

FIGURE 11.22   Basal area-volume regression for Teak trees at Aruligo site.

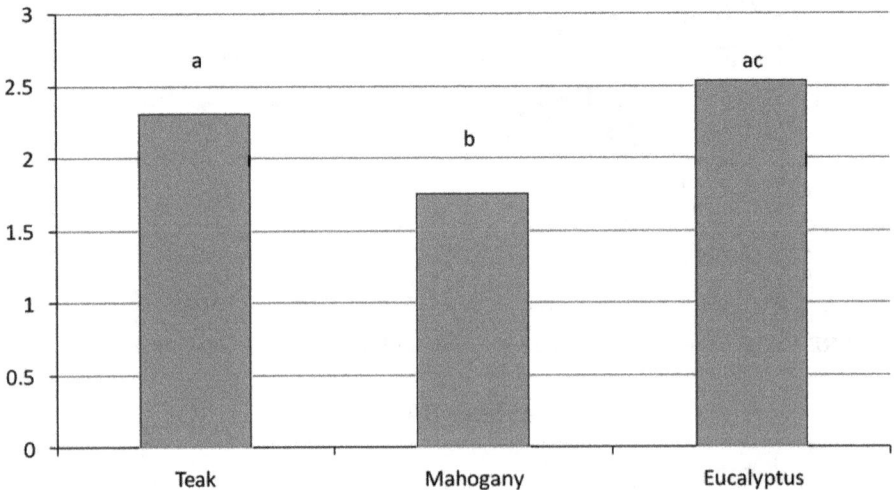

FIGURE 11.23   Tukey HSD comparisons of mean annual increment of total height (m/year) per species at Aruligo site. Bars with different letters denote significant differences ($p < 0.001$).

At the Golf site, eight Teak trees had a similar mean basal area as Eucalyptus trees (32 trees in total), but their mean total height was of around 5 m less than those of Eucalyptus trees. The Teak trees were planted in a single row at the border of the plantation providing them with more space for growth. Eucalyptus trees in average yielded twice as much timber volume per tree than the other three species (Figures 11.26–11.32).

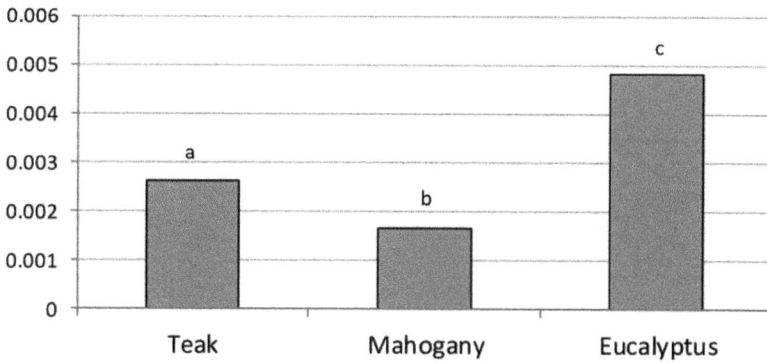

FIGURE 11.24 Tukey HSD comparisons of mean annual increment of basal area (m²/year) per species at Aruligo site. Bars with different letters denote significant differences (p < 0.001).

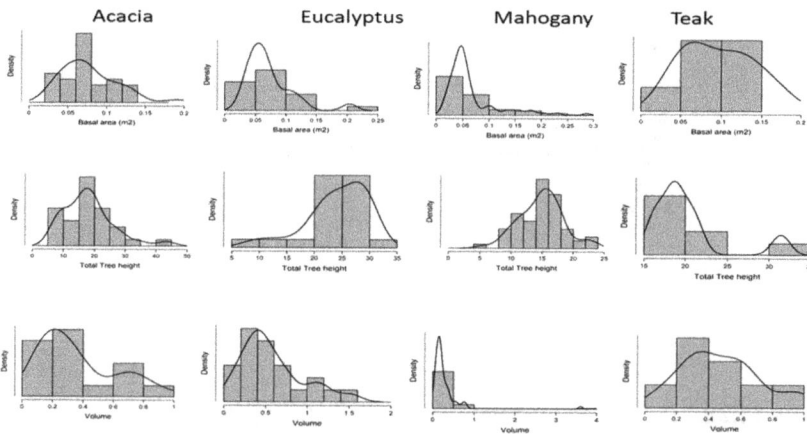

FIGURE 11.25 Histograms for basal areas, commercial heights, and volumes of Acacia, Eucalyptus, Mahogany, and Teak trees at Golf site.

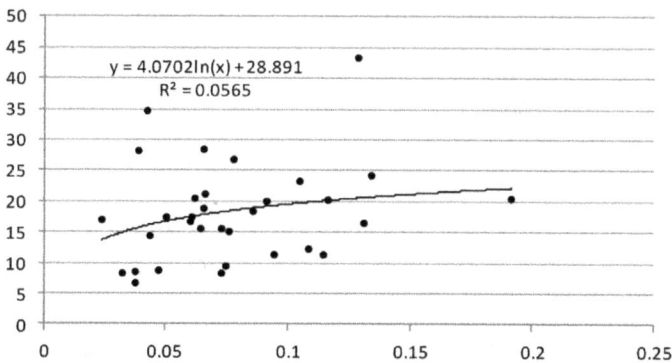

FIGURE 11.26 Basal area-total height regression for Acacia trees at the Golf site.

Diversity and Dynamics in Forest Ecosystems

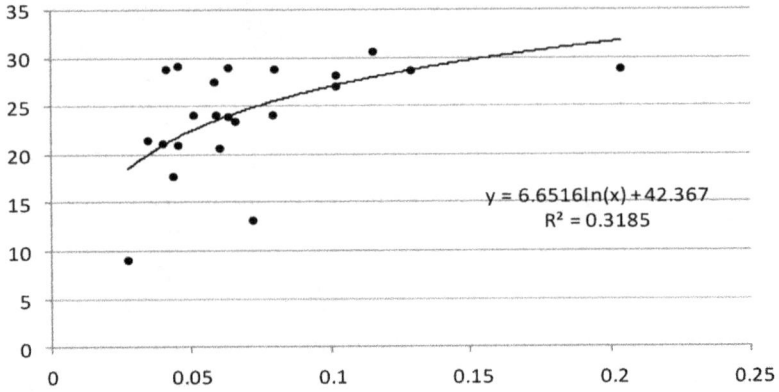

FIGURE 11.27   Basal area-total height regression for Eucalyptus trees at Golf site.

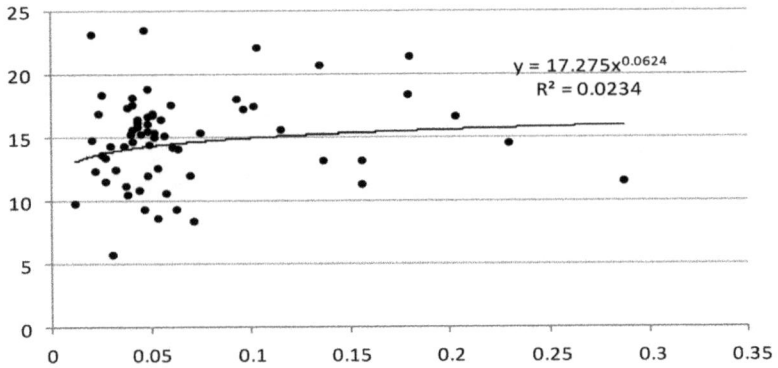

FIGURE 11.28   Basal area-total height regression for Eucalyptus trees at Golf site.

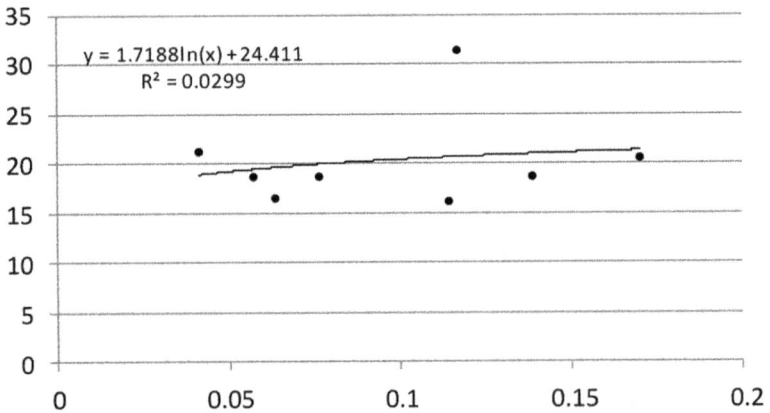

FIGURE 11.29   Basal area-total height regression for Teak trees at the Golf site.

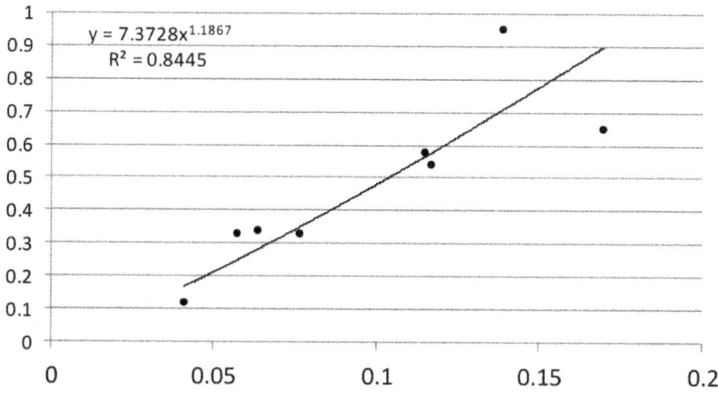

FIGURE 11.30   Basal area-volume regression for Teak trees at the Golf site.

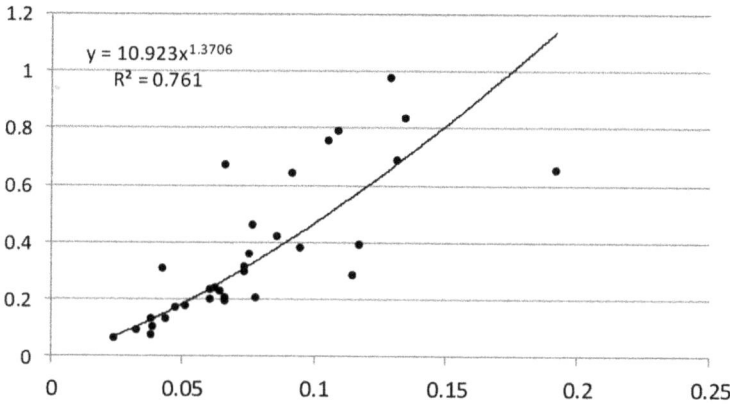

FIGURE 11.31   Basal area-volume regression for Acacia trees at the Golf site.

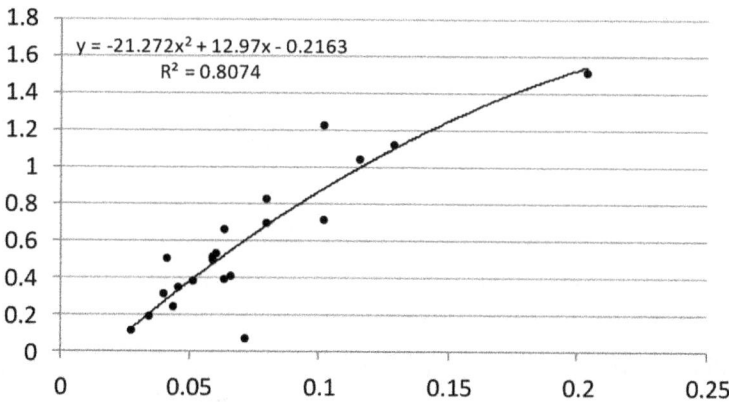

FIGURE 11.32   Basal area-volume regression for Eucalyptus trees at Golf site.

The regression curve between basal areas and heights at Tetere site does not fit as well as in the case of basal areas with commercial volumes for the three species (Figures 11.33–11.35).

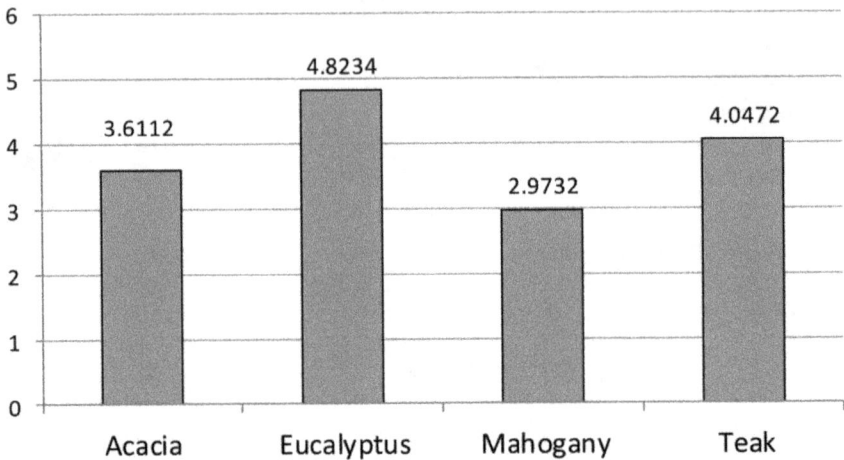

FIGURE 11.33   Mean annual increment of total heights (m/year) per species at the Golf site.

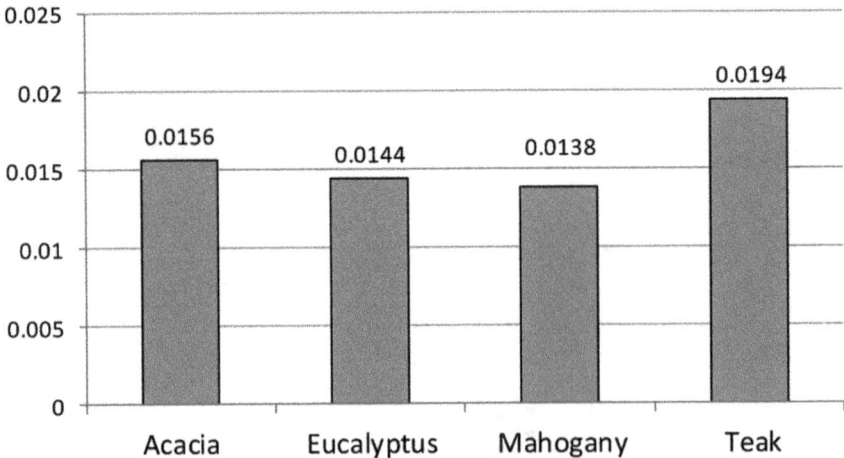

FIGURE 11.34   Mean annual increment of basal areas (m²/year) per species at the Golf site.

Acacia reached the highest basal area MAI in all the sites with a standard deviation (0.041) almost twice as the mean basal area MAI for the other tree species in all the sites (0.0274) (Figure 11.36).

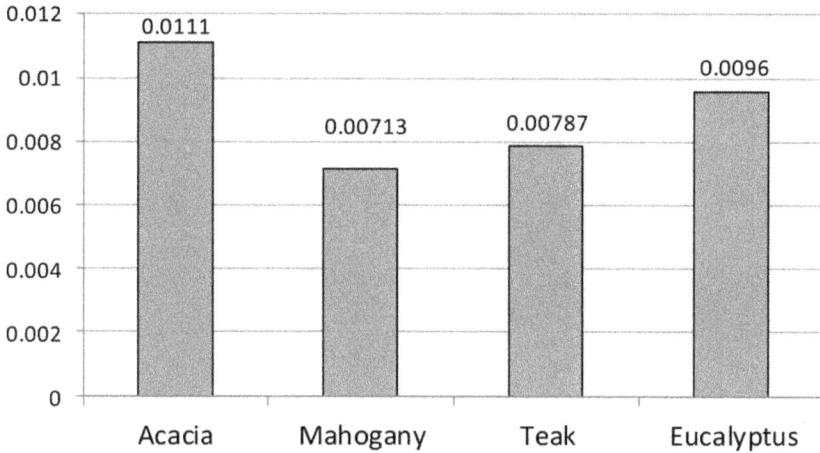

FIGURE 11.35 Overall mean annual increment of basal area (m²/year) per species at all sites.

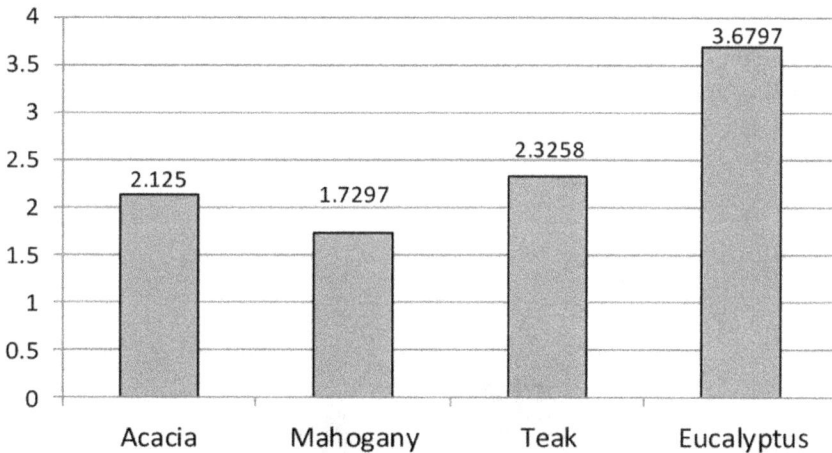

FIGURE 11.36 Overall mean annual increment of commercial tree heights (m/year) per species for all sites. At Aruligo site, only total heights were recorded (in average the trees were too young), at Tetere site only commercial heights (the lack of thinning made it difficult to detect the top end of the trees) and at Golf site both heights were recorded because most of the trees were planted along two widely spaced single lines.

Eucalyptus yielded higher commercial volumes in the three sites with a standard deviation of 3.3196 followed by Teak (2.0645), Acacia (2.4425), and Mahogany (2.684), this last with a poor performance at Aruligo site (Figure 11.37).

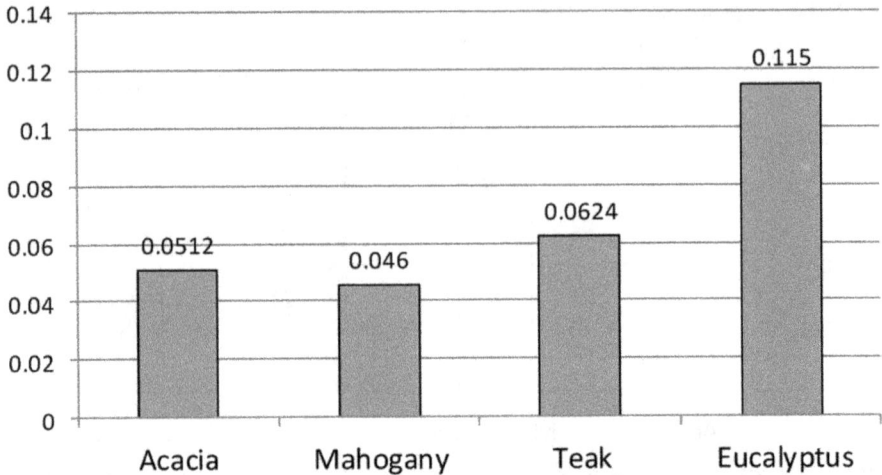

FIGURE 11.37   Overall mean annual increment of volume (m³/year) per species for all sites.

With the exception of Acacia, the diameter increments for Teak, Mahogany, and Eucalyptus at the three sites were higher that averages from overseas (Figures 11.38–11.41).

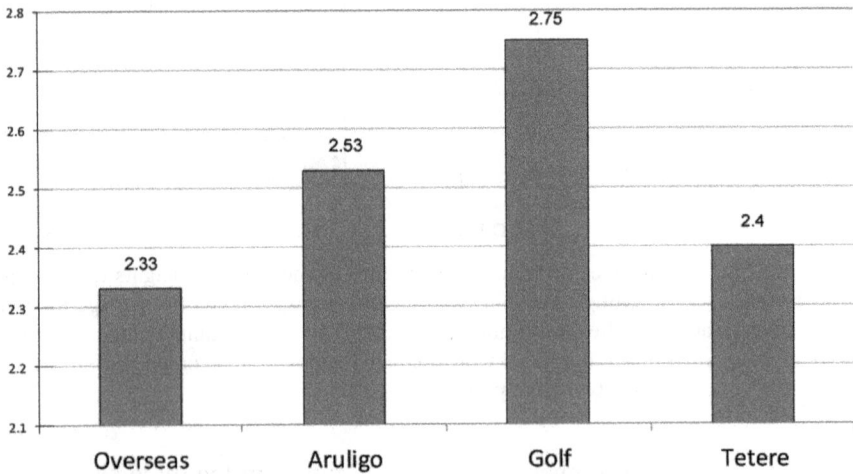

FIGURE 11.38   Comparison of diameter increments (cm/year) of Teak trees for all sites with overseas performance (White, 1991).

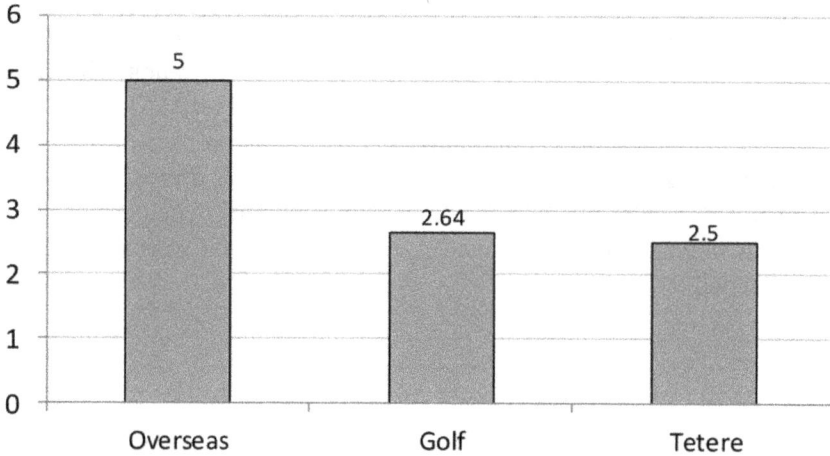

FIGURE 11.39 Comparison of diameter increments (cm/year) of Acacia trees for all sites with overseas performance. Overseas data for *Acacia mangium* from NRC (1983) and Gunn and Midgley (1991).

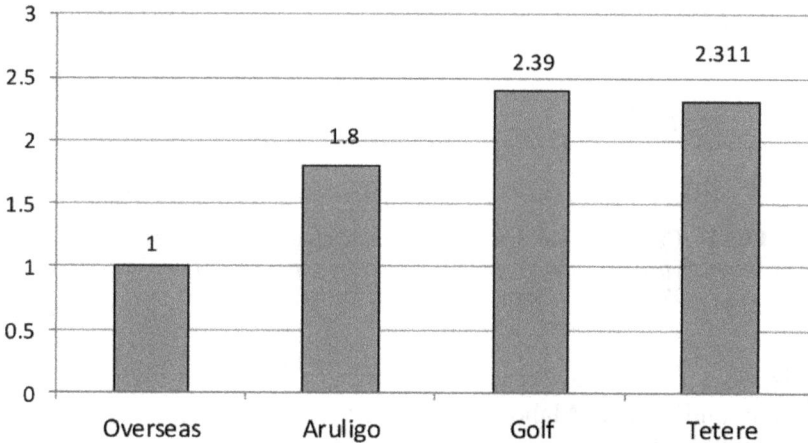

FIGURE 11.40 Comparison of diameter increments (cm/year) of Mahogany trees for all sites with overseas performance. Overseas data for Mahogany from Snook (2000), Bird (1998); Grogan et al. (2005); Lamb (1996); and Shono and Snook (2006).

11.3 CONCLUSION

The tree dimensions for the three species at Tetere site were similar; therefore the MAI values were also similar. Larger basal areas were achieved by trees in the "borders" due to less competition for nutrients and water and higher insolation. Results for Mahogany at Aruligo were not optimum, soils are drier in hilly terrain; however, it grows fast in lowlands with regular water retention or alongside drains as in Tetere. Diameter MAIs were higher for Teak, Eucalyptus, and Mahogany than in overseas; Acacia follows an opposite trend, although they reach comparatively big dimensions in short time, they are susceptible to termites and fungus attack. The performance of *Eucalyptus deglupta* is exceptionally good even in land were other planted species "failed."

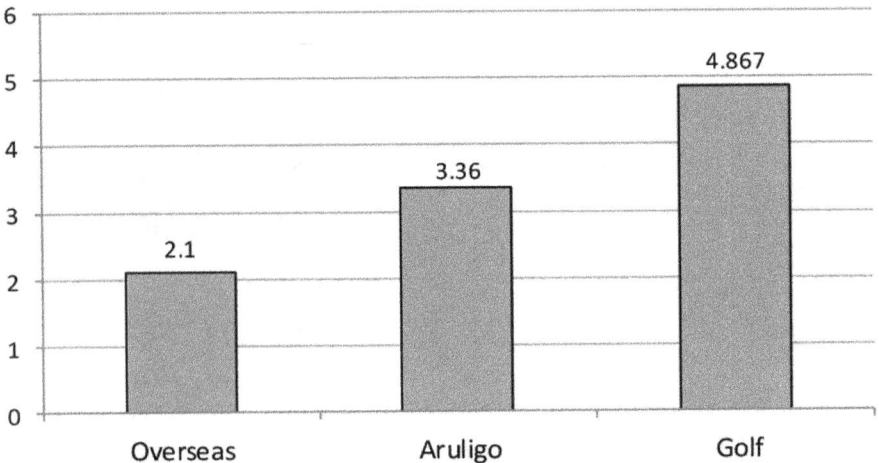

FIGURE 11.41 Comparison of diameter increments (cm/year) of Eucalyptus trees for all sites with overseas performance. Overseas data for the genre *eucalyptus* were obtained from Eldridge et al. (1993); Ugalde (1980); and Sanchez (1994).

It is recommended to set up trial plots of Eucalyptus interplanted with mahogany and/or teak. Mahogany (which is self-pruning) as an under crop for teak will facilitate heavy thinning of the latter without exposing the soil to desiccation and/or erosion. Intercropping with native hardwoods (such as *flueggea sp.* which can be harvested after few years), enable growers early returns, as well as facilitating thinning of the plantation. Maintenance works such as pruning and thinning on especially Teak plantations will increase the potential commercial volume of the tree and its shape quality. A similar survey in other provinces together with soil and tree canopy assessments will strengthen the conclusions.

ACKNOWLEDGMENTS

The indispensable fund provided by the Research Office at SINU is very appreciated, together with the participation in the field surveys of the following lecturers during the past four months at different times and locations: Mr. Ramon Polycarpio, Miss Inesha Mazini, and Mr. Henry Kaomara; the students Dorothy Oshea, Osborn Rigeo, Elsie Meesa, Israel Liva, Hackman Mela, Lawrence Rodoi, Christina Kangitagi, Lynne Robo, Joseph Tavuata, Alice Samo and Sereima Paramate; and lastly our main driver Moses Davis.

KEYWORDS

- *Acacia mangium*
- *Eucalyptus sp.*
- forest inventory
- mean annual increment
- Solomon Islands
- timber volume estimation

REFERENCES

Ball, J., Pandey, D., & Hirai, S., (1999). *Global Overview of Teak Plantations.* Paper presented at Chiang Mai, Thailand, Regional seminar on site, technology and productivity of teak plantations.

Bird, N. M. *Sustaining the Yield: Improved Timber Harvesting Practices in Belize 1992–1998.* Natural Resources Institute, University of Greenwich.

Boots, B., (1999). Spatial tessellations. In: Longley, P., Goodchild, M., Maguire, D., & Rhind, D., (eds.), *Geographical Information Systems.* New York, John Wiley & Sons.

Eldridge, K., Davidson, J., Hardwood, C., & Van, W. G., (1993). *Eucalyptus Domestication and Breeding* (p. 288). Oxford Science Publications. USA.

Grogan, J., Landis, R., Ashton, M., & Galva, J., (2005). Growth response by big-leaf mahogany (*Swietenia macrophylla*) advance seedling regeneration to overhead canopy release in southeast Para, Brazil. *Forest Ecol. Manage, 204,* 399–412.

Grogan, J., Schulze, M., & Galvao, J., (2010). Survival, growth and reproduction by big-leaf mahogany (*Swietenia macrophylla*) in open clearing vs. forested conditions in Brazil. *New For., 40*(3), 335–347.

Gunn, B., & Midgley, S., (1991). Genetic resources and tree improvement: Exploring and accessing the genetic resources of four selected tropical acacias. In: Turnbull, J. W., (ed.), *Advances in Tropical Acacia Research* (pp. 57–63). ACIAR Proceedings, No. 35, Australian Centre for International Agricultural Research, Canberra, Australia.

Hay, A., Kimberley, M., & Kampfraath, B., (1999). Monthly diameter and height growth of young *eucalyptus fastigata, E. regnans,* and *E. saligna. New Zealand Journal of Forestry Science, 29*(2), 263–273.

Krisnawati, H., Kallio, M., & Kanninen, M., (2011). *Acacia Mangium Willd: Ecology, Silviculture and Productivity.* CIFOR, Bogor, Indonesia.

Lamb, F., (1966). *Mahogany of Tropical America: Its Ecology and Management* (p. 220). University of Michigan Press, Ann Arbor, MI.

Lee, K., (1969). Some soils of the British Solomon Islands protectorate. Philosophical transactions of the royal society of London. *Series B, Biological Sciences, 255,* 211–257.

Meteorological Services Division, (2019). *Website of the Solomon Islands Ministry of Environment, Climate Change, Disaster Management, and Meteorology.* www.met.gov.sb (accessed on 5 December 2020).

Miller, J., Hay, A., & Ecroyd, C., (2000). *Introduced Forest Trees in New Zealand: Recognition, Role, and Seed Source* (Vol. 124, p. 18). Ash eucalyptus, *Eucalyptus fastigata, E. regnans, E. obliqua, E. delegatensis, E. fraxinoides, E. sieberi, E. oreades, E. pauciflora, E. dendromorpha, and E. paliformis.* FRI Bulletin.

Ministry of Forests and Research, (2012). *State of the Forest Genetic Resources in the Solomon Islands,* p. 126.

National Research Council, (1983). *Mangium and other Fast-Growing Acacias for the Humid Tropics.* National Academy Press, Washington, DC.

Negreiros-Castillo, P., & Mize, C., (2006). Stand and species growth of a tropical forest in Quintana Roo, Mexico. *J. Sust. For., 23*(6), 83–96.

Orwa, C., Mutua, A., Kindt, R., Jamnadass, R., & Anthony, S., (2009). *Agroforestry Database: A Tree Reference and Selection Guide Version 4.0.* http://www.worldagroforestry.org/sites/treedbs/treedatabases.asp (accessed on 5 December 2020).

Pauku, (2009). *Asia-Pacific Forestry Sector Outlook Study II Working Paper Series.* Working Paper No. APFSOS II/WP/2009/31 Solomon Islands Forestry Outlook Study.

Sánchez, S., (1994). *Crecimiento de Eucalyptus deglupta y E. grandis bajo tres sistemas de plantación a nivel de finca en la zona de Turrialba, Costa Rica* (p. 95). Master's Thesis, CATIE.

Shono, K., & Snook, L., (2006). Growth of big-leaf mahogany (*Swietenia macrophylla*) in natural forests in Belize. *Journal of Tropical Forest Science, 18,* 61–73.

Snook, L., (2000). Regeneration and growth of mahogany (*Swietenia macrophylla* King) in the forests of Quintana Roo, Mexico. *Cien. For. Mex., 25*(87), 59 –76.

Snook, L., (2003). Regeneration, growth, and sustainability of mahogany in Mexico's Yucatan forests. In: Lugo, A., Figueroa, C. J., & Alayo, M., (eds.), *Big-Leaf Mahogany: Genetics, Ecology, and Management* (pp. 169–192). Springer, New York.

Solomon Islands Historical Encyclopedia-1893–1978 (2013) http://www.solomonencyclopaedia.net/biogs/E000134b.htm (accessed on 5 December 2020).

Solomon Islands Initial National Communications, (2015). United Nations Framework Convention on Climate Change (UNFCCC), p. 64.

Ugalde, A., (1980). Rendimiento y aprovechamiento de dos intensidades de raleos selectivos en Eucalyptus deglupta Blume, Turrialba, Costa Rica *Tesis Magister Scientiae* (p. 127). CATIE. C. R.

Vanclay, J., (1994). *Modeling Forest Growth and Yield Applications to Mixed Tropical Forests.* CAB International, Wallingford UK as ISBN: 0851989136.

Webber, R., (2011). *Solomini: Times and Tales from Solomon Islands.* Troubador Publishing Ltd, Leicester, UK.

White, K., (1991). *Teak: Some Aspects of Research and Development* (p. 17). FAO/RAPA.

CHAPTER 12

Diversity and Regeneration of Tree Species in Western Himalayas: A Case Study from Kedarnath Wildlife Sanctuary

ZUBAIR A. MALIK,[1,2] JAHANGEER A. BHAT,[3] MUDASIR YOUSSOUF,[4] and A. B. BHATT[1]

[1]Ecology Laboratory, Department of Botany and Microbiology, HNB Garhwal University Srinagar, Garhwal, Uttarakhand–246174, India

[2]Department of Biology, Government HSS, Harduturoo, Anantnag, Jammu and Kashmir–192201, India, E-mail: malikmzubair081@gmail.com

[3]Department of Forestry, College of Agriculture, Fisheries, and Forestry, Koronivia, PO Box–1544, Nausori, Fiji National University, Republic of Fiji Islands, Fax: +679 340 0275. Presently at College of Horticulture and Forestry, Rani Lakshmi Bai Central Agricultural University, Jhansi - 284003 (U.P.), India

[4]Centre for Environmental Science and Technology, Central University of Punjab, Bathinda, Punjab–151001, India

ABSTRACT

The study explored the diversity and regeneration status of tree species at different altitudes of Kedarnath Wildlife Sanctuary (KWLS). A total of 39 tree species belonging to 33 genera and 22 families were reported in the study area. Species richness (SR) exhibited direct relation with altitude. Tree density ranged from 280 ± 12 (lower altitude) to 465 ± 13 (higher altitude), while total basal cover (TBC) varied from 13.67 ± 0.94 m²/ha (lower altitude) to 31.51 ± 1.86 m²/ha (higher altitude). Shannon Wiener index of diversity ($\bar{H}$) ranged between 2.67 (lower altitude) to 3.53 (higher altitude). Seedling and sapling diversity ($\bar{H}$) varied from 2.41 to 3.32 and 2.56 to 3.59, respectively.

Due to the highest IVI, *Quercus glauca* (51), *Litsea elongata* (50.84) and *Quercus floribunda* (45.61) were the dominant tree species at lower, middle, and higher altitudes respectively. Two different types of dominance-diversity (d-d) curves were reported from the present study. The tree species of lower altitude showed lognormal curves while those of middle and higher altitudes showed geometric curves. Maximum percentage of seedlings (64.05%), saplings (38%), and trees (5.54%) were reported from lower, higher, and higher altitudes, respectively, while their minimum values were recorded for higher (56.39%), lower (32.75%), and lower (3.38%) altitudes respectively. The regeneration status of tree species in the study area was fairly high (41–56% of tree species had "good" regeneration, 7–8% fair, 24–41% poor, while 11–14% exhibited "new" regeneration). However, many important tree species exhibited discontinuous regeneration due to the absence of some of their diameter classes. Such species would be in trouble in the future and may result in the change in composition of their respective communities.

12.1 INTRODUCTION

Biodiversity is crucial for human sustenance and the normal ecosystem function and stability (Singh, 2002). Global diversity is the sum of contributions from different organizational levels, i.e., from genetic diversity to ecosystem diversity. Biological diversity is the richness and evenness (relative abundance) of species amongst and within living organisms and ecological complexes (Polyakov et al., 2008).

The Himalaya is important and one of the richest mountain ecosystems on Earth. The unique physiography, climatic conditions, and soil characteristics of the area have resulted in a variety of habitats and a significant biological diversity. The Indian Himalayan Region (IHR) although occupies a small percentage (18%) of the national geographical area but it has about 50% of national forest cover and also harbors about 40% of the Indian endemic species (Gairola et al., 2009; Malik, 2014). Western Himalaya which includes three Indian states *viz.*, Jammu Kashmir, Himachal Pradesh, and Uttarakhand, is one of the biodiversity rich hotspots of IHR. Climatic and topographic variation has resulted in the unique assemblage of floristic composition in this region, distributed from warm-moist submontane zone to the dry-cold alpine region. Usually, *Pinus* and *Quercus* species dominate the vegetation of this region. *P. roxburghii* grows at mid altitudinal zones (1000–2200 m) in the form of pure patches. In valleys and it can also be found growing with some broad-leaved species depending upon the altitude like *Lyonia ovalifolia*, *Pyrus pashia*, *Mallotus philipensis* and *Shorea robusta* at lower altitude and

Cedrus deodara, Myrica esculenta, Q. leucotrichophora, Q. glauca and *Rhododendron arboreum* at higher altitudes. Banj oak (*Q. leucotrichophora*) flourish well in northern aspects below 2000 m as pure patches or in association with other broadleaved and conifer species (Semwal et al., 2007).

Vegetation, being an important component of biodiversity, is nature's one of the precious gifts because humans have been dependent on it since the beginning for most of their needs like food, fodder, fuel, medicine, timber, etc., (Gaur, 1999). Vegetation of any area is determined by its structure and species composition. Comprehension of species diversity patterns and vegetation as a whole is considered essential for the conservation of natural areas, and these have frequently been the focus of ecological studies (Zhang et al., 2013). The species is among the prime analytical attributes of the plant community (Odum, 1959). SR is an easily explicable index of biological diversity (Peet, 1974). The evolution and composition of plant communities in a region is governed by several factors; time, altitude, slope, and aspect, soil, humidity, and precipitation. Among these factors, change in altitude corresponds with changes in both climatic and site conditions, which in turn has significant influence on plant communities (Gauthier et al., 2000; Kharkwal et al., 2005). Comprehension of forest composition is essential to describe various ecological processes and also to model the functioning and dynamics of forests (Elourard et al., 1997). In the case of forest ecosystems, the diversity of trees is fundamental to total biodiversity, as these produce resources and habitat for almost all other forest species (Huang et al., 2003). Trees form and maintain general physical structure of habitats (Jones et al., 1997) and thus define fundamentally the templates for structural complexity and environmental heterogeneity. The nature of forests depends on the ecological characteristics of sites, species diversity, and regeneration status of species (Mishra et al., 2003; Singh et al., 2016).

Regeneration of tree species is an essential process during which new seedlings and saplings get successfully established, and hence it determines the desired species composition and stocking in the future. The potential regenerative status of tree species can clearly show the possible future changes in the composition of forests within a stand in space and time (Henle et al., 2004). So understanding the activities that have an influence on the regeneration of tree species is of much importance to both ecologists and forest managers (Slik et al., 2003). The temperate forests of Himalaya are the centers of high species diversity, but a dearth of adequate regeneration is a big problem of mountain forests (Krauchii et al., 2000). Successful management and conservation of natural forest requires reliable data on aspects such as the regeneration trends (Eilu and Obua, 2005).

Kedarnath Wildlife Sanctuary (KWLS) is one of the largest protected area and biodiversity-rich sites of the Western Himalayas; but sadly, the forests of this region have been degraded during the last few years mostly because of anthropogenic perturbations (Malik et al., 2016). KWLS has received little attention as far as floristic and ecological investigations are concerned. Therefore, keeping in mind the aforesaid facts, the present study was carried out to investigate the diversity and regeneration status of tree species at different altitudes KWLS. Ecological data obtained in this regard would be useful for the application of sound management and conservation practices in the Himalayan forests.

12.2   MATERIALS AND METHODS

12.2.1   STUDY AREA

The study was carried out at different altitudes in KWLS and its adjoining areas. The KWLS (30°25′–30°41′N, 78° 55′–79°22′E), situated in the Northern Indian state of Uttarakhand (in the North-eastern part of Garhwal Himalayas) is one of the largest (975 km^2) protected areas in the Western Himalaya located in Chamoli and Rudraprayag districts (Table 12.1). The sanctuary lies in the upper catchment of the Alaknanda and Mandakini Rivers, which are the major tributaries of the Ganges (Malik et al., 2014a). Besides the grandeur of Himalayan wilderness, the area also encloses many important shrines viz., Triyuginarayan (2300 m asl), Madhmeshwar (3200 m), Rudranath (3500 m), Tungnath (3750 m) and Kedarnath (3400 m). The present study was carried out on the south-western part of KWLS and its adjoining areas in Rudraprayag district. Following the preliminary survey, three sites at three different altitudes, i.e., lower (Kund, 900–1200 m asl), middle (Phata, 1600–1900 m asl) and higher (Triyuginarayan, 2300–2600 m asl) were selected. The study area falls in three different altitudinal zones of Garhwal Himalaya viz., sub-montane, montane, and subalpine. Triyuginarayan forests at higher altitude form the core zone of KWLS; those at middle altitude (Phata) form fringe areas, while at lower altitude (Kund) come under the adjoining areas of KWLS.

Most of the valleys of the study area (KWLS and adjoining areas) are in north-south directions and thus suitable for the reception of summer monsoon. The area receives 300 cm of annual rainfall of which the rainy months (June-September) contribute approximately 60%. The relative humidity varies from 35–85% annually. Generally, the monsoon arrives after mid-June and

TABLE 12.1 Characteristics of the Study Area[*]

Study Regions (Name)	Comparative Altitude	Altitudinal Range (m asl)	Geographic Coordinates	Aspect	Slope	Anthropogenic Disturbances[#]
Kund	Lower	(900–1100 m asl)	(3030'02.91 N, 079°05'22.12 E)	NWW	30° ± 10°	LTL, LG, HSC
Phata	Middle	(1650–1750 m asl)	(30 35' 07.75 N, 079°12'26.85 N)	SW	15° ± 8°	HTL, HG, LSC, CNTFP
Triyuginarayan	Higher	(2300–2600 m asl)	(3038'04.02 N, 078°58'49.90 E)	SSW	33° ± 6°	HG, HTL, CNTFP, HSC

[*]*Adapted from:* Malik et al. (2014b).

[#]*LTL = Low tree lopping, LG = low grazing, HTL = heavy tree lopping, HG = Heavy grazing, LSC = low stem cutting, HST = heavy stem cutting, CNTFP = collection of non-timber forest products.*

continues up to the end of September. July and August are the wettest months in this region. During these months, the rainfall occurs almost every day, and the whole area is usually covered with fog reducing the visibility.

During the study period, mean annual temperature ranged from $2.50 \pm 0.53°C$ to $31.50 \pm 7.18°C$, mean annual relative humidity was recorded from $35.42 \pm 6.86\%$ to $90.43 \pm 5.66\%$, mean annual rainfall fluctuated between 125.00 ± 20.43 mm to 640.00 ± 110.43 mm (Malik, 2014). The soil types found in the region are dark brown and black forest soils and podozolic soils. Soils are generally gravelly and large boulders are common in the area. Geology plays an important role in the distribution of forest types. The main rock types identified in the research area were quartzofeldspathic gneiss, granite, and biotite schists and quartz albitite (Malik and Nautiyal, 2016).

12.2.2 METHODOLOGY

12.2.2.1 PHYTOSOCIOLOGICAL ATTRIBUTES OF WOODY SPECIES

The quantitative information regarding the distribution and abundance of tree species was collected using randomly placed sampling plots (quadrats) as per Mishra (1968). Quadrats were laid down in a spatially distributed manner so as to minimize the autocorrelation among the vegetation. The size of the square plots was 100 m² (10 m × 10 m). A total of twenty 100 m² quadrats covering an area of 2000 m² were laid to analyze the trees (≥30 cm dbh) at different altitudinal zones. The collected plants were identified with the help of taxonomists, available literature, and regional floras (Naithani, 1984, 1985; Gaur, 1999).

The tree species were analyzed for SR, density, and diversity at each altitudinal zone following standard methods (Curtis McIntosh, 1950; Phillips, 1959; Whittaker, 1972). The ratio of abundance to frequency indicated the distribution pattern. According to Curtis and Cottam (1956), the abundance to frequency ratio signifies regular distribution (below 0.025), random distribution (0.025–0.05), and contagious distribution (> 0.05). The basal area was calculated by dividing the square of CBH (circumference at breast height, measured at 1.37 m height.) with 4π. The basal cover was multiplied with respective densities of the species to obtain total basal cover (TBC) (m²/ha). IVI is a statistical quantity, which gives an overall picture of the importance of the species in the plant community. It was calculated by summing up the three relative values viz., relative frequency, relative density, and relative dominance (Phillips, 1959).

The Shannon-Wiener diversity index (Shannon and Weaver, 1963), Simpson concentration of dominance (CD) (Simpson, 1949), Pielou equitability (Pielou, 1966), Margalef (1958) and Menheink (1964) indices of SR, beta diversity (Whittaker, 1972), Sorenson index of similarity (Sorenson, 1948) and maturity index (Pichi-Sermolli, 1948) were calculated with the following formulae:

$$\overline{H} = -\sum_{i=1}^{s}\left(\frac{N_i}{N}\right) \log_2\left(\frac{N_i}{N}\right) \tag{1}$$

$$Cd = \sum_{i=1}^{s} P_i^2 \tag{2}$$

$$MI = \frac{S-1}{\ln(N)} \tag{3}$$

$$MeI = \frac{S}{\sqrt{N}} \tag{4}$$

$$\beta - Div = \frac{Sc}{S} \tag{5}$$

$$I_s = \frac{2C}{A+B} \times 100 \tag{6}$$

$$\text{Degree of Maturity} = \frac{\text{Frequency of all species in a forest (stand)}}{\text{Total number of all the species}} \tag{7}$$

where; $\overline{H}$ is the Shannon-Wiener diversity index, Cd is Simpson concentration of dominance, MI, and MeI are Margalef and Menheink indices of SR respectively, lnS is natural log of S, N is total IVI values of all species, N = total number of individuals, ni = IVI value of a species, S = total number of species in forest, Sc = the total number of species occurring in a set of samples counting each species only once whether or not it occur more than once and Sa = the average number of species per individual sample, C = the common number of species in two comparable sites while A and B are the total number of species in site A and B respectively.

Following Khan et al. (1987) and Uma Shankar (2001), regeneration status of individual tree species was ascertained on the basis of densities of seedlings and saplings as: (i) good regeneration, if seedlings > saplings > adults; (ii) fair regeneration, if seedlings > or ≤ saplings ≤ adults; (iii) poor regeneration, if the species survives only in sapling stage, but no seedlings (saplings may be <, > or = adults) (iv) no regeneration, if a species is present only in adult form and (v) new regeneration, if the species has no adults but only seedlings or saplings.

12.3 RESULTS

During the present study, a total of 39 tree species belonging to 33 genera and 22 families were reported at different altitudes. At lower altitude 12 species (belonging to 12 genera and 10 families) were reported. Two families (Euphorbiaceae and Lauraceae) were represented by two species each while the rest were monotypic. The highest density was reported for Lauraceae (75 trees/ha), followed by Euphorbiaceae (50 trees/ha), Fagaceae (40 trees/ha) and so on (Table 12.2). At middle altitude 15 tree species (belonging to 14 genera and 9 families) were reported. Lauraceae with three species was the dominant family. Four families (Juglandaceae, Myricaceae, Oleaceae, and Pinaceae) were represented by one single species while the rest were represented by two species each (Table 12.2). The highest and lowest densities were reported for Lauraceae (120 trees/ha) and Oleaceae (10 trees/ha), respectively (Table 12.2). At higher altitude, 22 tree species (belonging to 18 genera and 15 families) were reported. Ericaceae, Fagaceae, and Lauraceae were represented by three species each. Aceraceae had two species while all the rest families were represented by a single species (Table 12.2). Fagaceae had highest density (120 trees/ha) followed by Lauraceae (80 trees/ha) and so on (Table 12.2).

12.3.1 FOREST CHARACTERISTICS AND REGENERATION PARAMETERS

Number of tree families was highest (15) at higher altitude, followed by lower (10) and middle altitudes (9). The number of genera was highest (18) at higher altitude, followed by middle (14) and lower altitudes (12). SR exhibited direct relation with altitude, i.e., highest at higher altitude, followed by middle and lower altitudes (Table 12.3). Tree density ranged from 280 ± 12 (lower altitude) to 465 ± 13 (higher altitude), while TBC varied from 13.67 ± 0.94 m^2/ha (lower altitude) to 31.51 ± 1.86 m^2/ha (higher altitude). Shannon Wiener index of diversity ($\bar{H}$) ranged between 2.67 (lower altitude) to 3.53 (higher altitude) (Table 12.3).

SR of seedlings and saplings varied along the altitudinal gradient (Table 12.3). Seedling and sapling densities ranged from 4535 ± 685 to 5330

± 563 and 2685 ± 283 to 3200 ± 155, respectively (Table 12.3). Seedling and sapling diversity ($\bar{H}$) varied from 2.41 to 3.32 and 2.56 to 3.59, respectively (Table 12.3). The reported tree species exhibited different types of 'regeneration status.' 41.17–56% of tree species exhibited "good" regeneration, 7.14–8% had "fair" regeneration, 24–41.17% showed "poor" regeneration while 11.76–14.28% exhibited "new" regeneration (Figure 12.1). No tree species was encountered in "no regeneration" category (Table 12.3).

TABLE 12.2 Number and Density of Tree Species of Commonly Found Families in the Study Area

Families	Lower Altitude		Middle Altitude		Higher Altitude	
	No. of Species	Density (Ind/ha)	No. of Species	Density (Ind/ha)	No. of Species	Density (Ind/ha)
Aceraceae	–	–	–	–	2	25
Aquifoliaceae	–	–	–	–	1	30
Betulaceae	1	15	2	80	1	5
Buxaceae	–	–	–	–	1	30
Caesalpinaceae	1	20	–	–	–	–
Celastraceae	–	–	–	–	1	20
Ericaceae	–	–	2	70	3	50
Euphorbiaceae	2	50	–	–	–	–
Fabaceae	1	20	–	–	–	–
Fagaceae	1	40	2	50	3	120
Hippocastanaceae	–	–	–	–	1	15
Juglandaceae	1	15	1	20	1	10
Lauraceae	2	75	3	120	3	80
Meliaceae	1	15	–	–	–	–
Moraceae	1	15	–	–	–	–
Myricaceae	–	–	1	20	–	–
Oleaceae	1	15	1	10	–	–
Pinaceae	–	–	1	15	1	10
Rhamnaceae	–	–	–	–	1	5
Rosaceae	–	–	2	25	1	25
Symplocaceae	–	–	–	–	1	15
Taxaceae	–	–	–	–	1	25
Total	**12**	**280**	**15**	**410**	**22**	**465**

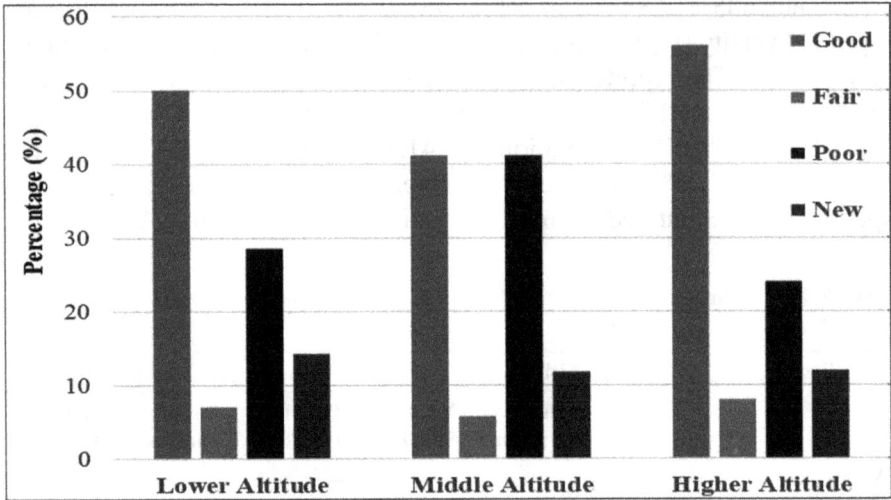

FIGURE 12.1   Regeneration status (%) of tree species along the altitudinal gradient.

TABLE 12.3   Summary of Forest Characteristics and Regeneration Parameters of the Study Area

Parameter	Lower Altitude	Middle Altitude	Higher Altitude
Number of families	10	9	15
Number of genera	12	14	18
Tree SR (species richness)	12	15	22
Tree density (Ind/ha)	280 ± 12	410 ± 20	465 ± 13
TBC (m² ha⁻¹)	13.67 ± 0.94	20.40 ± 1.24	31.51 ± 1.86
Margalef's index of SR	2.73	3.17	4.63
Shannon-Wiener diversity index ($\bar{H}$)	2.67	3.02	3.53
Maturity Index	15.41	20.66	15.68
Seedling species richness	10	10	18
Sapling species richness	14	17	25
Seedling diversity ($\bar{H}$)	2.64	2.41	3.32
Sapling diversity ($\bar{H}$)	2.56	3.15	3.59
Seedling density (Ind/ha)	5330 ± 563	4535 ± 685	4740 ± 259
Sapling density (Ind/ha)	2685 ± 283	2950 ± 279	3200 ± 155
Percentage of tree species with "good regeneration"	50%	41.17%	56%
Percentage of tree species with "fair regeneration"	7.14%	5.88%	8%
Percentage of tree species with "poor regeneration"	28.57%	41.17%	24%
Percentage of tree species with "no regeneration"	–	–	–
Percentage of tree species with "new regeneration"	14.28%	11.76%	12%

12.3.2   SPECIES RICHNESS (SR) AND DIVERSITY PATTERNS

In the present study, the SR and Shannon Wiener diversity index ($\bar{H}$) varied from 12 to 22 and 2.67 to 3.53 respectively (Table 12.3). Lowest SR (12) and lowest $\bar{H}$ (2.67) was recorded from lower altitude while their highest values were reported from higher altitude (SR = 22; $\bar{H}$ = 3.53). Middle altitude occupied the intermediate position with respect to SR and $\bar{H}$ (Table 12.3 and Figure 12.2).

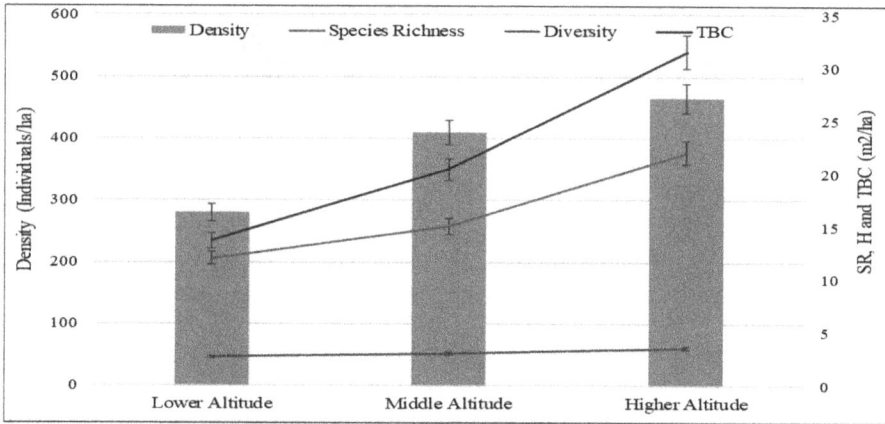

FIGURE 12.2   Variation of density, SR, diversity, and TBC along the altitudinal gradient.

12.3.3   DIVERSITY SCENARIO AT LOWER ALTITUDE

At this altitude, a total of 12 species belonging to 12 genera and 10 families were reported (Table 12.4). The highest values for density (40 trees/ha), TBC (3.18 m²/ha), IVI (51) and $\bar{H}$ (0.76) were reported for *Quercus glauca* (Table 12.4). The lowest value of TBA (0.22 m²/ha), IVI (13.91) and H (0.08) were reported for *Ficus semicardifolia, Sapium insigne* and *Ficus semicardifolia* respectively (Table 12.4). All the tree species had contagious distribution except *Neolitsea cuipala* that exhibited a random distribution pattern (Table 12.4).

12.3.4   DIVERSITY SCENARIO AT MIDDLE ALTITUDE

A total of 15 tree species belonging to 14 genera and 9 families were recorded for middle altitude. The highest values for density (70 trees/ha) and $\bar{H}$ (0.43) were reported for *Litsea elongata*, while the highest TBC (5.08 m²/ha) and IVI (50.84) were recorded for *Alnus nepalensis* (Table 12.5). The lowest value of density (5 tree/ha), TBA (0.31 m² ha⁻¹), IVI (6.66) and H̄ (0.08) were reported for *Prunus cerasoides, Fraxinus micrantha, Prunus cerasoides,* and *Pinus roxburghii,* respectively (Table 12.5). About 47% of tree species had random distribution while the rest 53% exhibited a contagious distribution pattern (Table 12.5).

TABLE 12.4 Phytosociological Attributes of Individual Tree Species of Lower Altitude

Tree Species	Family	D	TBA	IVI	H	A/F Ratio	DP
Alnus nepalensis D. Don	Betulaceae	15	0.64	15.46	0.15	0.15	C
Bauhinia variegata L.	Caesalpinaceae	20	0.50	18.94	0.26	0.08	C
Cinnamomum tamala (Buch.-Ham.) Nees and Eberm.	Lauraceae	30	1.97	33.24	0.14	0.13	C
Engelhardtia spicata Lechen ex. Blume	Juglandaceae	15	0.71	15.99	0.10	0.15	C
Ficus semicardifolia Buch.-Ham ex Sm.	Moraceae	15	0.22	15.11	0.08	0.06	C
Mallotus philippensis (Lam.) Mull.-Arg.	Euphorbiaceae	35	1.42	33.71	0.13	0.08	C
Neolitsea cuipala (D. Don) Kosterm.	Lauraceae	45	2.34	49.39	0.42	0.05	Ra
Olea glandulifera Wall. ex G. Don	Oleaceae	15	0.76	16.35	0.11	0.15	C
Ougeinia oogeinensis (Roxb.) Hoch.	Fabaceae	20	0.32	14.91	0.17	0.20	C
Quercus glauca Thunb.	Fagaceae	40	3.18	51.09	0.67	0.06	C
Sapium insigne (Royle) Benth. and Hook. f.	Euphorbiaceae	15	0.43	13.91	0.25	0.15	C
Toona hexandra M. Roemer	Meliaceae	15	1.15	21.91	0.19	0.06	C
Total		**280**	**13.67**		**2.67**		

12.3.5   *DIVERSITY SCENARIO AT HIGHER ALTITUDE*

A total of 22 tree species belonging to 18 genera and 15 families were reported from this altitude (Table 12.6). The highest values for density (55 trees/ha), TBC (6.99 m²/ha), IVI (45.61) and H̄ (0.36) were reported for *Quercus floribunda* of Fagaceae (Table 12.6). The lowest value of TBC (0.11 m²/ha) and IVI (2.87) were reported for *Rhododendron campanulatum,* while the lowest value of $\bar{H}$ (0.07) was recorded for *Rhamnus virgatus* (Table 12.6). All the tree species had contagious distribution except *Quercus floribunda* that exhibited a random distribution pattern (Table 12.6).

TABLE 12.5 Phytosociological Attributes of Individual Tree Species of Middle Altitude

Tree Species	Family	D	TBA	IVI	H	A/F Ratio	DP
Alnus nepalensis D. Don	Betulaceae	60	5.08	50.84	0.41	0.05	Ra
Betula alnoides Buch. Ham. ex. D. Don	Betulaceae	20	1.36	18.01	0.11	0.05	Ra
Cinnamomum tamala (Buch.-Ham.) Nees and Eberm.	Lauraceae	10	0.54	8.32	0.13	0.10	C
Fraxinus micrantha Lingelsh.	Oleaceae	10	0.31	7.19	0.09	0.10	C
Juglans regia L.	Juglandaceae	20	0.81	12.07	0.21	0.20	C
Lindera pulcherrima (Nees) Hook. f.	Lauraceae	40	0.95	24.09	0.39	0.04	Ra
Litsea elongata (Nees) Hook. f.	Lauraceae	70	1.89	42.46	0.43	0.03	Ra
Lyonia ovalifolia (Wall.) Drude	Ericaceae	45	1.85	31.34	0.27	0.04	Ra
Myrica esculenta Buch.-Ham. ex D. Don	Myricaceae	20	0.68	13.06	0.14	0.09	C
Pinus roxburghii Sarg.	Pinaceae	15	2.87	20.95	0.08	0.15	C
Prunus cerasoides Buch.-Ham. ex D. Don	Rosaceae	5	0.78	6.66	0.12	0.20	C
Pyrus pashia Buch.-Ham. ex. D. Don.	Rosaceae	20	0.42	11.77	0.16	0.09	C
Quercus glauca Thunb.	Fagaceae	15	0.36	8.65	0.17	0.15	C
Quercus leucotrichophora A. Camus	Fagaceae	35	1.46	25.38	0.12	0.04	Ra
Rhododendron arboreum Sm.	Ericaceae	25	1.03	19.22	0.19	0.04	Ra
Total		**410**	**20.40**		**3.02**		

12.3.6 DOMINANCE-DIVERSITY CURVE (D-D CURVE)

Dominance-diversity (d-d) curves were prepared to analyze species dominance and ascertain resource allocation among different tree species at different altitudinal zones. Two different types of d-d curves were reported from the present study. As depicted in Figure 12.3, the tree species of lower altitude showed lognormal curves. Among the three species of this altitude, *Quercus glauca* (IVI 51.09), and *Neolitsea cuipala* (IVI 49.39)

TABLE 12.6   Phytosociological Attributes of Individual Tree Species of Higher Altitude

Tree Species	Family	D	TBA	IVI	H	A/F Ratio	DP
Acer caesium Wall. ex. Brandis	Aceraceae	15	0.53	9.26	0.12	0.07	C
Acer cappadocicum Gled.	Aceraceae	10	0.29	5.96	0.09	0.10	C
Aesculus indica (Wall. ex. Cambess.) Hook.	Hippocastanaceae	15	2.69	16.10	0.13	0.07	C
Betula utilis D. Don.	Betulaceae	5	0.20	3.14	0.16	0.20	C
Buxus wallichiana Baill.	Buxaceae	30	0.77	13.23	0.15	0.13	C
Euonymus pendulus Wall. ex. Roxb.	Celastraceae	20	0.32	9.67	0.16	0.09	C
Ilex dipyrena Wall.	Aquifoliaceae	30	1.34	16.51	0.18	0.08	C
Juglanus regia L.	Juglandaceae	10	1.35	9.32	0.11	0.10	C
Lindera pulcherrima (Nees) Hook. f.	Lauraceae	30	0.60	14.14	0.26	0.08	C
Litsea elongata (Nees) Hook. f.	Lauraceae	40	1.19	19.64	0.11	0.06	C
Lyonia ovalifolia (Wall.) Drude	Ericaceae	30	1.34	16.51	0.19	0.08	C
Persea odoratissima (Nees) Kosterm.	Lauraceae	10	0.39	6.29	0.12	0.10	C
Picea smithiana (Wall.) Boiss.	Pinaceae	10	0.51	6.67	0.17	0.10	C
Pyrus pashia Buch.-Ham. ex. D. Don.	Rosaceae	25	0.48	12.69	0.13	0.06	C
Quercus floribunda Lindl. ex. A. Camus	Fagaceae	55	6.99	45.61	0.36	0.03	Ra
Quercus leucotrichophora A. Camus	Fagaceae	30	2.80	21.14	0.21	0.08	C
Quercus semicarpifolia Sm.	Fagaceae	35	6.51	32.53	0.28	0.16	C
Rhamnus virgatus Roxb.	Rhamnaceae	5	0.12	2.91	0.07	0.20	C
Rhododendron arboreum Sm.	Ericaceae	15	0.70	9.80	0.09	0.07	C
Rhododendron campanulatum D. Don	Ericaceae	5	0.11	2.87	0.14	0.20	C
Symplocos ramosissima Wall. ex. G. Don.	Symplocaceaea	15	0.53	9.26	0.11	0.07	C
Taxus baccata L.	Taxaceae	25	1.76	16.75	0.19	0.06	C
Total		**465**	**31.51**		**3.53**		

were dominating species and hence occupied the top positions of the d-d curve. Tree species of middle and higher altitudes showed geometric curves. In this case, only the highly dominating tree species covers larger space, exploit the major share of the resources, and occupies the top niche because of the highest IVI, leaving the remaining space and resources for other species for competition. In the present study, because of the highest IVI, *Alnus nepalensis* (IVI 50.84) and *Quercus floribunda* (IVI 45.61) occupied top positions and utilized major resources in middle and higher altitudinal zones, respectively.

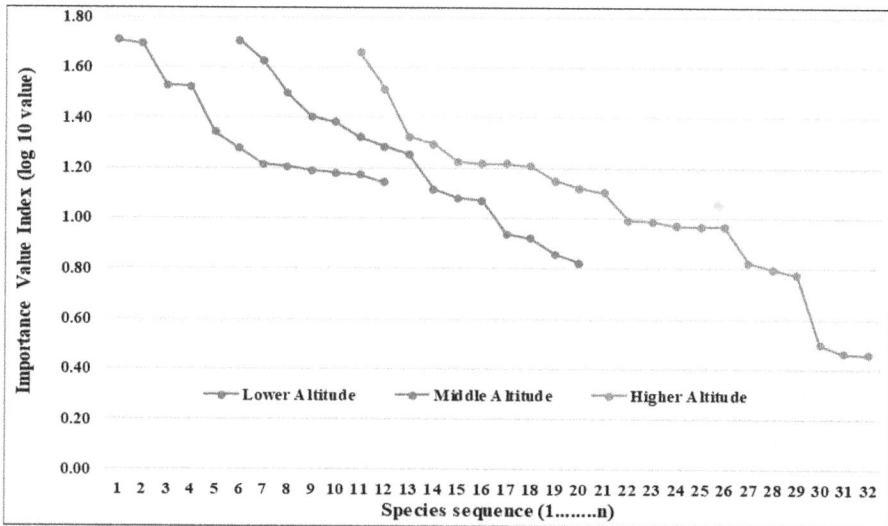

FIGURE 12.3 Dominance-diversity (d-d) curve of tree species along the altitudinal gradient.

12.3.7 DIAMETER-DENSITY DISTRIBUTION

12.3.7.1 LOWER ALTITUDE

Diameter wise stem density distribution of different species from lower altitude is represented in Table 12.7. Highest number of individuals were recorded in 0–30 cm diameter class (8015 ind/ha) that included seedlings and saplings, followed by 130 ind/ha in 31–60 cm category and so on (Table 12.7). No individual was recorded beyond 151–180 cm diameter class. *Acer oblongum* was represented by seedlings/saplings only (Table 12.7).

TABLE 12.7 Diameter Wise Stem Density (ind/ha) Distribution for Different Species of Lower Altitude

Tree Species	Different Diameter (cm) Classes with Respective Densities								
	0–30	31–60	61–90	91–120	121–150	151–180	181–210	>211	Total
Acer oblongum	50	–	–	–	–	–	–	–	50
Alnus nepalensis	465	–	15	–	–	–	–	–	480
Bauhinia variegata	110	15	5	–	–	–	–	–	130
Cinnamomum tamala	1220	5	10	15	–	–	–	–	1250
Engelhardtia spicata	150	5	5	5	–	–	–	–	165
Ficus auriculata	350	15	–	–	–	–	–	–	365
Ficus semicardifolia	735	15	–	–	–	–	–	–	750
Mallotus philippensis	355	20	10	–	5	–	–	–	390
Neolitsea cuipala	3235	10	15	5	–	5	–	–	3270
Olea glandulifera	45	–	15	–	–	–	–	–	60
Ougeinia oogeinensis	135	20	–	–	–	–	–	–	155
Quercus glauca	555	5	15	15	–	5	–	–	595
Sapium insigne	115	15	–	–	–	–	–	–	130
Toona hexandra	495	5	5	–	5	–	–	–	510
Total	**8015**	**130**	**95**	**40**	**10**	**10**	**0**	**0**	**8300**

12.3.7.2   MIDDLE ALTITUDE

Diameter wise stem density distribution of different species from middle altitude is represented in Table 12.8. Highest number of individuals were recorded in 0–30 cm diameter class (7485 ind/ha) that included seedlings and saplings, followed by 175 ind/ha in 31–60 cm category and so on (Table 12.8). No individual was recorded in 151–180 cm diameter classes. Only a few individuals of *Pinus roxburghii* and *Alnus nepalensis* were reported beyond 121–150 diameter class. *Carpinus faginea* and *Symplocos racemosa* were represented by seedlings/saplings only (Table 12.8).

TABLE 12.8 Diameter Wise Stem Distribution of Tree Species from Middle Altitude

Tree Species	Different Diameter (cm) Classes with Respective Densities								
	0–30	31–60	61–90	91–120	121–150	151–180	181–210	>211	Total
Alnus nepalensis	870	5	25	20	5	–	–	5	930
Betula alnoides	365	–	15	–	5	–	–	–	385
Carpinus faginea	175	–	–	–	–	–	–	–	175
Cinnamomum tamala	40	–	5	–	–	–	–	–	45
Fraxinus micrantha	295	5	5	–	–	–	–	–	305
Juglans regia	20	–	5	10	–	–	–	–	35
Lindera pulcherrima	500	30	10	–	–	–	–	–	540
Litsea elongata	3570	35	30	–	–	–	–	–	3635
Lyonia ovalifolia	85	20	20	5	–	–	–	–	130
Myrica esculenta	120	5	15	–	–	–	–	–	140
Pinus roxburghii	180	–	5	–	–	–	5	5	195
Prunus cerasoides	20	5	5	–	–	–	–	–	30
Pyrus pashia	115	15	5	–	–	–	–	–	135
Quercus glauca	85	10	5	–	–	–	–	–	100
Q. leucotri-chophora	195	25	10	5	–	–	–	–	235
R. arboreum	480	20	5	5	–	–	–	–	510
Symplocos racemosa	370	–	–	–	–	–	–	–	370
Total	**7485**	**175**	**165**	**45**	**10**	**0**	**5**	**10**	**7895**

12.3.7.3 HIGHER ALTITUDE

Diameter wise stem density distribution of tree species from higher altitude is given in Table 12.9. Highest number of individuals were recorded in 0–30 cm diameter class (7940 Ind/ha) that included only seedlings and saplings, followed by 250 ind/ha in 31–60 cm class and so on (Table 12.9). No individual was recorded in 181–210 cm diameter classes. The diameter class 151–180 was represented by only by a few trees of *Juglans regia*. *Abies pindrow* and *Fraxinus micrantha* were represented by seedlings/saplings only (Table 12.9).

TABLE 12.9   Diameter Wise Distribution of Stems at Higher Altitude

Tree Species	Different Diameter (cm) Classes with Respective Densities								
	0–30	31–60	61–90	91–120	121–150	151–180	181–210	>211	Total
Abies pindrow	70	–	–	–	–	–	–	–	70
Acer caesium	570	20	–	–	–	–	–	–	590
Acer cappadocicum	160	10	5	–	–	–	–	–	175
Aesculus indica	20	–	5	10	–	–	–	5	40
Betula utilis	30	10	5	–	–	–	–	–	45
Buxus wallichiana	720	40	–	–	–	–	–	–	760
Carpinus faginea	30	5	–	–	–	–	–	–	35
Euonymus pendulus	1250	20	–	–	–	–	–	–	1270
Fraxinus micrantha	50	–	–	–	–	–	–	–	50
Ilex dipyrena	290	10	10	5	–	–	–	–	315
Juglans regia	30	–	5	–	–	5	–	–	40
Lindera pulcherrima	620	20	5	–	–	–	–	–	645
Lyonia ovalifolia	150	15	–	5	–	–	–	–	170
Litsea elongata	1620	20	10	–	–	–	–	–	1650
Persea odoratissima	190	–	15	–	–	–	–	–	205
Picea smithiana	30	–	5	5	–	–	–	–	40
Pyrus pashia	55	15	–	–	–	–	–	–	70
Quercus floribunda	580	–	10	20	5	–	–	5	620
Q. leucotrichophora	235	5	15	5	5	–	–	–	265

TABLE 12.9 *(Continued)*

Tree Species	Different Diameter (cm) Classes with Respective Densities								
	0–30	31–60	61–90	91–120	121–150	151–180	181–210	>211	Total
Q. semecarpi-folia	90	5	10	10	–	–	–	10	125
Rhamnus virgatus	405	10	–	–	–	–	–	–	415
Rhododendron arboreum	135	5	5	5	–	–	–	–	150
R. campanula-tum	50	25	–	–	–	–	–	–	75
Symplocos ramosissima	210	5	5	–	–	–	–	–	220
Taxus baccata	350	10	5	–	5	–	–	5	375
Total	**7940**	**250**	**115**	**65**	**15**	**5**	**0**	**25**	**8415**

12.3.8 DIAMETER WISE TBC DISTRIBUTION

12.3.8.1 LOWER ALTITUDE

Diameter wise distribution of TBC of tree species at lower altitude is represented in Table 12.10. Maximum TBC value (4.03 m²/ha) was recorded for 61–90 cm diameter class, followed by 91–120 cm (3.05 m²/ha) and so on (Table 12.10). The least TBC was observed for 121–150 diameter class (1.67 m²/ha). The TBC of all the diameter classes including seedlings and saplings, was found to be 15.14 m²/ha. The lowest and highest TBC was recorded for *Acer oblongum* (0.004 m²/ha) and *Quercus glauca* (3.24 m²/ha), respectively (Table 12.10). *Acer oblongum* was represented by the lowest diameter class (0–30 cm) only (Table 12.10).

12.3.8.2 MIDDLE ALTITUDE

Diameter wise distribution of TBC values for different tree species of middle altitude are represented in Table 12.11. Maximum TBC value (6.20 m²/ha) was observed in 91–120 cm diameter class, followed by 61–90 cm diameter class (5.56 m²/ha) and so on (Table 12.11). The least TBC was observed for 181–210 diameter class (0.58 m²/ha). The TBC of all the diameter classes including seedlings and saplings, was found to be 22.55 m²/ha, that is higher than that of lower altitude by 7.41 m²/ha. The lowest and highest TBC was recorded for *Carpinus faginea* (0.005 m² ha⁻¹) and *Alnus nepalensis* (3.39 m²/ha), respectively (Table 12.11). *Carpinus faginea* was represented by the lowest diameter class only (Table 12.11).

TABLE 12.10   Diameter Wise Distribution of TBC (m²/ha) of Tree Species at Lower Altitude

Tree Species	Diameter Classes (cm) with Respective TBC (m²/ha)								
	0–30	31–60	61–90	91–120	121–150	151–180	181–210	>211	Total (m²/ha)
Acer oblongum	0.004	–	–	–	–	–	–	–	0.004
Alnus nepalensis	0.014	–	0.54	–	–	–	–	–	0.554
Bauhinia variegata	0.019	0.31	0.19	–	–	–	–	–	0.519
Cinnamomum tamala	0.09	0.06	0.6	0.99	–	–	–	–	1.74
Engelhardtia spicata	0.268	0.14	0.2	0.39	–	–	–	–	0.998
Ficus auriculata	0.035	–	–	–	–	–	–	–	0.035
Ficus semi-cardifolia	0.08	0.22	–	–	–	–	–	–	0.3
Mallotus philippensis	0.203	0.43	0.44	–	0.78	–	–	–	1.853
Neolitsea cuipala	0.433	0.39	0.58	0.57	–	0.98	–	–	2.95
Olea glandulifera	0.071	–	0.53	–	–	–	–	–	0.6
Ougeinia oogeinensis	0.121	0.32	–	–	–	–	–	–	0.44
Quercus glauca	0.194	0.06	0.67	1.13	–	1.22	–	–	3.244
Sapium insigne	0.110	0.37	–	–	–	–	–	–	0.48
Toona hexandra	0.113	0.14	0.287	–	0.89	–	–	–	1.43
Total	**1.75**	**2.44**	**4.037**	**3.05**	**1.67**	**2.2**	**0**	**0**	**15.147**

TABLE 12.11   Diameter Wise TBC Distribution in the Tree Species of Middle Altitude

Tree Species	Diameter Classes (cm) with Respective TBC (m²/ha)								
	0–30	31–60	61–90	91–120	121–150	151–180	181–210	>211	Total (m²/ha)
Alnus nepalensis	0.33	0.13	0.63	0.59	0.57	0.39	–	0.75	3.39
Betula alnoides	0.072	–	0.72	0.63	–	0.67	–	–	2.092
Carpinus faginea	0.005	–	–	–	–	–	–	–	0.005
Cinnamomum tamala	0.031	–	0.29	0.54	–	–	–	–	0.861
Fraxinus micrantha	0.019	0.07	0.21	0.25	–	–	–	–	0.549
Juglans regia	0.048	–	0.15	0.19	0.73	–	–	–	1.118
Lindera pulcherrima	0.28	0.56	0.36	0.4	–	–	–	–	1.6
Litsea elongata	0.62	0.65	0.76	0.81	–	–	–	–	2.84
Lyonia ovalifolia	0.049	0.57	0.69	0.73	0.35	–	–	–	2.389
Myrica esculenta	0.051	0.07	0.37	0.62	–	–	–	–	1.111
Pinus roxburghii	0.107	–	0.15	0.19	–	–	0.58	0.63	1.657
Prunus cerasoides	0.008	0.13	0.21	0.25	–	–	–	–	0.598
Pyrus pashia	0.07	0.23	0.15	0.19	–	–	–	–	0.64
Quercus glauca	0.018	0.16	0.15	0.19	–	–	–	–	0.518
Q. leucotri-chophora	0.197	0.52	0.41	0.37	0.39	–	–	–	1.887
R. arboreum	0.103	0.29	0.31	0.25	0.34	–	–	–	1.293
Symplocos racemosa	0.01	–	–	–	–	–	–	–	0.01
Total	**2.018**	**3.38**	**5.56**	**6.20**	**2.38**	**1.06**	**0.58**	**1.38**	**22.558**

12.3.8.3   HIGHER ALTITUDE

Diameter wise distribution of TBC values for different tree species of higher altitude is represented in Table 12.12. The highest TBC value (15.73 m²/ha)

was observed in >211 cm diameter class, followed by 61–90 cm diameter class (5.55 m²/ha) and so on (Table 12.12). The least TBC was observed for 121–150 and 151–180 diameter classes (1.21 m²/ha each). The TBC of all the diameter classes including seedlings and saplings was found to be 33.70 m²/ha that is higher than that of lower altitude (by 18.56 m²/ha) and middle altitude (by 11.15 m²/ha). The lowest and highest TBC was recorded for *Carpinus faginea* (0.003 m²/ha) and *Quercus floribunda* (8.92 m²/ha), respectively (Table 12.12). *Abies pindrow, Carpinus faginea,* and *Fraxinus micrantha* were represented by the lowest diameter class (0–30 cm) only (Table 12.12).

TABLE 12.12   Diameter Wise Distribution of TBC of Tree Species of Higher Altitude

Tree Species	Diameter Classes (cm) with Respective TBC (m²/ha)								
	0–30	31–60	61–90	91–120	121–150	151–180	181–210	>211	Total (m²/ha)
Abies pindrow	0.08	–	–	–	–	–	–	–	0.081
Acer caesium	0.1	0.39	–	–	–	–	–	–	0.491
Acer cappadocicum	0.1	0.15	0.11	–	–	–	–	–	0.361
Aesculus indica	0.07	–	0.26	0.81	–	–	–	3.02	4.160
Betula utilis	0.02	0.17	0.24	–	–	–	–	–	0.430
Buxus wallichiana	0.21	0.48	–	–	–	–	–	–	0.690
Carpinus faginea	0.003	–	–	–	–	–	–	–	0.003
Euonymous pendulus	0.19	0.23	–	–	–	–	–	–	0.421
Fraxinus micrantha	0.01	–	–	–	–	–	–	–	0.011
Ilex dipyrena	0.12	0.07	0.34	0.28	–	–	–	–	0.810
Juglans regia	0.01	–	0.2	–	–	1.21	–	–	1.420
Lindera pulcherrima	0.46	0.34	0.12	–	–	–	–	–	0.920
Lyonia ovalifolia	0.05	0.15	–	0.28	–	–	–	–	0.480
Litsea elongate	0.36	0.23	0.58	–	–	–	–	–	1.170
Persea odoratissima	0.02	–	0.65	–	–	–	–	–	0.670

TABLE 12.12   *(Continued)*

Tree Species	Diameter Classes (cm) with Respective TBC (m²/ha)								
	0–30	31–60	61–90	91–120	121–150	151–180	181–210	>211	Total (m²/ha)
Picea smithiana	0.03	–	0.21	0.37	–	–	–	–	0.610
Pyrus pashia	0.03	0.17	–	–	–	–	–	–	0.201
Quercus floribunda	0.23	–	0.75	1.68	0.41	–	–	5.85	8.921
Q. leucotricho-phora	0.09	0.05	0.72	0.37	0.47	–	–	–	1.730
Q. semecarpi-folia	0.05	0.05	0.49	0.72	–	–	–	5.78	7.090
Rhamnus virgatus	0.01	0.01	–	–	–	–	–	–	0.020
Rhododendron arboreum	0.02	0.06	0.21	0.28	–	–	–	–	0.570
R. campanu-latum	0.009	0.24	–	–	–	–	–	–	0.249
Symplocos ramosissima	0.008	0.05	0.17	–	–	–	–	–	0.228
Taxus baccata	0.028	0.15	0.5	–	0.33	–	–	1.08	2.088
Total	**2.308**	**2.99**	**5.55**	**4.79**	**1.21**	**1.21**	**0**	**15.73**	**33.708**

12.3.9   REGENERATION STATUS

Regeneration status of tree species at the three altitudes is illustrated in Figure 12.4.

12.3.9.1   REGENERATION STATUS AT LOWER ALTITUDE

At lower altitude, seedling density per hectare was highest (5330), followed by saplings (2685) and trees (280) (Figure 12.5(A)). As far as the regeneration status is concerned, maximum (50%) species showed good regeneration, 7.14% fair, 28.57% poor and 14.28% new regeneration in this forest. No tree species was encountered in the "No Regeneration" category (Figure 12.5(B)). Regeneration status of individual tree species at lower altitude is given in Table 12.13.

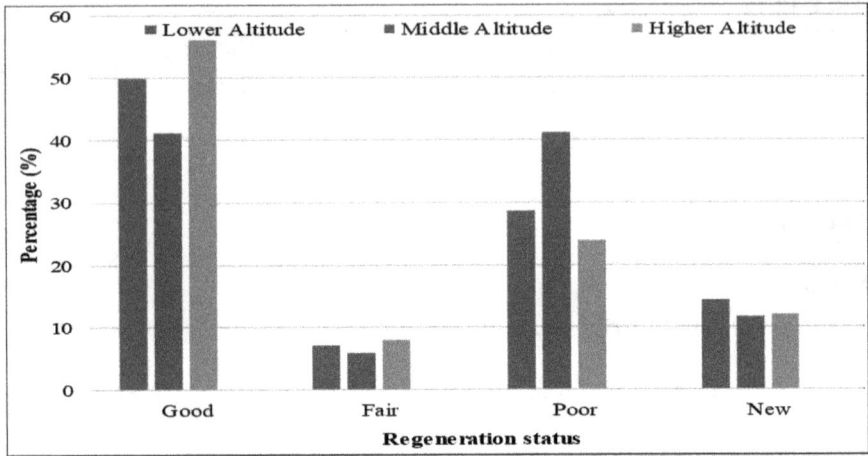

FIGURE 12.4 Overall regeneration status of all the forests in the study area.

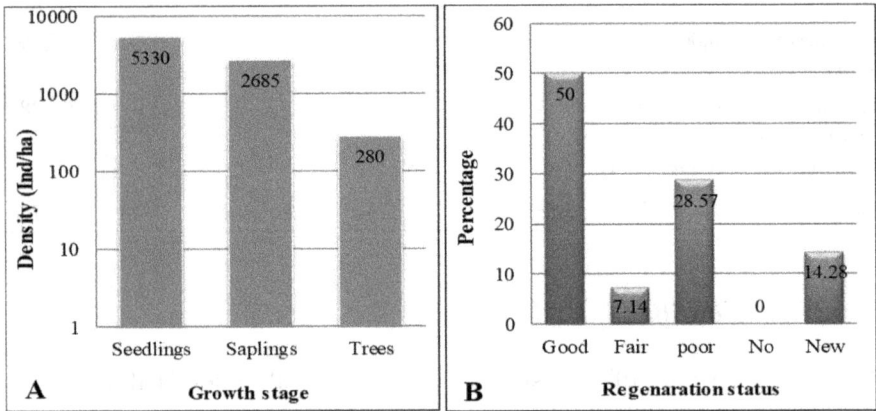

FIGURE 12.5 Regeneration pattern (A) and regeneration status (B) of tree species at lower altitude.

12.3.9.2 REGENERATION STATUS AT MIDDLE ALTITUDE

At mid-altitude, seedling density (per hectare) was highest (4535), followed by saplings (2950) and trees (410) as illustrated in Figure 12.6(A). Maximum species showed either good or poor regeneration (41.17% each), 5.88% fair and 11.76% new regeneration. No tree species was recorded in the "No Regeneration" Category (Figure 12.6(B)). Regeneration status of individual tree species of the mid-altitude is given in Table 12.14.

TABLE 12.13 Regeneration Status of Individual Tree Species of Lower Altitude

Lower Altitude	Density (Ind/ha)			
Species	Trees	Saplings	Seedlings	Regeneration Status
Acer oblongum	0	25	25	New
Alnus nepalensis	15	165	300	Good
Bauhinia variegata	20	65	45	Fair
Cinnamomum tamala	30	270	950	Good
Engelhardtia spicata	15	150	0	Poor
Ficus auriculata	0	80	270	New
Ficus semicardifolia	15	185	550	Good
Mallotus philippensis	35	120	235	Good
Neolitsea cuipala	45	1150	2085	Good
Olea glandulifera	15	135	0	Poor
Ougeinia oogeinensis	20	95	0	Poor
Quercus glauca	40	115	460	Good
Sapium insigne	15	85	0	Poor
Toona hexandra	15	45	410	Good

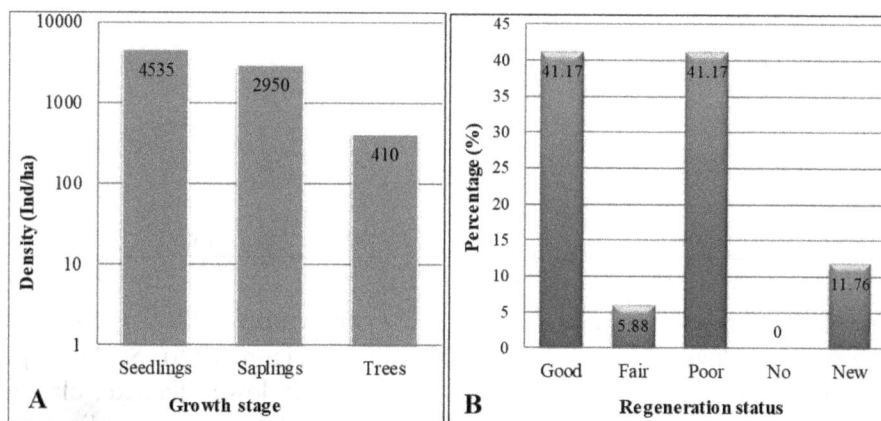

FIGURE 12.6 Regeneration pattern (A) and regeneration status (B) of tree species at mid-altitude.

12.3.9.3 REGENERATION STATUS AT HIGHER ALTITUDE

At higher altitude, seedling density (per hectare) was highest (4740), followed by saplings (3200) and trees (465) as illustrated in Figure 12.7(A).

Maximum species (56%) showed good regeneration, 8% fair, 24% poor and 12% new regeneration. No tree species was recorded in the "No Regeneration" category (Figure 12.7(B)). Regeneration status of individual tree species of higher altitude is worked out in Table 12.15.

TABLE 12.14 Regeneration Status of Individual Tree Species at Middle Altitude

Middle Altitude	Density (Ind/ha)			
Species	Trees	Saplings	Seedlings	Regeneration Status
Alnus nepalensis	60	270	600	Good
Betula alnoides	20	130	235	Good
Carpinus faginea	0	60	115	New
Cinnamomum tamala	10	40	0	Poor
Fraxinus micrantha	10	85	210	Good
Juglans regia	20	20	0	Poor
Lindera pulcherrima	40	260	240	Fair
Litsea elongata	70	1210	2360	Good
Lyonia ovalifolia	50	85	0	Poor
Myrica esculenta	20	50	70	Good
Pinus roxburghii	15	45	135	Good
Prunus cerasoides	10	20	0	Poor
Pyrus pashia	20	115	0	Poor
Quercus glauca	15	85	0	Poor
Q. leucotrichophora	40	195	0	Poor
R. arboreum	30	210	270	Good
Symplocos racemosa	0	70	300	New

All the forests, irrespective of altitude, exhibited Inverse-J (i-J) curve on the basis of diameter wise density distribution (Figure 12.8). Such a type of population curve is obtained when the density in lower diameter classes is highest that declines gradually or abruptly with increase in diameter class which is further due to high mortality of young seedlings, saplings, and juvenile trees in the lower diameter classes. As depicted in Figure 12.8, the number of individuals declined abruptly with the increase of diameter class. It indicates continuous regeneration. On the basis of individual-level also, maximum species showed i-J curve although there were some species that showed sporadic (S) curves. The sporadic type of population structure indicates that the adjacent classes were poorly represented; some diameter classes are not represented at all, and frequency rises more or less sharply in intermediate classes.

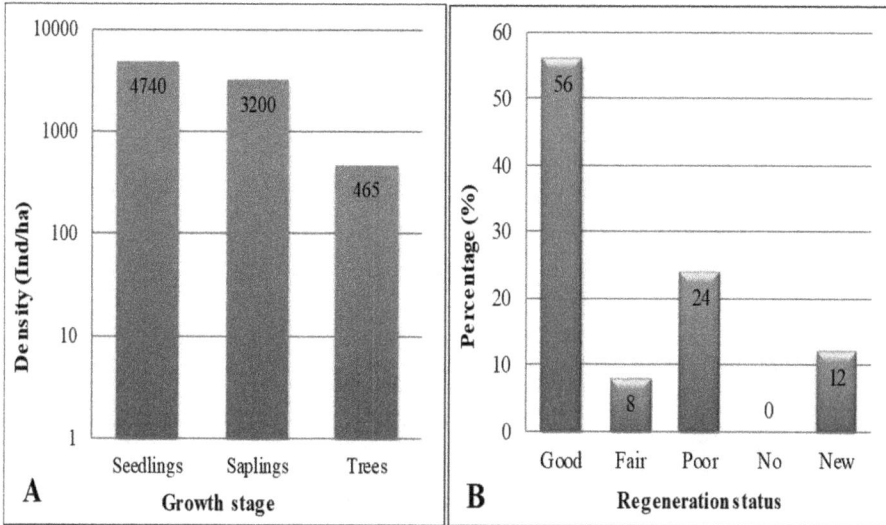

FIGURE 12.7 Regeneration pattern (A) and regeneration status (B) of tree species at high altitude.

TABLE 12.15 Regeneration Status of Individual Tree Species at Higher Altitude

Higher Altitude	Density (Ind/ha)			
Species	Trees	Saplings	Seedlings	Regeneration Status
Abies pindrow	0	35	35	New
Acer caesium	20	170	400	Good
Acer cappadocicum	15	50	110	Good
Aesculus indica	20	20	0	Poor
Betula utilis	15	30	0	Poor
Buxus wallichiana	40	500	220	Fair
Carpinus faginea	0	30	0	New
Euonymous pendulus	20	450	800	Good
Fraxinus micrantha	0	20	30	New
Ilex dipyrena	25	90	200	Good
Juglans regia	10	30	0	Poor
Lindera pulcherrima	20	300	320	Good
Lyonia ovalifolia	20	50	100	Good
Litsea elongata	30	550	1070	Good

TABLE 12.15 *(Continued)*

Higher Altitude			Density (Ind/ha)	
Species	Trees	Saplings	Seedlings	Regeneration Status
Persea odoratissima	15	75	115	Good
Picea smithiana	10	30	0	Poor
Pyrus pashia	15	55	0	Poor
Quercus floribunda	40	180	400	Good
Q. leucotrichophora	25	85	150	Good
Q. semecarpifolia	35	50	40	Fair
Rhamnus virgatus	10	155	250	Good
Rhododendron arboreum	15	35	100	Good
R. campanulatum	15	50	0	Poor
S. ramosissima	10	60	150	Good
Taxus baccata	25	100	250	Good

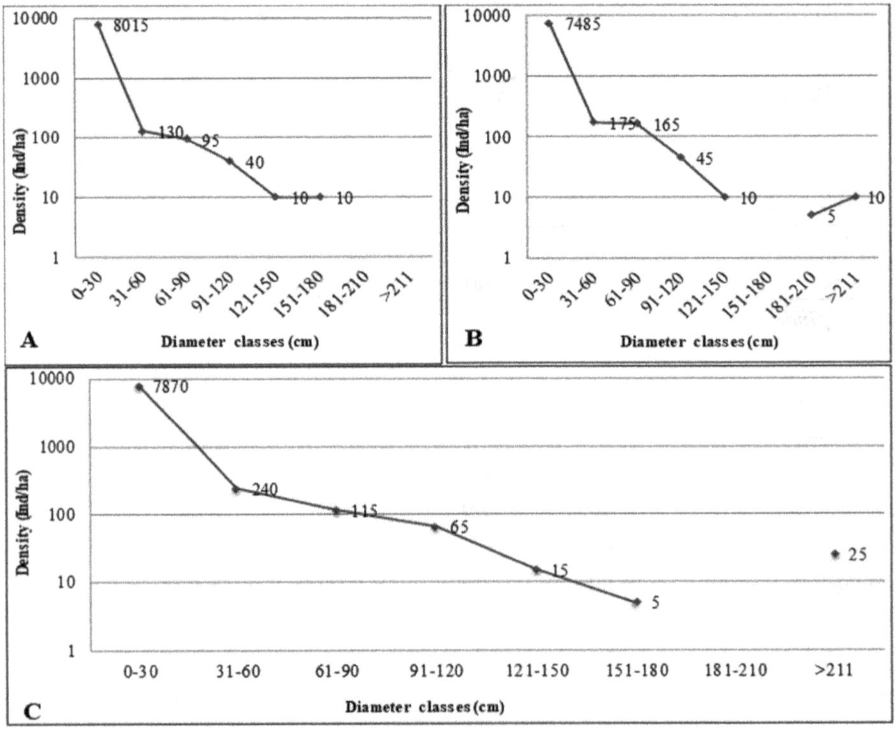

FIGURE 12.8 Population structure based on the diameter at breast height (1.37 m) class distribution of the tree species at different altitudes (A-lower, B-middle, and C-higher altitude) in the study area.

Overall seedling density ranged between 4535 ind/ha (middle altitude) and 5330 ind/ha (lower altitude), whereas sapling density varied between 2685 ind/ha (lower altitude) and 3200 ind/ha (higher altitude). Maximum percentage of seedlings (64.05%) was recorded from lower altitude and minimum (56.39%) at higher altitude (Figure 12.9). Highest percentage of saplings (38%) was recorded at higher altitude and lowest (32.57%) from lower altitude. Maximum percentage of trees (5.54%) was recorded at higher altitude and minimum (3.38%) at lower altitude (Figure 12.9).

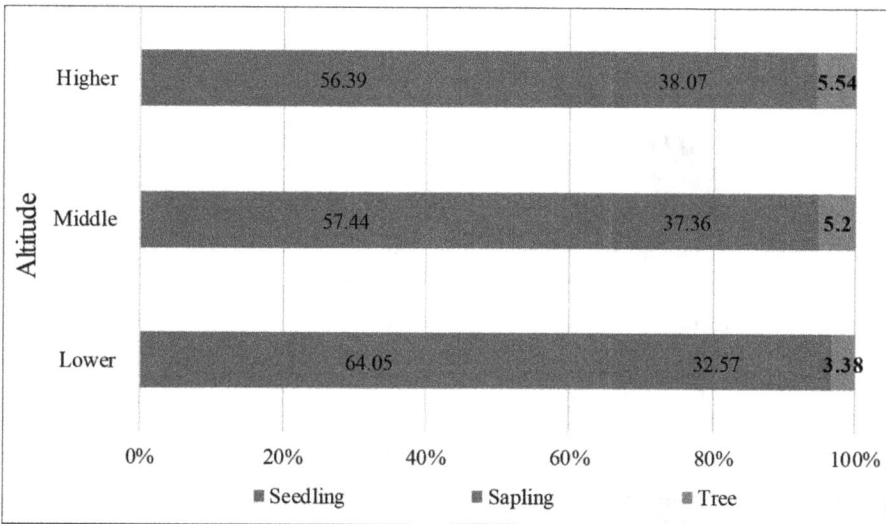

FIGURE 12.9   Percentage of seedling, sapling, and trees in different forests.

12.4   DISCUSSION

The values obtained during the current investigation for different phyto-sociological and diversity indices are well within those reported earlier from various parts of Western Himalaya like Kumaon, Garhwal, Himachal Pradesh, Nepal, and Pakistan.

12.4.1   SPECIES RICHNESS (SR)

Communities under different environmental conditions differ in the number of species they contain. Margalef's index of SR for trees ranged from

2.37 (lower altitude) to 4.63 (higher altitude). Earlier some authors have reported similar values of SR like Uniyal et al. (2010) (2.21–7.00); Gairola et al. (2011) (1.36–2.17); Bhat (2012) (1.17–3.43). Along the altitude, the geographic and climatic conditions change sharply (Kharkwal et al., 2005). The maximum SR at higher altitude may be due to the presence of favorable conditions like high moisture, humidity, rainfall, etc. The SR increases with increase in altitude till a considerable altitudinal level is reached, and tree richness increases with increasing moisture in the Indian Himalaya (Rikhari et al., 1989). According to Champion and Seth (1968), more than 60% of plant species are generally present at middle altitudes, where the temperature covers a range from 10°C to 24 °C.

12.4.2 DIVERSITY

Shannon-Wiener index (H) of tree diversity, which is the most popular measures of species diversity, was recorded highest (3.53) for higher altitude and the lowest (2.67) for lower altitude. The difference in species diversity at different altitudes generally results from variation in site quality (Denslow, 1980). The higher diversity at higher altitude may be due to the interaction of different species on these sites. A higher number of species with overlapping niches may occur in the same habitat leading to higher diversity and higher stability (Kharkwal, 2009). Tree species diversity exhibits great variation from region to region largely due to variation in biogeography, habitat, and intensity of disturbance (Hubbell et al., 1999; Sagar et al., 2003), which are important factors for structuring the forest communities. Diversity is likely to increase as the environment becomes favorable and predictable (Putman, 1994).

The values of species diversity are comparable to those reported for other temperate forests by Ghildiyal et al. (1998) [1.86–2.73], Uniyal et al. (2010) [0.70–3.08], Gairola et al. (2011) [2.43–3.33], Raturi (2012) [0.78–3.45] and Bhat (2012) [2.09–3.37] from different parts of Western Himalaya. The values are also comparable to the values reported from Nepal and Pakistan Himalayas. For instance, Kunwar and Sharma (2004) reported Shannon-Wiener Index value for trees ranging from 2.36–3.02 while Koirala (2004) reported a value of 2.40–2.61 from Nepal Himalaya. Khan et al. (2012) studied temperate forests of Narran Valley, Western Himalaya Pakistan and found $\bar{H}$ values fluctuating between 3.30–4.00, while Shaheen et al. (2012) reported species diversity between 0.75–2.27 from moist temperate forests of Pakistan Himalaya. Pant and Samant (2007) reported tree diversity from

0.99 to 2.93 in Mornaula Reserve Forests of Central Himalaya. Pande et al. (2001) computed diversity values ranging from 1.80 to 2.33. The values rehearsed in the present study are however, higher than those reported by Srivastava et al. (2005) (1.33–2.01), Kumar and Ram (2005) (1.2), Sanjeev et al. (2006) (0.65–1.71), and Rawat and Rawat (2010) (0.98–1.88) from different areas of Uttarakhand Himalaya. The reason behind the higher diversity of trees may be that the present investigation was carried out in a protected area.

12.4.3 BASAL COVER

The TBC ranged from 13.67 m²/ha (lower altitude) to 31.51 m²/ha (higher altitude). These values are best fitted within the values reported earlier for temperate forests. The highest TBC (31.51 m²/ha) was recorded from the higher altitude, which is similar to values recorded by Saxena and Singh (1982) and Singh and Singh (1986) for Kumaon Himalaya, Ghildiyal et al. (1998), Gairola et al. (2011), Raturi (2012), Bhat (2012); Singh (2013), etc., for Garhwal Himalaya, Koirala (2004) for Nepal Himalaya and Shaheen et al. (2012) for Pakistan Himalaya. On the other hand, lowest TBC value (13.67 m²/ha) was recorded from lower altitude, which is similar to the values reported by Kusumlata and Bisht (1991), Sharma et al. (2001), Rawat and Rawat (2010) for Garhwal Himalaya, Pant and Samant (2012) for N-W (Himachal) Himalaya. The value (13.67 m²/ha) is however lower than the values recorded by Singh et al. (1994) for Kumaun Himalaya, Pande et al. (2001, 2002) for Garhwal Himalaya and Kunwar and Sharma (2004) for Nepal Himalaya. It is higher than the values reported by Khera et al. (2001), Dhaulkhandi et al. (2008), Raturi (2012), and Bhat (2012). As vegetation matures, total stand density tends to decrease, and the stand increases in height, basal area, and volume (Kunwar and Sharma, 2004). The possible reason for a higher value of TBC at higher altitude could be that the forests of this altitude constitute the buffer zone of KWLS and hence are protected and mature, which appears to have attained their highest limit of productivity while the forest of lower altitude forms the adjoining areas of KWLS that is unprotected from anthropogenic disturbances. Saxena et al. (1978) had earlier stated that the trees having higher TBC signify their best performance in the specific environmental conditions while their lower TBC indicate either the chance occurrence of the species or presence of the anthropogenic disturbances in the past.

12.4.4   *MATURITY INDEX*

Maturity index varied from 15.41(lower altitude) to 20.66 (middle altitude) for trees. The maturity index is an important tool for the representation of quantitative and qualitative characteristics of a community. It helps in assessing the overall biodiversity status as well as the conservation of intact habitat and plant life in a specific area. Shaheen et al. (2012) reported values of maturity index between 38 and 53 from Western Himalayan moist temperate forests of Pakistan. The lower values of maturity index in the present study may be different forms of natural (landslides, cloudburst) and anthropogenic disturbances (lopping, grazing, and fuelwood collection by local people) in these forests. Mature communities consist of a few species that are well acclimatized and consistently distributed, occupy maximum space, and out-compete the sporadic flora. In well-established and mature communities, species manage to acclimatize with the local conditions and obtaining an equilibrium state with other species (Nautiyal and Kaechele, 2007). Reduced values of maturity index means the communities are heterogeneous, which is further due to the low acclimatization to the local ecological conditions. The higher levels of perturbations frequently disturb the natural equilibrium of forest communities, thus hindering them to attain a climax stage of community maturity (Saxena and Singh, 1984). Moreover, most of the tree species showed contagious distribution in the study area, and according to Whitford (1949), in pioneer communities' plants are likely to be aggregated, but as the community advances towards climax, their dispersal become more random or even regular. The reduced values of maturity index and contiguous dispersal of species reveal the early successional status of the studied forests.

The regeneration of a forest is the key activity during which aged trees perish and are replaced by juveniles (Malik and Bhatt, 2016). The potential regenerative status of tree species constituting the forest stand is the single factor on which the forest wealth and health depends (Jones et al., 1994). Knowing the regeneration potential/status and understanding of the various factors, including disturbances that influence the regeneration of forests is of key significance for their maintenance. In the current study, an effort was made to investigate the tree regeneration status at three different altitudes in KWLS and its adjoining areas. The overall regeneration status was fairly high in the study area (Figure 12.10). Since it is a reserve forest, there is a curb to some level on anthropogenic activities like forest clearing, collection of firewood and litter, etc. This eventually promoted the regeneration of most tree species. Besides, sufficient precipitation, adequate temperature, and vast differences in altitude and soil attributes provided a beneficial environment for the vigorous growth of many tree species.

FIGURE 12.10 Seedlings of (A) *Quercus semecarpifolia*; (B) *Ilex dipyrena*; (C) *Rhododendron arboreum*; (D) *Quercus floribunda*.

Source: Photographs by Zubair A. Malik.

The survival and/or mortality of seedlings over relatively shorter time periods may give essential indication regarding the future composition of a forest stand (Deb and Sundriyal, 2008); and results of the present study indicated important differences among seedling groups. Of the total number of individuals (seedling + sapling + tree), 64.05%, 32.57% and 3.38% individuals at lower altitude; 57.44%, 37.36% and 5.20% individuals at middle altitude and 56.39%, 38.07% and 5.54% individuals at higher altitude were recorded in respective stages. The current study reported a net loss of

18–31% seedlings during the developmental stage from seedling to sapling, and a further loss of 29–32% individuals in developmental stage from sapling to tree stage. The Inverse-J diameter distribution of stem density, as shown by maximum tree species (Figure 12.8), indicate that a small percentage of the seedlings and saplings survived to the larger tree classes. A general reason for seedling mortality could be infections caused by microbes and small beetles, etc. Increased mortality is generally ascribed to gap size and light accessibility, physical injury, soil desiccation and herbivory (Nunez-Farfan and Dirzo, 1989). Seedling survival is further significantly influenced by the surrounding environment, biotic, and abiotic factors (Harper, 1977). Higher seedling mortality during the winter season could be ascribed to the low soil moisture and frost conditions. The significance of soil moisture in affecting the seedling survival and growth is well known (Mcleod and Murphy, 1977; Mueller-Dombois et al., 1980; Schulte and Marshall, 1983; Deb and Sundriyal, 2008). The tree seedling recruitment is frequently restricted by reduced and uncertain seed supply and establishment. It is further limited by the absence of favorable microsites and factors, which influence initial seedling growth and mortality (Clark et al., 1999).

The reproductive status of a population is regulated by the relative proportions of different age groups (seedlings, saplings, and trees). In the current investigation, seedling density ranged from 4535 ± 685 Ind/ha (middle altitude) to 5330 ± 563 Ind/ha (lower altitude) while sapling density varied from a minimum of 2685 ± 283 Ind/ha (lower altitude) to a maximum of 3200 ± 155 Ind/ha (higher altitude). These figures are nearly similar to those reported earlier by various authors from different areas of Garhwal (Uttarakhand) Himalayas. Dhaulkhandi et al. (2008) studied the regeneration potential of natural forest site in Gangotri (Uttarakhand) and recorded the seedling and sapling density respectively as 5200 and 1880 Ind/ha, respectively. While investigating the regeneration status of tree species in different temperate forests of Garhwal Himalaya, Gairola (2010) reported seedling density of 600 to 30,000 Ind/ha and sapling density 96 to 9792 Ind/ha. Pokhriyal et al. (2010) reported seedling density ranging between 2329 and 3320 Ind/ha; while sapling density varied from 370 to 701 Ind/ha in forests of Phakot and Pathri Rao watersheds in Garhwal Himalaya. Ballabha et al. (2013) reported that the seedling density ranging from 520–1240 Ind/ha whereas sapling density ranges from 400–800 ind/ha for sub-tropical forest in Alaknanda Valley of Garhwal Himalaya. Bhat et al. (2015) reported seedling density 155 Ind/ha to 695 Ind/h and sapling density ranged from 160 Ind/ha to 330 Ind/ha for the higher altitudes of Garhwal Himalaya. Singh et al. (2016) reported seedling and sapling densities from 1367–9600 and 167–1296 ind/ha respectively for Oak dominated forests of Garhwal Himalaya, and finally, Haq et al. (2019) reported density of seedling and sapling as 1200–3200 and 800–2300 ind/ha respectively from Kashmir Himalaya. The variation in seedling/sapling density among the forests and/or species may be due to change in climatic conditions along the altitudinal gradient (low temperature, rainfall, and high humidity) that restrict the distribution of some species as well as germination and establishment of seedlings (Vera, 1997; Bharali et al., 2012).

The population structure of the forests is determined by the density-diameter (d-d) distribution of stems (UNESCO/UNEP/FAO, 1978; Khan et al., 1987). The population structures characterized by the presence of sufficient seedlings, saplings, and adults, indicate a successful regeneration. The presence of seedlings and saplings under the canopies of adult trees also indicates the future composition of a community (Saxena and Singh, 1984). In the current investigation, an inverse relation was reported between the density of individuals and diameter classes. Most of the tree species had the highest number of individuals in the lower girth classes that reduced progressively as the girth classes increased. Variations in the density at different altitudes can be due to many factors, including the existing environmental conditions and level of disturbance. The species exhibiting 'inverse J-shaped' distribution are supposed to be the dominant in a stand. The 'inverse J-shaped' distribution depicts higher density of individuals in the lower diameter classes that decreases progressively as the diameter class increases. Reverse J type of distribution is an indication of good regeneration status (Vetaas, 2000; Tesfaye et al., 2010). The 'J' and 'Inverse-J' distributions determine the pioneer and successional status of the species, respectively. If the tree individuals are dominant in lower and middle diameter classes, it means that the forest is still evolving (Campbell et al., 1992). The 'i-J' diameter-distribution curve depicts that a low fraction of seedling and saplings persisted up to the adult tree classes. The population structure of any tree species in any forest stand is supposed to be on the verge of extinction if it has a high density of higher diameter classes as compared to lower diameter classes (Benton and Werner, 1976). If the population of a tree species has adequate seedlings, saplings, young poles and adult/mature trees, it indicates successful regeneration (Khan et al., 1987). Further, the regeneration of a tree species can be continuous (or discontinuous) depending upon whether the tree species is represented by all the diameter classes (or some of them are missing). According to Khan et al. (1987), in a natural forest, the dominant tree species usually exhibit continuous regeneration (i.e., represented by all diameter classes). In the present study, we encountered a few tree species at each altitude that had discontinuous regeneration (Table 12.16). Although dominant at present, these tree species may be in trouble in the future (Malik and Bhatt, 2016). The cause for discontinuous regeneration and deviation from 'Inverse-J' curve can be many, including most importantly, the anthropogenic disturbances (and/or sometimes-natural calamities). At mid-altitude forest, the main reasons for the absence of various diameter classes are the power project tunnels that are dug beneath these forest stands due to which trees do not have sufficient support to bear heavy rains or winds and thus get uprooted easily. The other reason for their absence at any altitude is the cutting of trees for domestic uses, especially fuelwood and fodder. At higher altitude, heavy snowfall during the winter season is also responsible for uprooting and breakage of trees. *Bauhinia variegata, Quercus floribunda, Q. leucotrichophora, Q. semecarpifolia*, etc., are frequently used as fodder as well as fuel and hence are under tremendous pressure throughout the area. *Aesculus indica, Mallotus philippensis, Neolitsea cuipala, Olea glandulifera, Quercus glauca, Toona hexandra*, etc., are cut or lopped for fuel. *Fraxinus micrantha, Neolitsea cuipala, Quercus glauca, Quercus floribunda, Q. leucotrichophora, Q. semecarpifolia, Lyonia ovalifolia, Juglans regia*, etc., are regularly exploited for making agricultural implements. *Betula alnoides, Persea odoratissima, Q. semecarpifolia, Taxus baccata, Fraxinus micrantha, Picea smithiana* and *Pinus roxburghii*, etc., are exploited as timber for construction purposes. Illegal extraction of bark, leaves, and roots of *Taxus baccata* is one of the main reasons for its discontinuous regeneration. In nature, this species (*T. baccata*) has lower regeneration rates (Gairola et al., 2012).

According to Ghimire et al. (2010), the deviation from 'Inverse-J' shaped distribution to any other type (unimodal, sporadic, etc.), indicates that the forest is in distress because of significant variations in the state and pattern of forest regeneration. A perturbed forest that has impeded regeneration usually exhibits bell-shaped class distribution (Saxena et al., 1984). During the current study, we found many tree species showing 'Inverse-J' shaped population structure (i.e., good regeneration) that had sufficient small tree individuals, a significant number of medium-sized individuals, and few large tree individuals. Some tree species had poor regeneration, while the others were found in seedling and sapling stages only, which seem to be the new intruders in the studied forests and may form sub-canopy in the future (Malik and Bhatt, 2016).

TABLE 12.16 List of Tree Species Showing Discontinuous Regeneration at Different Altitudes Due Absence of the Respective Diameter Class(es)

Tree Species/Altitude	
Lower Altitude	**Absent Diameter Classes (cm)**
Alnus nepalensis	31–60
Mallotus philippensis	91–120
Neolitsea cuipala	121–150
Olea glandulifera	31–60
Quercus glauca	121–150
Toona hexandra	91–120
Middle Altitude	
Alnus nepalensis	151–180, 181–210
Betula alnoides	31–60, 91–120
Cinnamomum tamala	31–60
Juglans regia	31–60
Pinus roxburghii	31–60, 91–120, 121–150, 151–180
Higher Altitude	
Aesculus indica	31–60, 121–150, 151–180, 181–210
Juglans regia	31–60, 91–120, 121–150
Lyonia ovalifolia	61–90
Persea odoratissima	31–60
Picea smithiana	31–60
Quercus floribunda	31–60, 150–180, 181–210
Quercus semecarpifolia	121–150, 151–180, 181–210
Taxus baccata	91–120, 151–180, 181–210

12.5 CONCLUSION

The study area possesses a rich diversity of woody species. However, it (diversity) varied at different altitudes. The forests of lower altitudes were recorded to be less diverse than those of mid and higher altitudes. Density and TBC (m²/ha) also followed the same trend. The unprotected status of these forests seems to be the only reason, due to which a number of anthropogenic disturbances are frequent that have resulted in the degradation of these forests. Hence the importance of the establishment of protected areas for the conservation of the whole biodiversity is revealed. However, the overall regeneration status of tree species in the study area was fairly high. However, many important tree species exhibited discontinuous regeneration due to the absence of some of their diameter classes. These species will be in trouble in the future and may result in a change in the composition of their respective communities in the future. On the other hand, in the forests where dominant species showed excellent regeneration, there is no possibility of major changes in their composition in the near future.

KEYWORDS

- **population structure**
- **regeneration**
- **saplings**
- **seedlings**
- **species richness**
- **total basal cover**
- **western Himalaya**

REFERENCES

Ballabha, R., Tiwari, J. K., & Tiwari, P., (2013). Regeneration of tree species in the sub-tropical forest of Alaknanda Valley, Garhwal Himalaya, India. *For. Sci. Pract., 15*(2), 89–97.

Benton, A. H., & Werner, W. E., (1976). *Field Biology and Ecology*. McGraw-Hill, New York.

Bharali, S., Paul, A., Khan, M. L., & Singha, L. B., (2012). Impact of altitude on population structure and regeneration status of two *Rhododendron* species in a temperate broad-leaved forest of Arunachal Pradesh, India. *Int. J. Ecosys., 2*(1), 19–27.

Bhat, J. A., (2012). *Diversity of Flora Along an Altitudinal Gradient in Kedarnath Wildlife Sanctuary.* PhD thesis, HNB Garhwal University Srinagar, Garhwal.

Bhat, J. A., Negi, A. K., & Todaria, N. P., (2015). Regeneration status of woody species in a protected area of Western Himalaya. *Acta Ecolog. Sini., 35*(3), 51–58.

Campbell, D. G., Stone, J. L., & Rosas, Jr. A., (1992). A comparison of the phytosociology and dynamics of three floodplain (Varzea) forests of known ages, Rio Jurua, western Brazilian Amazon. *Bot. J. Linn. Soc., 108*(3), 213–237.

Clark, J. S., Beckage, B., Camill, P., Cleveland, B., Hillerislambers, J., Lichter, J., Mclachlan, J., et al., (1999). Interpreting recruitment limitation in forests. *Am. J. Bot., 86*(1), 1–16.

Curtis, J. T., & Cottam, G., (1956). *Plant Ecology Work Book: Laboratory Field References Manual* (p. 193). Burgus Publ. Co., Minnesota.

Curtis, J. T., & McIntosh, R. P., (1950). The Interrelation of certain analytic and synthetic phytosociological characters. *Ecol., 31*, 434–455.

Diversity and Regeneration of Tree Species 317

Deb, P., & Sundriyal, R. C., (2008). Tree regeneration and seedling survival patterns in old-growth lowland tropical rainforest in Namdapha National Park, north-east India. *For. Ecol. Manag., 255*, 3995–4006.

Denslow, J. S., (1980). Gap partitioning among tropical rain forests. *Biotrop., 12*, 47–55.

Dhaulkhandi, M., Dobhal, A., Bhatt, S., & Kumar, M., (2008). Community structure and regeneration potential of natural forest site in Gangotri, India. *J. Basic Appl. Sci., 4*(1), 49–52.

Eilu, G., & Obua, J., (2005). Tree condition and natural regeneration in disturbed sites of Bwindi impenetrable forest National Park, Southwestern Uganda. *Trop. Ecol., 46*(1), 99–101.

Elourard, C., Pascal, J. P., Pelissier, R., Ramesh, B. R., Houllier, F., Durand, M., Aravajy, S., et al., (1997). Monitoring the structure and dynamics of a dense moist evergreen forest in the Western Ghats (Kodagu District, Karnataka, India). *Trop Ecol., 38*(2), 193–214.

Gairola, S., (2010). *Phytodiversity (Angiosperms and Gymnosperms) in Mandal-Chopta Forest of Garhwal Himalaya, Uttarakhand, India.* PhD thesis, HNB Garhwal University Srinagar, Garhwal.

Gairola, S., Sharma, C. M., Ghildiyal, S. K., & Suyal, S., (2012). Regeneration dynamics of dominant tree species along an altitudinal gradient in moist temperate valley slopes of the Garhwal Himalaya. *J. For. Res., 23*(1), 53–63.

Gairola, S., Sharma, C. M., Ghildiyal, S. K., Suyal, S., Rana, C. S., & Butola, D. S., (2009). Biodiversity conservation and sustainable rural development in the Garhwal Himalaya. *Rep. Opin., 1*(4), 6–12.

Gairola, S., Sharma, C. M., Suyal, S., & Ghildiyal, S. K., (2011). Species composition and diversity in mid-altitudinal moist temperate forests of the Western Himalaya. *J. For. Sci., 27*(1), 1–15.

Gaur, R. D., (1999). *Flora of the District Garhwal North West Himalaya (with Ethanobotanical Notes).* Transmedia Publication, Srinagar (Garhwal) India.

Gauthier, S., Grandpre, L. D., & Bergeron, Y., (2000). Differences in forest composition in two boreal forest ecoregions of Quebec. *J. Veg. Sci., 11*, 781–790.

Ghildiyal, S., Baduni, N. P., Khanduri, V. P., & Sharma, C. M., (1998). Community structure and composition of Oak forests along altitudinal gradient in Garhwal Himalaya. *Ind. J. For., 21*(3), 242–247.

Ghimire, B., Mainali, K. P., Lekhak, H. D., Chaudhary, R. P., & Ghimeray, A. K., (2010). Regeneration of *Pinus wallichiana* AB Jackson in a trans-Himalayan dry valley of north-central Nepal. *Himal. J. Sci., 6*(8), 19–26.

Haq, S. M., Rashid, I., Khuroo, A. A., Malik, Z. A., & Malik, A. H., (2019). Anthropogenic disturbances alter community structure in the forests of Kashmir Himalaya. *Trop. Ecol., 60*(1), 6–15.

Harper, J. L., (1977). *Population Biology of Plants.* Academic Press, London.

Henle, K., Saree, S., & Wiegand, K., (2004). The role of density regulation in extinction processes and population viability analysis. *Biodiv. Conser., 13*(1), 9–52.

Huang, W., Pohjonen, V., Johansson, S., Nashanda, M., Katigula, M. I. L., & **Luvkkanen, O.,** (2003). Forest structure, species composition, and diversity of Tanzanian rain forest. *For. Ecol. Manag., 173*, 11–24.

Hubbell, S. P., Foster, R. B., O'Brien, S. T., Harms, K. E., Condit, R., Wechsler, B., Wright, S. J., & Loode, L. S., (1999). Light gap disturbance, recruitment limitation, and tree diversity in a neotropical forest. *Sci., 283*, 554–557.

Jones, C. G., Lawton, J. H., & Shachak, M., (1997). Positive and negative effects of organisms as physical ecosystem engineers: Positive interactions in communities. *Ecol., 78*(7), 1946–1957.

Jones, R. H., Sharitz, R. R., Dixon, P. M., Segal, P. S., & Schneider, R. L., (1994). Woody plant regeneration in four flood plants forests. *Ecol. Monogr., 64*, 345–367.

Kennedy, D. N., & Swaine, M. D., (1991). Germination and growth of colonizing species in artificial gaps of different sizes in Dipterocarp rain forest. In: Marshall, A. G., & Swain, M. S., (eds.), *Tropical Rain Forest: Disturbance and Recovery* (pp. 357–367). Royal Society, London.

Khan, M. L., Rai, J. P. N., & Tripathi, R. S., (1987). Population structure of some tree species in disturbed and protected sub-tropical forests of North East India. *Acta Oecol: Oecol. Appl., 8*, 247–255.

Khan, S. M., Page, S., Ahmad, H., & Harper, D., (2012). Anthropogenic influences on the natural ecosystem of the Naran Valley in the western Himalayas. *Pak. J. Bot., 44*, 231–238.

Kharakwal, G., Mehrortra, P., Rawat, Y. S., & Pangtey, Y. P. S., (2005). Phytodiversity and growth form in relation to altitudinal gradient in the Central Himalayan (Kumaun) region of India. *Curr. Sci., 89*(5), 873–878.

Kharkwal, G., (2009). Qualitative analysis of tree species in evergreen forests of Kumaun Himalaya, Uttarakhand, India. *Afr. J. Pl. Sci., 3*(3), 49–52.

Khera, N., Kumar, A., Ram, J., & Tewari, A., (2001). Plant biodiversity assessment in relation to disturbances in mid-elevational forest of Central Himalaya, India. *Trop. Ecol., 42*(1), 83–95.

Koirala, M., (2004). Vegetation composition and diversity of Piluwa micro-watershed in Tinjure-Milke region, East Nepal. *Himal. J. Sci., 2*(3), 29–32.

Krauchii, N., Brang, P., & Schonenberger, W., (2000). Forests of mountainous regions: Gaps in knowledge and research needs. *For. Ecol. Manag., 132*, 73–82.

Kumar, A., & Ram, J., (2005). Anthropogenic disturbances and plant biodiversity in forests of Uttaranchal, Central Himalaya. *Biodiv. Conser., 14*, 309–331.

Kunwar, R. M., & Sharma, S. P., (2004). Quantitative analysis of tree species in two community forests of Dolpa district, mid-west Nepal. *Himal. J. Sci., 2*(3), 23–28.

Kusumlata, & Bisht, N. S., (1991). Quantitative analysis and regeneration potential of moist temperate forest in Garhwal Himalaya. *Ind. J. For., 14*(2), 98–106.

Malik, Z. A., & Bhatt, A. B., (2016). Regeneration status of tree species and survival of their seedlings in Kedarnath Wildlife Sanctuary and its adjoining areas in Western Himalaya, India. *Trop. Ecol., 57*(4), 677–690.

Malik, Z. A., & Nautiyal, M. C., (2016). Species richness and diversity along the altitudinal gradient in Tungnath, the Himalayan benchmark site of Himadri. *Trop. Pl. Res., 3*(2), 396–407.

Malik, Z. A., Bhat, J. A., & Bhatt, A. B., (2014b). Forest resource use pattern in Kedarnath wildlife sanctuary and its fringe areas (a case study from Western Himalaya, India). *En. Pol., 67*, 138–145.

Malik, Z. A., Hussain, A., Iqbal, K., & Bhatt, A. B., (2014a). Species richness and diversity along the disturbance gradient in Kedarnath Wildlife Sanctuary and its adjoining areas in Garhwal Himalaya, India. *Int. J. Curr. Res., 6*(12), 10918–10926.

Malik, Z. A., Pandey, R., & Bhatt, A. B., (2016). Anthropogenic disturbances and their impact on vegetation in Western Himalaya, India. *J. Mt. Sci., 13*(1), 69–82.

Margalef, D. R., (1958). Information theory in ecology. *Gen. Sys., 3*, 36–71.

Mcleod, K. W., & Murphy, P. G., (1977). Establishment of *Ptelea trifoliate* on lake Michigan sand dunes. *Am. Midl. Nat., 97*, 350–362.

Menhinick, E. F., (1964). A comparison of some species diversity indices applied to samples of field insects. *Ecol., 45*, 859–861.

Mishra, B. P., Tripathi, R. S., Tripathi, O. P., & Pandey, H. N., (2003). Effect of disturbance on the regeneration of four dominant and economically important woody species in a broad-leaved subtropical humid forest of Meghalaya, northeast India. *Curr. Sci., 84,* 1449–1453.

Mishra, R., (1968). *Ecology Workbook.* Oxford and IBH Publication Co., Calcutta.

Mueller-Dombios, D. J., Jucobi, D., Cooray, R. G., & Balakrishnan, N., (1980). *Ohia Rain Forest Study: Ecological Investigation of the Ohia Dieback Problem in Hawaii* (p. 183). Hawaii Institute of Tropical Agriculture and Human Resource, Honolulu HI, Miscellaneous Pub.

Naithani, B. D., (1984, 1985). *Flora of Chamoli, I-II.* Botanical Survey of India, Calcutta.

Nautiyal, S., & Kaechele, H., (2007). Conserving the Himalayan forests: Approaches and implications of different conservation regimes. *Biodiv. Conser., 16,* 3737–3754.

Nunez-Farfan, J., & Divzo, R., (1989). Leaf survival in relation to herbivory in two tropical pioneer species. *Oikos, 54,* 71–74.

Odum, E. P., (1959). *Fundamentals of Ecology* (2nd edn.). Sounders, Philadelphia.

Pande, P. K., Negi, J. D. S., & Sharma, S. C., (2001). Plant species diversity and vegetation analysis in moist temperate Himalayan forest. *Ind. J. For., 24*(4), 456–470.

Pande, P. K., Negi, J. D. S., & Sharma, S. C., (2002). Plant species diversity, composition, gradient analysis and regeneration behavior of some tree species in a moist temperate western Himalayan forest ecosystem. *Ind. For., 128*(8), 869–886.

Pant, S., & Samant, S. S., (2007). Assessment of plant diversity and prioritization of communities for conservation in Mornaula reserve forest. *Appl. Ecol. Environ. Res., 5*(2), 123–138.

Pant, S., & Samant, S. S., (2012). Diversity and regeneration status of tree species in Khokhan wildlife sanctuary, North-Western Himalaya. *Trop. Ecol., 53*(3), 317–331.

Peet, R. K., (1974). The measurement of species diversity. *Ann. Rev. Ecol. Evol. Syst., 5,* 285–307.

Phillips, E. A., (1959). *Methods of Vegetation Study* (p. 107). New York: Henry Holt and Co. Inc.

Pichi-Sermolli, (1948). An index for establishing the degree of maturity in plant communities author(s), source. *J. Ecol., 36*(1), 85–90.

Pielou, E. C., (1966). The measurement of diversity in different types of biological collections. *J. Theor. Bio., 13,* 131–144.

Pokhriyal, P., Uniyal, P., Chauhan, D. S., & Todaria, N. P., (2010). Regeneration status of tree species in forest of phakot and Pathri Rao watersheds in Garhwal Himalaya. *Curr. Sci., 98*(2), 171–175.

Polyakov, M., Majumdar, I., & Teeter, L., (2008). Spatial and temporal analysis of the anthropogenic effects on local diversity of forest trees. *For. Ecol. Manag., 255,* 1379–1387.

Putman, R. J., (1994). *Community Ecology.* Chapman and Hall, London.

Raturi, G. P., (2012). Forest community structure along an altitudinal gradient of district Rudraprayag of Garhwal Himalaya, India. *Ecologia, 2*(3), 76–84.

Rawat, Y. S., & Rawat, V. S., (2010). Van panchayats as an effective tool in conserving biodiversity at the local level. *J. Environ. Protec., 1,* 278–283.

Rikhari, H. C., Chandra, R., & Singh, S. P., (1989). Pattern of species distribution and community characters along a moisture gradient within an oak zone of Kumaon Himalaya. *Proc. Ind. Nat. Sci. Acad., 55*(B), 431–438.

Sagar, R., Raghubanshi, A. S., & Singh, J. S., (2003). Tree species composition, dispersion and diversity along a disturbance gradient in a dry tropical forest region of India. *For. Ecol. Manag., 186,* 61–71.

Salick, J., Zhendong, F., & Byg, A., (2009). Eastern Himalayan alpine plant ecology, Tibetan ethnobotany, and climate change. *Glob. Env. Chang., 19*, 147–155.

Sanjeev, Gera, M., & Sankhayan, P. L., (2006). Phytosociological analysis of Arnigad micro-watershed in Mussorie hills of Garhwal Himalayas. *Ind. For., 132*(1), 19–29.

Saxena, A. K., & Singh, J. S., (1982). A phytosociological analysis of woody species in forest communities of a part the Kumaon Himalaya. *Vegetatio., 50*, 3–22.

Saxena, A. K., & Singh, J. S., (1984). Tree population structure of certain Himalayan forest associations and implications concerning the future composition. *Vegetatio., 58*(2), 61–69.

Saxena, A. K., Pandey, U., & Singh, J. S., (1978). On the ecology of Oak forest in Nainital hills, Kumaun Himalaya. In: Sigh, et al., (eds.), *Glimpses of Ecology* (pp. 167–180). Jaipur International Scientific Publication.

Saxena, A. K., Singh, S. P., & Singh, J. S., (1984). Population structure of forest of Kumaon Himalaya: Implications for management. *J. Environ. Manag., 19*, 307–324.

Schulte, P. J., & Marshall, P. E., (1983). Growth and water relations of black locust and Pine seedlings exposed to controlled water-stress. *Canad. J. For. Res., 13*, 334–338.

Semwal, D. P., Saradhi, P. P., Nautiyal, B. P., & Bhatt, A. B., (2007). Current status, distribution and conservation of rare and endangered medicinal plants of Kedarnath Wildlife Sanctuary, Central Himalaya, India. *Curr. Sci., 92*, 1733–1738.

Shaheen, H., Ullah, Z., Khan, S. M., & Harper, D. M., (2012). Species composition and community structure of Western Himalayan moist temperate forests in Kashmir. *For. Ecol. Manag., 278*, 138–145.

Shannon, C. E., & Weaver, W., (1963). *The Mathematical Theory of Communication* (p. 117). Urbana, USA: University of Illinois Press.

Sharma, C. M., Khanduri, V. P., & Goshwami, S., (2001). Community composition and population structure in temperate mixed broad-leaved and coniferous forest along an altitudinal gradient in a part of Garhwal Himalaya. *J. Hill. Res., 14*(1), 32–43.

Simpson, E. H., (1949). Measurement of diversity. *Nature, 163*, 688–690.

Singh, D., (2013). *Forest Structure, Diversity, Growing Stock Variation and Regeneration Status of Different Forest Cover Types in Dudatoli Area of Garhwal Himalaya.* PhD Thesis submitted to HNB Garhwal University Srinagar Garhwal, Uttarakhand.

Singh, J. S., & Singh, S. P., (1986). Structure and function of central Himalayan Oak forests. *Proc. Ind. Acad. Sci., 96*, 156–189.

Singh, J. S., (2002). The biodiversity crisis: A multifaceted review. *Curr. Sci., 82*, 638–647.

Singh, S. P., Adhikari, B. S., & Zobel, D. B., (1994). Biomass productivity, leaf longevity and forest structure in central Himalaya. *Ecol. Monogr., 64*, 401–421.

Singh, S., Malik, Z. A., & Sharma, C. M., (2016). Tree species richness, diversity, and regeneration status in different oak (Quercus spp.) dominated forests of Garhwal Himalaya, India. *J. As-Pac. Biodiv., 9*(3), 293–300.

Slik, J. W. F., Kebler, P. J. A., & Van, W. P. C., (2003). *Macaranga* and *Mallotus* species (Euphorbiaceae) as indicators for disturbance in the mixed lowland dipterocarp forest of East Kalimantan (Indonesia). *Ecol. Ind., 2*, 311–324.

Sorensen, T., (1948). A method of establishing groups of equal amplitude in plant sociology based on similarity of species content and its application to analyses of the vegetation on Danish commons. *Viden. Sels. Biol. Skr., 5*, 1–34.

Srivastava, R. K., Khanduri, V. P., Sharma, C. M., & Kumar, P., (2005). Structure, diversity, and regeneration potential of oak dominant conifer mixed forest along an altitudinal gradient in Garhwal Himalaya. *Ind. For., 131*(12), 1537–1553.

Tesfaye, G., Teketay, D., Fetene, M., & Beck, E., (2010). Regeneration of seven indigenous tree species in a dry Afromontane forest, southern Ethiopia, flora-morphology, distribution. *Func. Ecol. Pl., 205*(2), 135–143.

Uma, S., (2001). A case study of high tree diversity in a Sal (*Shorea robusta*)-dominated lowland forest of Eastern Himalaya: Floristic composition, regeneration and conservation. *Curr. Sci., 81*, 776–786.

UNESCO/UNEP/FAO, (1978). *Tropical Forest Ecosystems: A State-of-Knowledge Report.* UNESCO, Paris, France.

Uniyal, P., Pokhriyal, P., Dasgupta, S., Bhatt, D., & Todaria, N. P., (2010). Plant diversity in two forest types along the disturbance gradient in Dewalgarh Watershed, Garhwal Himalaya. *Curr. Sci., 98*(7), 938–943.

Vera, M. L., (1997). Effects of altitudes and seed size in germination and seedling survival of heathland plants in North Spain. *Pl. Ecol., 133*(1), 101–106.

Verma, R. K., Kapoor, K. S., Subramani, S. P., & Rawat, R. S., (2004). Evaluation of plant diversity and soil quality under plantation raised in surface-mined areas. *Ind. J. For., 27*(2), 227–233.

Vetaas, O. R., (2000). The effect of environmental factors on the regeneration of *Quercus semecarpifolia* Sm. in Central Himalaya, Nepal. *Pl. Ecol., 146*, 137–144.

Whitford, P. B., (1949). Distribution of woodland plants in relation to succession and clonal growth. *Ecol., 30*, 199–208.

Whittaker, R. H., (1972). Evolution and measurement of species diversity. *Tax., 21*, 213–215.

Zhang, J. T., Xu, B., & Li, M., (2013). Vegetation patterns and species diversity along elevational and disturbance gradients in the Baihua mountain reserve, Beijing, China. *Mt. Res. Dev., 33*(2), 170–178.

CHAPTER 13

Seed Handling of Tropical Forestry Species

P. K. CHANDRASEKHARA PILLAI

Ex-Head, Silviculture Department and Scientist-in-Charge, Kerala Forest Seed Centre, Kerala Forest Research Institute, Peechi, Thrissur–680653, Kerala, India, E-mail: pkcpillai@gmail.com

ABSTRACT

Seeds are unique in regeneration since they are resulting from parental genetic materials. The success of tree improvement programs depends on the availability of quality seeds and seed handling information. The use of quality seeds ensures better survival and growth. Factors like seed source, collection time, and techniques of harvesting, processing, and storage practices influence the seed quality. Quality of seeds can ensure through a collection of seeds from genetically superior stands/trees and scientific handling practices. The right stage for seed collection, processing, pre-sowing treatments, and storage are significant for efficient utilization of seeds. Knowledge on different stages of seed handling procedure will be beneficial for the production of superior planting stock, which helps to achieve a high survival rate. This chapter provides a glimpse on handling of forestry seeds in scientifically.

13.1 INTRODUCTION

Seeds are the primary source of planting material all over the world. They are unique in regeneration since they developed from mixing parental genetic materials and results in genetic variation of the offspring, which helps ecological adaptability, resistant to physical, chemical, and biological constraints (Chacko et al., 2002). Integral components in tree improvement programs are the availability of quality seeds and seed handling information. Each seed has the potential for a full-grown tree; however, most of them may be perished due to dispersal failure, predation, pest infestation, deterioration, failure of germination, etc. The use of superior and quality seeds ensure better survival and growth of plantations. Seed quality depends on factors like source, time, and techniques of harvest, processing, and storage practices. Quality of seeds can be ensured through a collection of seeds from genetically superior stands/ trees and scientific handling practices (Chacko, 2009).

The right stage for seed collection, processing, pre-sowing treatments, and storage are vital for efficient utilization of seeds. Knowledge on different stages of seed handling procedure will be beneficial for the production of superior planting stock (Chacko et al., 2002). Systematic handling of seeds helps to obtain a high rate of survival. Seed handling is a chain of procedures starting from selection of seed source, collection of seeds, processing of seeds, storage of seeds and pre-sowing treatment to enhance germination of seeds. Each component in the chain is significant since there is a potential risk of losing seed quality (Schmidt, 2000).

Seeds are broadly classified as orthodox and recalcitrant group according to storage physiology (Schmidt, 2000). Orthodox seeds (e.g., *Tectona grandis*) usually contain lower moisture content (MC) at maturity and survive drying up to 2–5% moisture level. This enables them to store for a long period. However, recalcitrant seeds (e.g., *Dysoxylum malabaricum*) are short-lived, commonly surviving only for a few days after maturity. They do not undergo a drying phase during maturation but continue germination throughout their short life. They require special attention during collection, transportation, and storage. Seeds of several tropical evergreen species show recalcitrant nature. Seeds shed from mother plants with high MC and are sensitive to desiccation and low temperature; hence, difficult to store for longer periods (Roberts, 1973; Ellis et al., 1991). However, considerable variation in tolerance to desiccation has been reported (Roberts, 1973). Planting activities have been limited to such species due to desiccation-sensitivity and non-storability (Gunn, 1991). Longevity of such seeds shall extend if store under optimum environmental condition (Parimalam et al., 2013). Another group described by Ellis et al. (1990) as intermediate type (e.g., *Swietenia macrophylla*) having characteristics of both the groups (orthodox and recalcitrant). Such seeds can be desiccated to low moisture level (12–14%) and tolerant of low temperature (2–5°C).

Desiccation of seeds to the lowest possible moisture level helps for safe storage and maintenance of viability. The actual germination potential of seeds can be fully attained through pre-sowing treatments. Storage of seeds at the lowest temperature commensurate with their chilling tolerance helps to improve seed longevity (Chacko et al., 2002). This topic is dealt with a general discussion on seed handling procedure of forestry species.

13.2   SEED HANDLING PROCEDURES

Seed handling procedures include selection of seed source, seed collection, processing, viability testing, pre-sowing treatments, and storage.

13.2.1   *SELECTION OF SEED SOURCE*

Performance of a progeny is determined by the genetic constitutions of parent trees carried by seeds. Therefore, while selecting seed sources of unknown genetic constitution, it is advisable to avoid seeds from inbred populations and phenotypically inferior trees.

13.2.1.1   *MANAGED SEED SOURCES*

Seed production areas, seeds stand and seeds orchards are the managed seed sources. Seed production areas and seed stands are interim sources of the seeds of non-proven genetic superiority. However, they are phenotypically upgraded by removal of undesired trees. Seed orchards are the source of the seeds of proven superior performance through progeny trials. They are raised from seeds (Seedling seed orchard) or vegetative materials (Clonal seed orchard). Provenance seed orchards are the seed source, where no progeny trials have been carried out.

13.2.2   *SEED COLLECTION*

Following measures are considered to avoid deleterious genetic effects during the collection of seeds. (i) prefer stands with heavily fruiting trees in close proximity to each other, (ii) within the preferred stand, collect seeds from 15 trees which are preferably 100 m apart, and (iii) collect seeds from vigorous trees of good form (Chacko et al., 2002). Seed collection should be during a large quantity of viable seeds available on trees. Usually, most of the seeds will be matured during this period. There is a chance of inbreeding during early or late flowering, which influences the quality of seeds (Chacko et al., 2002). Moreover, the fruits develop during this period may have underdeveloped seeds and likely to mature or dehisce more quickly than those of normally developed seeds. Consequently, physiological and genetic quality of the seed crop is often of poor.

Maturity index is significant for seed collection and other handling procedures. Knowledge of structural changes in fruits or seeds during the maturity stage important to determine the best seed collection time. Following are some of the maturity indices:

1. Dried calyx (*Tectona grandis*);
2. Color change (greenish yellow-*Azadirachta indica*, green to deep brown-*Acacia mangium*, greenish-yellow to dark brown-*Shorea robusta*);
3. Dehydration (*Ailanthus triphysa*);
4. Longitudinally furrowed capsules attain bright yellow color-*Dysoxylum malabaricum*);
5. Hydration (*Calamus* spp.).

Seeds of *T. grandis*, Bamboos, etc., shall collect from ground after dispersal by sweeping floor and *Melia dubia, Terminalia bellirica* by handpicking. Mature pods of *Caesalpinia sappan, Cassia fistula*, mature fruits of *Artocarous hirsutus, Calophyllum inophyllum, D. malabaricum*, etc., shall collect from trees using long pole. Fruits of *A. hirsutus, Syzygium cumini, S. travancoricum*, etc., shall collect by shaking branches and *Eucalyptus* capsules by pruning branch-lets with the help of a tree climber. In the case of *Gluta travancorica*, freshly fallen mature fruits shall collect from the ground in cloth bags. Optimum collection period of *A. hirsutus* is May-April, *C. inophyllum* during December, *D. malabaricum* during June-August, *G. travancorica* during May-July, *S. cumini*, and *S. travancoricum* during May-June (Pillai and Pandalai, 2015).

Prior to collection, seed source information such as details of species (Scientific name, common name, provenance-if known, seed source classification-unclassified, seed zone, seed production areas, etc.), location, site, stand, how to reach the site, flowering, and fruiting period (seed production), labourer availability, etc., must be recorded in a prescribed format. Similarly, collection methods (from trees, ground after shaking or natural fall, etc.), genetic representation (number of parent trees from which the seeds collected), average spacing between parent trees, phenotypic selection of trees, quantity of fruits/seeds collected, container used, weight before processing, etc., also shall be recorded after seed collection in the prescribed format.

13.2.3   TRANSPORTATION

Fruits/seeds are to be transported to the Processing Centre soon after collection, since it is difficult to maintain a proper environment during transportation as bags and containers may have to be piled up. Ensure optimal ventilation for orthodox seeds and avoid desiccation of recalcitrant seeds.

13.2.4 SEED PROCESSING

Seed processing is done to get clean and pure seeds of high physiological quality, which can store and easily handle during further process. Primary processing (pre-cleaning) can be done at the collection site in order to reduce bulk for easy transportation. Following processing are carried out at the processing centre:

1. **Seed Extraction:**
   - Beating sack/bag containing pods with stick until the fruits open or disintegrated (dry indehiscent fruits);
   - Soak in water for a while, macerate, and wash (*S. cumini* and *S. travancoricum*);
   - Split open the fruits using a sharp knife (*D. malabaricum*).
2. **Pre-Cleaning:** It is the removal of non-seed materials from the seed lot.
3. **Pre-Curing:** Fruits are kept moist for a while, which help to promote after-ripening of premature or immature fruits, and to ease extraction of seeds.
4. **De-Winging:** Remove dry appendages (wings, spines, aril, etc.), from seeds (*Swietenia macrophylla*)
5. **De-Pulping:** It is the extraction of seeds from fruit pulp (*A. hirsutus*, *S. cumini* and *S. travancoricum*).
6. **De-Coating:** It is the removal of seed coat (*D. malabaricum* and *C. inophyllum*).
7. **Cleaning:** It is the removal of remaining inert materials from the extracted seed lot to make sure viable seed lot.
8. **Grading:** It is the separation of below-average sized seeds from the seed lot to improve physiological quality.
9. **Adjustment of Moisture Content (MC%):** Seeds for storage have to be desiccated to the critical moisture level and can measure by oven-dry method as per the ISTA rules using the following formula:

$$MC\% = \frac{\text{Fresh weight} - \text{Dry weight}}{\text{Fresh weight}} \times 100$$

Fungal contamination is ubiquitous in recalcitrant seeds, and is a major cause of deterioration; hence, surface sterilization of the seeds immediately after removal from the fruit is critical.

13.2.5   SEED DRYING

High MC of seeds creates an ideal environment for infestation of mycoflora (fungi). Orthodox seeds have low MC% at the time of harvest and possible to dry up to 2–5% MC, which help for long term storage. Whereas, recalcitrant seeds (often >30–50% MC) pose a problem as they are intolerant to desiccation (below 12–30% MC) depending on species. Seeds with high MC are more prone to heat damage; hence, direct sun drying is avoided. Seeds shall dry to the critical moisture level (lowest safe MC) under open desiccation (spreading out seeds on concrete floor and dry by air circulation using fan) for safe storage and maintenance of viability. Seeds that dry below the critical moisture level shall be lost their viability (Sasaki, 1980). Most recalcitrant seeds can't survive drying below the critical MC, i.e., 12–50% (Roberts, 1973; Chin, 1988). Seeds shall treat with systemic fungicide (Captan 50% WP @ 4 g/kg) to prevent fungal infection (Chacko, 2009).

13.2.6   VIABILITY TESTING

Viability indicates the potential germinability of seeds, which reflect expected germination in the nursery (ISTA, 2004). Seed viability shall check through rapid viability test (cutting test, TTZ test, H_2O_2 test, etc.), as well as germination test.

13.2.6.1   CUTTING TEST

It is a simple test used for visual interpretation of the quality of seeds (emptiness, insect-damaged, underdeveloped, etc.), during their collection and processing.

13.2.6.2   TOPOGRAPHICAL TETRAZOLIUM (TTZ) TEST

This is one of the widely accepted rapid tests for assessing seed viability. Live tissues will be stained during the test. Fully stained seeds are considered as viable seeds and thus estimate the viability of the seed lots.

13.2.6.3   HYDROGEN PEROXIDE TEST (H_2O_2)

It is a transition to germination test (Bhodthipuks et al., 1996). H_2O_2 increases oxygen supply and stimulate seed germination during the test.

13.2.6.4   GERMINATION TEST

It is the direct measurement of seed quality-ability of seeds to germinate under suitable conditions. Seeds shall test for viability through germination trials in the laboratory. Sterilized vermiculite is used as a germination medium. Record daily observations up to the culmination of germination and calculate germination percentage.

13.2.7   PRE-SOWING TREATMENT

It is the treatment that carries out to enable rapid and uniform germination of seeds sown in the nursery, field, or seed quality testing. In some cases, pretreatment is the mere acceleration of the natural processes of dormancy release; in others, it is a simulation of this process. Even under favorable conditions, viable seeds may fail to germinate due to dormancy. Pre-sowing treatment can terminate dormancy, speed up germination, reduce germination inhibitors, etc. Mainly, seed dormancy is in three types, i.e., physical (hard and impermeable seed coat/pericarp), mechanical (hard seed coat), and morphological (underdeveloped embryo). Morphological dormancy can overcome through after-ripening process by storing seeds for a while (e.g., *T. grandis*).

13.2.7.1   WATER SOAKING

Soaking seeds in tap water for 12–48 hours, which helps in softening seed coat and leach out chemical inhibitors. It is applied to most medium-sized dry seeds to overcome physical, mechanical, or chemical dormancy (Chacko et al., 2002). About 31% increase in seed germination of *S. macrophylla* shall obtain under laboratory condition (Table 13.1).

13.2.7.2   HOT WATER TREATMENT

Soaking seeds in hot water (boiled, and then cooled for about 5 minutes to 80°C) for 1 to 45 minutes (depending on the hardness of seed coat) followed by soaking in tap water for 12 to 24 hours. It can overcome physiological dormancy with hard, thick, and waxy coated seeds by creating tension which consequently possess cracking of the macro sclerid layer or by affecting strophiolar plug (Schmidt, 2002). Hot water treatment tremendously improved seed germination of *C. sappan* when compared to tap-water soaking (Table 13.1).

13.2.7.3   ACID SCARIFICATION

Scarify the seeds with thick impermeable seed coat in concentrated H_2SO_4 for 1 to 30 minutes depending on the nature of seed coat, and thoroughly rinse under running water for 10 minutes to remove traces of acid and soak in water for 12 hours to enhance imbibition. Acid scarification increases seed germination of *Adenanthera pavonina* and *C. fistula* (Table 13.1). Acid causes wet combustion of seed coat, which results in disruption on the seed coat and increases permeability or lowers the mechanical resistance (Schmidt, 2002).

13.2.7.4   DE-WINGING

Remove seed appendages like wings, spines, hairs, and arils. De-winging creates a crack on the seed coat that helps to imbibe water. Wings usually increase the surface area of the seed, tend to accumulate moisture and promote fungal attack. De-winging helps to reduce bulk and ease in handling (Chacko et al., 2002), e.g., *S. macrophylla, Butea monosperma*

13.2.7.5   DE-COATING

Remove fleshy or hard seed coat to promote germination. Seed coat may setback seed germination due to the presence of chemical inhibitors. De-coating also helps to prevent fungal infestation in fleshy seeds (Chacko and Pillai, 1997). De-coating enhances about three to ten times germination in seeds of *Embelia Ribes, Garcinia gummi-gutta,* and *D. malabaricum* when compared to seeds with seed coat (Table 13.1).

13.2.7.6   WEATHERING

Weathering (wetting and drying) of seeds helps to soften hard seed coat. Wetting during night and drying under the sun for 7 days has a great improvement in seed germination of *T. grandis* (Table 13.1). It causes

TABLE 13.1   Comparison of Germination (Treated Seeds with Control-Untreated Seeds)

Species	Treatment	Treatment Duration	Germination (%)
Adenanthera pavonina	Control	–	30%
	Hot water soaking	5 minutes in hot water and overnight in tap water	46%
	Acid scarification	Acid scarification for 10 minutes followed by overnight tap water soaking	76%
Caesalpinia sappan	Control	–	32%
	Hot water soaking	5 minutes in hot water and overnight in tap water	56%
Cassia fistula	Control	–	25%
	Hot water soaking followed by tap water	5 minutes in hot water and overnight in tap water	30%
	Acid scarification	Acid scarification for 10 minutes followed by overnight tap water soaking	83%
Dysoxylum malabaricum	Control	–	17%
	De-coating	–	97%
Embelia ribes	Control	–	19%
	De-coating	Grinding using mortar and pestle	40%
Garcinia gummi-gutta	Control	–	7%
	De-coating	–	70%
Mimusops elenji	Control	–	65%
	Water soaking	12 hours	75%
Phyllanthus emblica	Control	–	40%
	Water soaking	12 hours	44%
Santalum album	Control	–	30%
	Cow dung slurry	48 hours	63%
	GA$_3$ (225 ppm)	16 hours	84%
Simarouba glauca	Control	–	66%
	De-pulping and water soaking	12 hours	71%
Spondias pinnata	Control	–	40%
	Clipping	–	60%
Swietenia macrophylla	Control	–	73%
	Water soaking	48 hours	96%
Tectona grandis	Control	–	18%
	Weathering	7 days	34%

Note: *Control = No treatment.*

softening of pericarp (due to expansion and contraction) and to crack the hard seed coat, which helps to imbibe water and gas exchange resulting in increased germination (Chacko et al., 2002).

13.2.7.7   CLIPPING

Chop at distal ends or micropylar end will create a crack on seed coat, through which moisture absorption and gaseous exchange take place (Chacko et al., 2002). Clipping enhances about 20% increase in seed germination of *Spondias pinnata* (Table 13.1).

13.2.7.8   SPLITTING ENDOCARP

It improves imbibition and promotes germination (Chacko et al., 2002), e.g., *Melia dubia*.

13.2.7.9   ABRASION

Seeds mix with sand (equal or double the volume of seeds) and grind gently in a mortar. Seeds rotate in drums that lined with sandpaper or barbed wire is another method for abrasion. Soak the abraded seeds in water for overnight prior to sowing.

13.2.7.10   MECHANICAL SCARIFICATION

Tumbling in cement mixer with sand/gravel (*M. dubia*). However, it is time consuming or labor-intensive.

13.2.7.11   STRATIFICATION

It is a process of treating seeds to stimulate natural conditions and break dormancy. Breaking thermo-dormancy in temperate species (Fagus, Quercus, etc.) and highland tropical species (pines and eucalypts) is possible through exposing seeds to cold and moist condition by stratification. It is done by spreading seeds in layers alter with layers of sand, or charcoal in boxes or baskets and keep them in cold condition.

13.2.7.12 COW-DUNG SLURRY TREATMENT

It is prepared by mixing cow-dung with equal quantity of water and soaks seeds for 24 hours. Cow-dung contains growth hormones like indole acetic acid (IAA), gibberellic acid (GA_3), kinetin, etc. It can overcome physiological dormancy. Soaking seeds in cow-dung slurry for 48 hours improves seed germination of *Santalum album* (Table 13.1).

13.2.7.13 BIOLOGICAL SCARIFICATION

Seeds germinate quickly if they are passed through the digestive system of birds/animals, e.g., *Acacia arabica*, *S. album*, etc. Seeds of *Gmelina arborea*, *A. nilotica*, *M. dubia*, etc., shows improved germination by the action of strong digestive chemicals in the guts of the organisms. Seeds coat dormancy of teak shall be removed by termite-aided treatment (Chacko, 1998).

13.2.7.14 HORMONAL TREATMENT

Treat seeds with GA_3 promotes cell elongation, cell division and thus helps in the growth and development (Chacko et al., 2002). Soaking seeds in GA_3 (225 ppm) for 16 hours enhances seed germination of *S. album* (Thahseen, 2018).

13.2.8 SEED STORAGE

Large variation in storability is encountered among species. General storage conditions should aim at: (i) reduce metabolism of seeds; (ii) keep pest-free condition; and (iii) reduce general seed aging. General mandates for seed storage are: (i) store seeds at the low temperature and relative humidity; (ii) store seeds under critical seed MC; (iii) protect seeds from pest infestation; and (iv) store seeds in suitable container-airtight container (orthodox seeds), container that permeable to gasses and retention of moisture (recalcitrant seeds). Recalcitrant seeds are to be stored at the lowest temperature commen-surate with their chilling tolerance. Storage conditions and seed longevity of some species are given in Table 13.2. Major seed characteristics of selected species in the Western Ghats are presented in Table 13.3.

TABLE 13.2   Storage Conditions and Seed Longevity of Selected Forestry Species

SL. No.	Species Scientific Name	Storage Container	Conditions	Longevity
1.	*Adenanthera pavonina*	Air-tight bin	16°C + 45% RH	> 1 year
2.	*Ailanthus triphysa*	Air-tight bin	16°C + 45% RH	6 months
3.	*Albizia lebbeck*	Air-tight bin	16°C + 45% RH	> 1 year
4.	*Bamboosa bambos*	Plastic bag	4°C + 45% RH	> 3 years
5.	*Barringtonia acutangula*	Soil pit	Ambient condition	2–3 months
6.	*Bauhinia variegata*	Air-tight bin	16°C + 45% RH	1 year
7.	*Butea monosperma*	Sealed container	16°C + 45% RH	1 year
8.	*Caesalpinia sappan*	Air-tight bin	16°C + 45% RH	2 years
9.	*Calamus* spp.	Moist sawdust	4°C + 45% RH	2 years
10.	*Calophyllum inophyllum*	Earthen pot inside wet vermiculite	16°C + 45% RH	14 months
11.	*Cassia fistula*	Air-tight bin	16°C + 45% RH	2 years
12.	*Dalbergia latifolia*	Air-tight bin	16°C + 45% RH	1 year
13.	*Dendrocalamus strictus*	Plastic bag	4°C + 45% RH	> 3 years
14.	*Dysoxylum malabaricum*	Earthen pot inside wet vermiculite	20°C + 45% RH	12 weeks
15.	*Mimuspos elengi*	Air-tight bin	16°C + 45% RH	2 years
16.	*Neolamarkia cadamba*	Air-tight bin	16°C + 45% RH	1 year
17.	*Oroxylum indicum*	Air-tight bin	16°C + 45% RH	1 year
18.	*Pterocarpus marsupium*	Air-tight bin	16°C + 45% RH	1 year
19.	*Santalum album*	Air-tight bin	4°C + 45% RH	2 years
20.	*Saraca asoca*	Air-tight bin	16°C + 45% RH	6 months
21.	*Strychnos nux-vomica*	Gunny bag	16°C + 45% RH	1 year
22.	*Swietenia macrophylla*	Air-tight bin	16°C + 45% RH	1 year
23.	*Syzygium cumini*	Earthen pot inside saw-dust	16°C + 45% RH	6 months
24.	*Tectona grandis*	Air-tight bin	4°C + 45% RH	5 years
25.	*Terminalia arjuna*	Air-tight bin	16°C + 45% RH	2 years
26.	*Terminalia bellirica*	Air-tight bin	16°C + 45% RH	1 year
27.	*Terminalia chebula*	Air-tight bin	16°C + 45% RH	2 years
28.	*Terminalia crenulata*	Air-tight bin	16°C + 45% RH	1 year
29.	*Wrightia tinctoria*	Air-tight bin	16°C + 45% RH	1 year
30.	*Xylia xylocarpa*	Air-tight bin	16°C + 45% RH	1 year

TABLE 13.3 Seed Characteristics of Selected Species in the Western Ghats

SL. No.	Species	Seed Weight (Seeds/kg)	Pretreatment	Germination %	Germination Period (Days)	Storage Physiology	References
1.	*Acacia nilotica*	5,500–11,600 at 10% MC	Hot water, acid scarification	50–95	15–30	Orthodox	Tjelele et al. (2015)
2.	*Acacia sinuata*	5,000–6,000 at 7% MC	Hot water	70	4–25	Orthodox	Pillai et al. (2020)
3.	*Adenanthera pavonina*	3,440–3,510 at 8% MC	hot-water, acid scarification	62–82	2–15	Orthodox	Vineeta et al. (2018)
4.	*Aegle marmelos*	10,200–10,860 at 6% MC	Hydro priming	84–97	9–23	Intermediate	Venudevan and Srimathi (2013)
5.	*Ailanthus triphysa*	7,500–10,000 at 8% MC	water soaking	80–99	3–20	Orthodox	Jesna (2018)
6.	*Albizia lebbeck*	7,000–10,000 at 8% MC	Hot water, mechanical/acid scarification	80–90	8–20	Orthodox	Jean Paul et al. (2017)
7.	*Albizia odoratissima*	About 23,250	Hot water, mechanical/acid scarification	70–99	4–15	Orthodox	Kannan et al. (1996)
8.	*Albizia saman*	4,000–6,000 at 8% MC	hot-water, acid scarification	70	20–40	Orthodox	Fathurrahman et al. (2015)
9.	*Alstonia scholaris*	About 3.57 lakh	overnight water soaking	85–95	12–90	Orthodox	Joker (2000)
10.	*Anamirta cocculus*	3,600–3,800 (fresh) at 30% MC	Water soaking for 24 hr	93–99	12–30	Intermediate	Un-published
11.	*Aphanamixis polystachya*	575–1,100 at 35% MC	De-coating + 24 hr water soaking	95	7–21	Recalcitrant	Rai (2014)
12.	*Artocarpus hirsutus*	2,200–2,300 at 40% MC	De-pulping	>97	11–18	Recalcitrant	Pillai and Pandalai (2015)

TABLE 13.3 (Continued)

SL. No.	Species	Seed Weight (Seeds/kg)	Pretreatment	Germination %	Germination Period (Days)	Storage Physiology	References
13.	Asparagus racemosus	21,000–23,000 at 8% MC	Overnight water soaking, soaking in cow-urine for 6 hr	90–100	15–90	Orthodox	Chacko and Pillai (1997a); Asha and Kasera (2012)
14.	Barringtonia acutangula	640–950 at 12% MC	De-coating + overnight water soaking	96	23–36	Orthodox	Nath et al. (2016)
15.	Bauhinia malabarica	7,320–7,480 at 7% MC	Hot water + 24 hr water soaking, acid scarification (30 mint)	33–100	2–28	Orthodox	Un-published
16.	Bauhinia variegate	6,400–6,800 at 10% MC	Hot water + overnight water soaking, acid scarification (20 mint	75–80	3–14	Orthodox	Martinelli-Seneme et al. (2006)
17.	Berrya javanica	21,000–23,500 at 10% MC	Hot water, acid scarification	68	12–40	Orthodox	Un-published
18.	Borassus flabellifer	5–7 at 47% MC	Not required	90	20–50	Recalcitrant	Pillai (2016)
19.	Butea monosperma	500–1,500 at 8% MC	Overnight water soaking	87	5–11	Orthodox	Un-published
20.	Caesalpinia coriaria	33,333–34,750 at 7% MC	Soaking in boiled water (1–2 mint) + 24 hr water soaking,	50–70	11–80	Orthodox	Un-published
21.	Caesalpinia sappan	1,500–1,800 at 10% MC	Hot water, acid scarification	85	5–20	Orthodox	Sutheesh et al. (2015)
22.	Calophyllum inophyllum	135–215 at 44% MC	De-coating	95	18–81	Recalcitrant	Pillai and Pandalai (2015)

TABLE 13.3 *(Continued)*

SL. No.	Species	Seed Weight (Seeds/kg)	Pretreatment	Germination %	Germination Period (Days)	Storage Physiology	References
23.	Calophyllum polyanthum	166–175 (fresh fruits) at 45% MC	De-pulping	45–47	10–45	Recalcitrant	Nair et al. (2002)
24.	Careya arborea	2,140–2,500 at 10% MC	Overnight water soaking, soaking in 500 ppm IBA/ GA$_3$ for 10 minute	86–98	11–46	Orthodox	Archana and Krishnamurthy (2016)
25.	Caryota urens	1,000–1,200 at 8% MC	50% HNO$_3$, mechanical scarification, oven heating	>80	25–110	Orthodox	Hitinayake et al. (2018)
26.	Cassia fistula	7,880–8,400 at 7% MC	Hot water, acid scarification (2 mint) + hot water for 6 mint	66–96	5–10	Orthodox	Sh. Soliman and Abbas (2013)
27.	Cassia grandis	2,250–2,500 at 7% MC	Mechanical/acid scarification, hot-water	65	11–46	Orthodox	Srivastava et al. (2012)
28.	Castanosper-mum australe	30–44 at 47% MC	Overnight water soaking	65–70	11–30	Recalcitrant	Un-published
29.	Casuarina equisetifolia	7.8–8.45 lakh at 10% MC	Petroleum flotation, treat with 0.1 mM GA$_3$, overnight water soaking	62–83	6–20	Orthodox	Sivakumar et al. (2007)
30.	Cerbera odollam	10–12 at 60% MC	Not required	98	7–25	Recalcitrant	Un-published
31.	Coscinium fenestratum	124–132 (fruits), 505–520 (seeds) at 22% MC	De-pulping + 3000 ppm GA$_3$ (24 hr)	40–95	33–81	Intermediate	Anilkumar et al. (2010)
32.	Couropita guianensis	5,200–5,500 at 10% MC	Overnight water soaking	90	12–42	Orthodox	Un-published

TABLE 13.3 *(Continued)*

SL. No.	Species	Seed Weight (Seeds/kg)	Pretreatment	Germination %	Germination Period (Days)	Storage Physiology	References
33.	Crescentia cujete	55,000–63,000 at 10% MC	Overnight water soaking	60	4–20	Orthodox	Un-published
34.	Croton malabaricus	2,700–2,900 at 10% MC	Overnight water soaking	99	14–92	Orthodox	Un-published
35.	Dalbergia latifolia	7,100–8,000 (pods); 20,000–22,700 (seeds) at 9% MC	Overnight water soaking	70	7–21	Orthodox	Udaykumar and Chavan (2018)
36.	Diospyros discolor	200–350 at 48% MC	Not required	90	15–30	Recalcitrant	Un-published
37.	Diospyros ebenum	6,500–7,000 at 15% MC	Overnight water soaking/ hot water (15 mint), acid scarification (30 mint)	69–85	7–30	Probably intermediate	Jeyavanan et al. (2016)
38.	Diospyros peregrine	7,000–8,500 at 29% MC	Overnight water soaking	95	11–28	Recalcitrant	Un-published
39.	Drypetes confertiflora	185–580 at 16.5% MC	Overnight water soaking	20	10–25	Recalcitrant	Un-published
40.	Dysoxylum malabaricum	125–130 at 45% MC	De-coating	97	14–45	Recalcitrant	Pillai and Pandalai (2015)
41.	Embelia ribes	9,000–9,100 (fruits); 44,000–45,000 (seeds) at 30% MC	cow-dung slurry (24 hr)	83	32–123	Recalcitrant	Raghu et al. (2016)
42.	Enterolobium cyclocarpum	1,350–1,500 at 4% MC	Soaking in boiling water and continue for 30 second	53	4–14	Orthodox	Chacko and Pillai (1997b)

TABLE 13.3 *(Continued)*

SL. No.	Species	Seed Weight (Seeds/kg)	Pretreatment	Germination %	Germination Period (Days)	Storage Physiology	References
43.	*Ficus benghalensis*	21–30 lakh at 24% MC	Soaking in dilute NH_3 for 12 hr	90	7–60	Orthodox	Mathew et al. (2011)
44.	*Ficus racemosa*	36–40 lakh at 16% MC	Hot water	19–28	6–15	Orthodox	Rai et al. (1988)
45.	*Ficus religiosa*	24.59–27.97 lakh at 14% MC	Hot water	52	8–35	Orthodox	Mathew et al. (2011)
46.	*Garcinia gummi-gutta*	1,160–1,300 at 33% MC	De-coating	90	25–164	Intermediate	Chacko and Pillai (1997c);
47.	*Garcinia indica*	1,500–1,800 at 21% MC	De-coating + overnight water soaking	>80	7–60	Recalcitrant	Malik et al. (2005)
48.	*Gluta travancorica*	55–65 at 45% MC	24 hr water soaking	99.7 ± 3.36	18–143	Recalcitrant	Pillai and Pandalai (2015)
49.	*Gmelina arborea*	950–1,250 at 6% MC	Water soaking (24–48 hr), 200 ppm GA_3 (12 hr)	34–97	6–30	Orthodox	Maharana et al. (2018)
50.	*Grewia tiliifolia*	16,600–22,250 at 13% MC	Acid scarification (10 mint), H_2O_2 (30 mint)	32–56	7–42	Orthodox	Seena and Udayan (2015)
51.	*Haldina cordifolia*	90–130 lakh	Water soaking for 24 hr	60–98	6–55	Orthodox	Jeena et al. (2012)
52.	*Hopea erosa*	350–380 at 47% MC	Not required	60	6–21	Recalcitrant	Pillai et al. (2018)
53.	*Hopea parviflora*	2,600–2,800 (winged seeds), 4,150–4,500 (de-winged seeds) at 40% MC	De-winging	75	2–27	Recalcitrant	Dayal and Kaveriappa (2000)

TABLE 13.3 *(Continued)*

SL. No.	Species	Seed Weight (Seeds/kg)	Pretreatment	Germination %	Germination Period (Days)	Storage Physiology	References
54.	*Hopea ponga*	2,800–3,000 at 52% MC	De-winging	69–99	2–10	Recalcitrant	Sukesh and Chandrashekar (2011)
55.	*Hopea racophloea*	936 (de-winged seeds) at 35% MC	De-winging	77	5–25	Recalcitrant	Pillai et al. (2018)
56.	*Indigofera tinctoria*	2–2.5 lakh at 7% MC	Acid scarification	76	3–50	Orthodox	Neema et al. (2018)
57.	*Kigelia africana*	9,000–10,000 at 12% MC	Soaking in water (24 hr)	50	6–49	Orthodox	Rønne and Jøker (2005)
58.	*Lagerstroemia microcarpa*	3.10–3.25 lakh at 10% MC	24 hr water soaking	9–17	6–35	Orthodox	Nair et al. (2002)
59.	*Lagerstroemia speciose*	91,000–2,74,000 at 10% MC	Hot water, acid scarification	20–79	14–56	Orthodox	Azad et al. (2010)
60.	*Lannea coromandelica*	8,500 to 9,000 at 11% MC	Water soaking for 24 hr	50–60	3–17	Orthodox	Hong et al. (1996)
61.	*Leucaena leucocephala*	13,000–14,300 at 9% MC	Hot water (20 mint) + water soaking (24 hr)	60–95	5–20	Orthodox	Tadros et al. (2011)
62.	*Melia dubia*	494–552 at 10% MC	Cow-dung slurry, seeds from animal excreta	20–78	30–70	Orthodox	Ravi et al. (2012); Singh et al. (2018)
63.	*Memecylon randerianum*	8,000–9,000 at 19% MC	De-pulping + overnight water soaking	89	2–29	Probably intermediate	Pillai and Nimisha (2017)
64.	*Muntingia calabura*	About 5.57 crore at 15% MC	Not required	82–87	7–47	Orthodox	Figueiredo et al. (2008)

TABLE 13.3 *(Continued)*

SL. No.	Species	Seed Weight (Seeds/kg)	Pretreatment	Germination %	Germination Period (Days)	Storage Physiology	References
65.	*Murraya koenigii*	4,500–6,000 at 35% MC	De-pulping	90	7–20	Recalcitrant	Un-published
66.	*Neolamarckia cadamba*	About 3.65 crore at 12% MC	Not required	43	18–110	Orthodox	Chacko et al. (2002)
67.	*Oroxylum indicum*	8,100–10,600 at 7% MC	Overnight water soaking, panjagavyam	95	3–25	Orthodox	Deepa et al. (2016)
68.	*Phyllanthus emblica*	55,333–64,444 at 7.54% MC	water soaking (24 hr), 200–500 ppm GA₃ (24 hr)	32–86	6–15	Orthodox	Lilabati and Sahoo (2015)
69.	*Pongamia pinnata*	550–700 at 10% MC	Water soaking (24 hr)	99	4–16	Orthodox	Kundu and Schmidt (2015a)
70.	*Pterocarpus marsupium*	1,400–1,500 pods at 10% MC	soaking in water or cow-dung slurry (24 hr)	60–85	7–15	Orthodox	Thanuja et al. (2018)
71.	*Pterocarpus santalinus*	875–975 pods at 10% MC	cow-dung slurry (3 days)	83	7–32	Orthodox	Naidu (2001)
72.	*Radermachera xylocarpa*	2–2.55 lakh at 12% MC	Overnight water soaking	43–62	5–35	Orthodox	Trivedi et al. (2016)
73.	*Salacia fruticose*	1,100–1,200 at 28% MC	De-pulping	100	30–50	Intermediate	Anaz (2018)
74.	*Salacia gambleana*	1,053 ± 35 at 28% MC	De-pulping and water soaking for 24 hour	26–37	27–49	Intermediate	Anaz (2018)
75.	*Santalum album*	5,550–5,750 at 7% MC	225 ppm GA₃ (16 hr)	84–90	15–60	Orthodox	Thahseen (2018)
76.	*Sapindus emarginattus*	1,400–1,500 at 10% MC	Water soaking (24 hr)	46	8–15	Orthodox	Swaminathan and Revathy (2013)

TABLE 13.3 (Continued)

SL. No.	Species	Seed Weight (Seeds/kg)	Pretreatment	Germination %	Germination Period (Days)	Storage Physiology	References
77.	Saraca asoca	69–76 at 42% MC	Overnight water soaking	85–90	20–60	Recalcitrant	Un-published
78.	Simarouba glauca	1,360–1,500 at 10% MC	water soaking (24 hr)	60–77	1130	Orthodox	Patil and Gaikwad (2011)
79.	Sterculia guttata	2,000–2,200 at 9% MC	Water soaking (24 hr)	88	10–20	Orthodox	Un-published
80.	Stereosper-mum colais	80,000–83,000 at 12% MC	Water soaking, panjagavyam	16–20	7–30	Orthodox	Deepa et al. (2016)
81.	Strobilanthes kunthianus	1,13,000–1,68,000 at 15% MC	Water soaking (24 hr)	85–97	3–70	Probably orthodox	Un-published
82.	Strychnos nux-vomica	542–563 at 8% MC	14-day alternate water soaking and drying, cow-dung slurry (24 hr)	32–92	38–80	Orthodox	Behera et al. (2017)
83.	Swietenia macrophylla	2,212–2,264 at 7% MC (de-winged seeds)	De-winging + water soaking (48 hr)	85–96	11–36	Intermediate	Jesna (2018)
84.	Syzygium cumini	1,410–1,500 at 35% MC	De-pulping	98	11–60	Recalcitrant	Pillai and Pandalai (2015)
85.	Syzygium samarangense	1,100–1,800 at 40% MC	Overnight water soaking	70–75	10–40	Recalcitrant	Un-published
86.	Syzygium travancoricum	12,500–12,600 at 38% MC	De-pulping	58	4–41	Recalcitrant	Pillai and Pandalai (2015)
87.	Tamarindus indica	1,220–1,265 at 9% MC	Water soaking (24 hr), acid scarification (30 mint)	80–95	4–30	Orthodox	Bello and Gada (2015)

TABLE 13.3 *(Continued)*

SL. No.	Species	Seed Weight (Seeds/kg)	Pretreatment	Germination %	Germination Period (Days)	Storage Physiology	References
88.	*Tectona grandis*	1,550–1,650 (≥ 9 mm size) at 9% MC	Termite-aided pericarp removal, weathering, dry heating, pit method	25–40	6–40	Orthodox	Chacko (1998); Rachmawati et al. (2002); Omokhua et al. (2015)
89.	*Terminalia bellirica*	158–162 fruits at 8% MC	Water soaking (7 days), de-coating + 48 hr water soaking	80–90	13–57	Orthodox	Kuniyal et al. (2015)
90.	*Terminalia chebula*	246–274 (fruits) at 9% MC	De-pulping + water soaking (7 days), remove hard endocarp + soaking in 500 ppm GA_3 (1 hr)	7–80	24–70	Orthodox	Hossain et al. (2013)
91.	*Terminalia cuneate*	280–300 (fruits) at 10% MC	Water soaking, IAA (500 ppm)	36–81	5–30	Orthodox	Kundu and Schmidt (2015b)
92.	*Terminalia elliptica*	800–870 (fruits) at 9% MC	De-winging + water soaking (24 hr)	56	6–26	Orthodox	Jesna (2018)
93.	*Terminalia paniculata*	11,909–28,908 (fruits) at 5% MC	Water soaking (24 hr)	2.63 ± 0.58	14–22	Orthodox	Pillai and Hrideek (2018)
94.	*Terminalia travancorensis*	About 400 (fruits) at 10% MC	Weathering for seven days	35	21–101	Orthodox	Pillai and Chandr-shekara (2011)
95.	*Thespesia populnea*	6,100 at 12% MC	Sandpaper rubbing + overnight water soaking, acid scarification (20 mint)	65–80	4–28	Orthodox	Friday and Okano (2006); Warrier (2010)
96.	*Trewia nudiflora*	5,550–5,880 at 10% MC	Treat with cow milk for one day	78	15–25	Orthodox	Un-published

TABLE 13.3 *(Continued)*

SL. No.	Species	Seed Weight (Seeds/kg)	Pretreatment	Germination %	Germination Period (Days)	Storage Physiology	References
97.	*Vateria indica*	23–24 (fresh fruits)	Not required	77–95	5–30	Recalcitrant	Nair et al. (2002)
98.	*Wrightia tinctoria*	39,100–40,100 at 6.91% MC	Overnight water soaking	82	4–15	Orthodox	Un-published
99.	*Xylia xylocarpa*	3,300–4,400 at 8% MC	Overnight water soaking	78	3–20	Orthodox	Schmidt (2004)
100.	*Ziziphus mauritiana*	855–893 at 8% MC	De-pulping, kernel extraction, 200 ppm GA$_3$, acid scarification (45 mint)	31–95	9–20	Orthodox	Jøker (2003); Boora (2016)

KEYWORDS

- **gibberellic acid**
- **pre-sowing treatments**
- **regeneration**
- **seed germination**
- **seed processing**
- **seed viability**

REFERENCES

Anaz, K. M., (2018). *Systematic Studies, Utilization and Conservation of the Genus Salacia (Celastraceae) in South India* (p. 166). PhD thesis, University of Calicut, Kerala.

Anilkumar, C., Chitra, C. R., Bindu, S., Prajith, V., & Mathew, P. J., (2010). Dormancy and germination of *Coscinium fenestratum* (Gaertn.) Colebr. Seeds. *Seed Science and Technology, 38*, 585–594.

Archana, S., & Krishnamurthy, G., (2016). Species conservation through applied seed research with special reference to *Careya arborea* (Kumbhi), A high value lesser known medicinal tree species of central India. *International Journal of Advances in Science Engineering and Technology, 4*(4), 62–66.

Asha, R., & Kasera, P. K., (2012). Seed germination behavior of *Asparagus racemosus* (Shatavari) under *in-vivo* and *in-vitro* conditions. *Asian J. Plant Sci. and Res., 2*(4), 409–413.

Azad, S., Paul, N. K., & Matin, A., (2010). Do pre-sowing treatments affect seed germination in *Albizia richardiana* and *Lagerstroemia speciosa*? *Frontiers of Agriculture in China, 4*(2), 181–184.

Behera, M. C., Mohanty, T. L., & Paramanik, B. K., (2017). Silvics, phytochemistry and ethnopharmacy of endangered poison nut tree (*Strychnos nux-vomica* L.): A review. *Journal of Pharmacognosy and Phytochemistry, 6*(5), 1207–1216.

Bello, A. G., & Gada, Z. Y., (2015). Germination and early growth assessment of *Tamarindus indica* L. in Sokoto State, Nigeria. *International Journal of Forestry Research*, 5. Article ID: 634108.

Bhodthipuks, J., Pukittayacamee, P., Saelim, S., Wang, B. S. P., & Yu, S. L., (1996). *Rapid Viability Testing of Tropical Tree Seeds*. Training Course Proceedings No. 4., ASEAN Forest Tree Seed Centre Project. Muak-Lek, Saraburi, Thailand.

Boora, R. S., (2016). Effect of various treatments on seeds germination in Indian jujube (*Ziziphus mauritiana* Lamk). *Agricultural Science Digest, 36*(3), 237–239.

Chacko, K. C., & Pillai, P. K. C., (1997a). Effect of pre-treatments on germination of *Asparagus racemosus* seeds. *Journal of Non-Timber Products, 4*(1/2), 23–25.

Chacko, K. C., & Pillai, P. K. C., (1997b). Storage and hot water treatments enhance germination of gunacaste (*Enterolobium cyclocarpum*) seeds. *Int. Tree Crops Journal, 9*(2), 103–107.

Chacko, K. C., & Pillai, P. K. C., (1997c). Seed characteristics and germination of *Garcinia gummi-gutta* (L) Robs. *Indian Forester, 123*(2), 123–126.

Chacko, K. C., (1998). Termite-aided mesocarp removal of teak (*Tectona grandis* L.F.) fruits for enhanced germination and cost-effective seed handling. *Indian Forester, 124*(2), 134–140.

Chacko, K. C., (2009). *Development of Protocols for Processing and Testing of Forest Seeds* (p. 38). KFRI Research Report No. 321.

Chacko, K. C., Pandalai, R. C., Seethalakshmi, K. K., Mohanan, C., Mathew, G., & Sasidharan, N., (2002). *Manual of Seeds of Forest Trees, Bamboos and Rattans* (p. 331). Kerala Forest Research Institute, Peechi, Kerala.

Chin, H. F., (1988). *Recalcitrant Seeds: A Status Report*. International Board for Plant Genetic Resource. FAO, Rome.

Dayal, B. R., & Kaveriappa, K. M., (2000). Effect of desiccation and temperature on germination and vigour of the seeds of *Hopea parviflora* Beddome and *H. ponga* (Dennst.) Mabb. *Seed Science and Technology, 28*(2), 497–506.

Deepa, K., Raghu, A. V., Daisy, M. I., Pillai, P. K. C., & Hrideek, T. K., (2016). Indigenous pre-sowing treatment for selected medicinal plants. *In: Proc of National Seminar on Trends and Innovations in Biological Research* (p. 24). (Abstract), organized by MES Kalladi College, Mannarkkad, Kerala.

Ellis, R. H., Hong, T. D., & Roberts, E. H., (1990). An intermediate category of seed storage behavior. *Journal of Exp. Bot., 41*, 1167–1174.

Ellis, R. H., Hong, T. D., & Roberts, E. H., (1991). Effect of storage temperature and moisture on the germination of papaya seeds. *Seed Science Research, 1*, 69–72.

Fathurrahman, Mohd, S. M. N., Wan, A. W. J., Doni, F., & Che, M. Z. C. R., (2015). Germination and seedling response of rain tree plants (*Albizia saman* Jacq. Merr) to seed priming using hot water. *Ecol., Envir. and Conserv., 21*(3), 1183–1187.

Figueiredo, R. A. D., Oliveira, A. A. D., Zacharias, M. A., Barbosa, S. M., Pereira, F. F., Cazela, G. N., Viana, J. P., & Camargo, R. A. D., (2008). Reproductive ecology of the exotic tree *Muntingia calabura* L. (Muntingiaceae) in southeastern Brazil. *Revista Árvore, 32*(6), 993–999.

Friday, J. B., & Okano, D., (2006). *Thespesia populnea* (milo). *Species Profiles for Pacific Island Agroforestry, Ver. 2.1* (p. 19). www.traditionaltree.org (accessed on 5 December 2020).

Gunn, S., (1991). *Banking on the Future* (pp. 16–21). Kew Spring, United Kingdom.

Hitinayake, G., Munasinghe, M., & Ranasinghe, C., (2018). Identifying methods for rapid and uniform germination and storage conditions for seeds of Kithul (*Caryota urens* L.). *International Journal of Agronomy and Agricultural Research, 12*(6), 94–100.

Hong, T. D., Linington, S., & Ellis, R. H., (1996). *Seed Storage Behavior: A Compendium* (p. 104). Handbooks for Gene banks: No. 4, IPGRI.

Hossain, M. A., Uddin, M. S., Rahman, M. M., & Ab Shukor, N. A., (2013). Enhancing seed germination and seedling growth attributes of a medicinal tree species *Terminalia chebula* through depulping of fruits and soaking the seeds in water. *Journal of Food, Agriculture and Environment, 11*(3/4), 2573–2578.

ISTA, (2004). *International Rules for Seed Testing*. International Seed Testing Association, Bassersdorf, CH-Switzerland.

Jean, P. T. M., Steve, M. L., Brunhel, V. N., & Désiré, N. M., (2017). Germination capacity stimulation of the "*Albizzia lebbeck*" seeds under Boma conditions in the democratic republic of Congo. *Int. J. Agriculture, Environment and Bioresearch, 2*(5), 272–281.

Jeena, L. S., Kaushal, R., Dhakate, P. M., & Tewari, S. K., (2012). Seed maturation indices for better regeneration and multiplication of Haldu (*Adina cordifolia*). *Indian Journal of Agricultural Sciences, 82*(4), 381–383.

Jesna, A. A., (2018). *Study on Seed Characteristics of Winged Seeds of Four Commercially Important Tropical Tree Species.* MSc thesis, Calicut University.

Jeyavanan, K., Sivachandiran, S., Vinujan, S., & Pushpakumara, D. K. N. G., (2016). Effect of different pre-sowing treatments on seed germination of *Diospyros ebenum* Koenig. *World Journal of Agricultural Sciences, 12*(6), 384–392.

Jøker, D., (2000). *Alstonia scholaris (L.)* R.Br. *Seed Leaflet, No. 9* (p. 2). Danida Forest Seed Centre, Denmark.

Jøker, D., (2003). *Ziziphus mauritiana* Lam. *Seed Leaflet No. 85* (p. 2). Danida Forest Seed Centre, Denmark.

Kannan, C. S., Sudhakara, K., Augustine, A., & Ashokan, P. K., (1996). Seed dormancy and pre-treatments to enhance germination in selected *Albizia* species. *J. Tropical For. Sci., 8*(3), 369–380.

Kundu, M., & Schmidt, L. H., (2015a). *Pongamia pinnata* (L.) Pierre. *Seed Leaflet* (p. 165). Danida Forest Seed Centre, Denmark.

Kundu, M., & Schmidt, L. H., (2015b). *Terminalia arjuna* (Roxb. ex DC) Wight & Arn. *Seed Leaflet* (p. 166). Danida Forest Seed Centre, Denmark.

Kuniyal, C. P., Purohit, V., Butola, J. S., & Sundriyal, R. C., (2015). Seed size correlates seedling emergence in *Terminalia bellerica*. *South African Journal of Botany, 87,* 92–94.

Lilabati, L., & Sahoo, U. K., (2015). Effect of pre-treatments on seed germination and seedling vigor of *Emblica officinalis* Gaertn. *Global Journal of Advanced Research, 2*(10), 1520–1526.

Maharana, R., Dobriyal, M. J., Behera, L. K., Gunaga, R. P., & Thakur, N. S., (2018). Effect of pre seed treatment and growing media on germination parameters of *Gmelina arborea* Roxb. *Indian Journal of Ecology, 45*(3), 623–626.

Malik, S. K., Chaudhury, R., & Abraham, Z., (2005). Desiccation - freezing sensitivity and longevity in seeds of *Garcinia indica, G. cambogia* and *G. xanthochymus. Seed Science and Technology, 33*(3), 723–732.

Martinelli-Seneme, A., Possamai, E., Schuta, L. R., & Vanzolin, S., (2006). Germination and sanity of seeds of *Bauhinia variegata. R. Árvore, Viçosa-MG, 30*(5), 719–724.

Mathew, G., Skaria, B. P., & Joseph, A., (2011). Standardization of conventional propagation techniques for four medicinal species of genus *Ficus* Linn. *Indian Journal of Natural Products and Resources, 2*(1), 88–96.

Naidu, C. V., (2001). Improvement of seed germination in red sanders (*Pterocarpus santalinus* Linn. f.) by plant growth regulators. *Indian Journal of Plant Physiology, 6*(2), 205–207.

Nair, K. K. N., Mohanan, C., & Mathew, G., (2002). *Plantation Technology for Nine Indigenous tree Species of Kerala* (p. 110). KFRI Research Report No. 231.

Nath, S., Nath, A. J., & Das, A. K., (2016). Seed germination in *Barringtonia acutangula*: A floodplain tree from North East India. *Int. J. Ecology and Environmental Sci., 42*(1), 47–53.

Neema, M., Aparna, V., Krishna, P., & Reghunath, B. R., (2018). Pre-sowing seed treatments for Indian indigo (*Indigofera tinctoria*). *Int. J. Innovative Horticulture, 7*(1), 54–55.

Omokhua, Godwin, E., & Alex, A., (2015). Improvement on teak (*Tectona Grandis* Linn. F.) germination for large scale afforestation in Nigeria. *Nature and Science, 13*(3), 68–73.

Parimalam, K., Subramanian, K., Mahalinga, K. S., & Vijayalakshmi, K., (2013). *Seed Storage Techniques: A Primer.* Centre for Indian knowledge systems.

Patil, M. S., & Gaikwad, D. K., (2011). Effect of plant growth regulators on seed germination of oil yielding plant *Simarouba glauca* DC. *Plant Sciences Feed, 1*(5), 65–68.

Pillai, P. K. C., & Chandrashekara, U. M., (2011). *Regeneration Study of Selected Terminalias in Kerala* (p. 62). KFRI Research Report No. 414.

Pillai, P. K. C., (2016). Cost effective method for planting stock production of Palmyra palm. In: *Proceedings of National Conference on "Palmyra Palm"* (pp. 27–31,). Organized by and ASPEE College of Horticulture and Forestry, Navsari Agricultural University.

Pillai, P. K. C., & Hrideek, T. K., (2018). *Study on Reproductive Constrains and Seed Characteristics of Terminalia Paniculata Roth* (p. 39). KFRI Research Report No. 551.

Pillai, P. K. C., & Nimisha, J., (2017). *Seed Characteristics of Memecylon Randerianum, an Endemic Species of Southern Western Ghats* (pp. 5–7). Evergreen, KFRI Newsletter No. 77 & 78.

Pillai, P. K. C., & Pandalai, R. C., (2015). *Storage Practices in Recalcitrant Tropical Forest Seeds of Western Ghats* (p. 33). KFRI Research Report No. 496.

Pillai, P. K. C., Jose, P. A., Sujanapal, P., Jayaraj, R., & Hrideek, T. K., (2018). *Population Analysis, Seed Biology, and Restoration of Hopea erosa and H. racophloea, Two Critically Endangered Trees of Western Ghats* (p. 43). KFRI Research Report No. 548.

Pillai, P. K. C., Sanal, C V., Hrideek, T. K., & Jiji, A. H., (2020). Seed Characteristics and Germination Behaviour of Bauhinia malabarica Roxb. (pp. 1–10), Horticultural crops, IntechOpen, Open Access book publisher, London. DOI: http://dx.doi.org/10.5772/intechopen.84970.

Rachmawati, H., Iriantono, D., & Hansen, C. P., (2002). *Tectona grandis* L.f. *Seed Leaflet. No. 61*. (p. 2). Danida Forest Seed Centre, Denmark.

Raghu, A. V., Deepa, K., Daisy, N. J., & Pillai, P. K. C., (2016). Effect of pre-germination treatments and storage conditions on germination of *Embelia ribes* Burm f. (Bidanga) with special reference to Vrikshayurveda. *Journal of Traditional and Folk Practices, 2–4*(1), 160–163.

Rai, S. N., Nagaveni, H. C., & Padmanabha, H. S. A., (1988). Germination and nursery technique of four species of *Ficus*. *Indian Forester, 114*(2), 63–68.

Rai, Y., (2014). Growth and development of medicinal endangered tree species *Aphanamixis polystachya* (Wall.) Parker in district Meerut, (U.P.) India. *International Journal of Multidisciplinary and Current Research, 2,* 1–4.

Ravi, R., Kalaiselvi, T., & Tilak, M., (2012). Effect of combined treatment of microbial consortia and microwave energy on germination of *Melia dubia* seeds. *International Journal of Scientific and Engineering Research, 3*(9), 1–4.

Roberts, E. H., (1973). Predicting storage life of seeds. *Seed Science and Technology, 1,* 499–514.

Rønne, C., & Jøker, D., (2005). *Kigelia africana* (Lam.) Benth. *Seed Leaflet No. 108* (p. 2). Danida Forest Seed Centre, Denmark.

Sasaki, S., (1980). Storage and germination of dipterocarp seed. *Malaysian Forester, 43,* 292–302.

Schmidt, L., (2000). *Guide to Handling of Tropical and Subtropical Forest Seed.* Danida Forest Seed Centre.

Schmidt, L., (2002). *Guide to Handling Tropical and Subtropical Forest Seed* (p. 532). Danida Forest Seed Centre, Humlebaek, Denmark.

Schmidt, L., (2004). *Xylia xylocarpa* (Roxb.) taub. *Species Leaflet No. 101*. Vietnam Tree Seed Project.

Seena, M., & Udayan, P. S., (2015). Seed germination studies of *Grewia tiliifolia* Vahl (Tiliaceae): A need for conservation. In: Buvaneswaran et al., (eds.) *Advances in Tree Seed*

Science and Silviculture (pp. 133–137). Institute of Forest Genetics and Tree Breeding (IFGTB), Prdag Print, Coimbatore.

Singh, R., Dhillon, G. P. S., Kumar, A., & Gill, R. I. S., (2018). Influence of pre-sowing seed and pre-planting cutting treatments on germination and growth of *Melia composita* Willd. under nursery conditions. *Indian Journal of Agroforestry, 20*(2), 94–99.

Sivakumar, V., Anandalakshmi, R., Warrier, R. R., Singh, B. G., Tigabu, M., & Oden, P. C., (2007). Petroleum flotation technique upgrades the germinability of *Casuarina equisetifolia* seed lots. *New Forests, 34*, 281–291.

Soliman, A. S., & Abbas, M. S., (2013). Effects of sulfuric acid and hot water pre-treatments on seed germination and seedlings growth of *Cassia fistula* L. *American-Eurasian J. Agric. and Environ. Sci., 13*(1), 07–15.

Srivastava, S., Kumar, K., & Malvika, S., (2012). Effect of different pre-sowing treatments on seed germination of *Cassia occidentalis*. *Res in Environment and Life Sci., 5*(3), 153–155.

Sukesh, & Chandrashekar, K. R., (2011). Biochemical changes during the storage of seeds of *Hopea ponga* (Dennst.) Mabberly: An endemic species of Western Ghats. *Research Journal of Seed Science, 4*(2), 106–116.

Sutheesh, V. K., Jijeesh, C. M., & Jiji, A. H., (2015). The influence of seed size variation and pre-sowing treatments on germination and early seedling growth of *Caesalpinia sappan* Linn. *J. Env. Bio-Sci., 29*(1), 161–166.

Swaminathan, C., & Revathy, R., (2013). Improving seed germination in *Sapindus emarginatus* vahl. *Pinnacle Agricultural Research and Management,* 1–3.

Tadros, M. J., Samarah, N. H., & Alqudah, A. M., (2011). Effect of different pre-sowing seed treatments on the germination of *Leucaena leucocephala* (Lam.) and *Acacia farnesiana* (L.). *New Forests, 42*, 397–407.

Thahseen, K., (2018). *Study on Influence of Pre-Sowing Treatments and Medium on Seed Germination of Santalum album L.* MSc thesis, Calicut University.

Thanuja, P. C., Nadukeri, S., Kolakar, S., Hanumanthappa, M., Ganapathi, M., & Vasudev, K. L., (2018). Enhancement of germination and seedling growth attributes of a medicinal tree species *Pterocarpus marsupium* Roxb. through pre sowing seed treatments. *Journal of Pharmacognosy and Phytochemistry (SP3),* 165–169.

Tjelele, T. J., Ward, D., & Dziba, L. E., (2015). The effects of passage through the gut of goats and cattle, and the application of dung as a fertilizer on seedling establishment of *Dichrostachys cinerea* and *Acacia nilotica*. *The Rangeland Journal, 37*(2), 147–156.

Trive, Joshi, A. G., & Nagar, P. S., (2016). Seed germination studies of tree species: *Radermarchera xylocarpa* and *Dolicandrone falcate*. *Bangladesh Journal of Scientific and Industrial Research, 51*(1), 41–46.

Udaykumar, & Chavan, R. L., (2018). Effect of pre-sowing treatments on germination parameters of *Dalbergia latifolia* Roxb. *J. Farm Sci., 31*(1), 99–101.

Venudevan, B., & Srimathi, P., (2013). Conservation of endangered medicinal tree Bael (*Aegle marmelos*) through seed priming. *Journal of Medicinal Plants Research, 7*(24), 1780–1783.

Vineeta, Shukla, G., Pala, N. A., Dobhal, S., & Chakravarty, S., (2018). Influence of seed priming on germination and seedling growth of *Adeanthera pavonina* in sub-humid region of West Bengal, India. *Indian Journal of Tropical Biodiversity, 26*(1), 87–91.

Warrier, K. C. S., (2010). *Thespesia populnea* (L.) Soland ex Coruea. In: Krishnakurnar, et al., (eds.), *Manual of Econornically Important Forestry Species in South India* (pp. 495–505). Institute of Forest Genetics and Tree Breeding, Coimbatore.

CHAPTER 14

Household Economic Dependence on Home Garden Forest Resources in Kashmir Himalaya, India

M. A. ISLAM, P. A. SOFI, G. M. BHAT, A. A. WANI, A. A. GATOO,
A. R. MALIK, NAZIR A. PALA, SHAH MURTAZA, and
KHUSHNOODA ANJUM

Faculty of Forestry, Sher-e-Kashmir University of Agricultural Sciences and Technology of Kashmir, Benhama, Ganderbal–191201, Jammu and Kashmir, India, E-mail: ajaztata@gmail.com (M. A. Islam)

ABSTRACT

The study reports the household economic dependence on home garden forest resources and the role of determinant socioeconomic factors modeling the income contribution in the Budgam district of Kashmir. The study administered a multistage random sampling procedure to withdraw the sample of 106 households from the sample villages having 10% sampling intensity. Secondary data were collected from all possible sources, and primary data were generated through field surveys. The primary data were collected by the personal interviews employing interview schedules and non-participant observations. The data were analyzed using descriptive statistics and an analytical regression model. Results indicated that the home garden forest resources generated an annual income of ₹ 15031.35/ household; of which fruits contributed maximum share (45.33%) followed by wicker (17.88%), fodder (13.34%), fuelwood (10.93%), timber (7.98%), vegetable (2.66%), and medicine (1.88%). The average gross annual income was ₹ 95921.73/household which is differentiated as agriculture (37.60%), livestock (16.09%), business (20.75%), home garden forest resources (15.67%), service (6.90%), wage labor (2.38%), and others (0.61%). Nonetheless, the home garden forest resources are the 4th major contributor of household economy. Regression analysis showed that the socioeconomic factors, namely, education, family size, family labor, farm size, livestock possession, main occupation, wealth status, and gross annual income are the key determinants influencing significantly the home garden forest resources based income and dependency. The study suggested that the home garden forest resources are the key option for socioeconomic development, poverty reduction, and livelihood security; hence, policy must be directed towards the livelihood diversification through sustainable production/collection, extraction, and commercialization of these resources.

14.1   INTRODUCTION

Cultivating multipurpose trees (MPTs), shrubs, herbs, lianas, and ornamentals in home gardens is an age-old farming practice in the Indian sub-continent, largely prevalent among the people inhabiting in hills and valleys (Tynsong and Tiwari, 2010). By and large, the home gardens are characterized by complex multi-layered vegetation that combines almost all the farming components including forest trees, fruit trees, agricultural crops, and medicinal herbs together with livestock within the homestead surroundings (Devi and Das, 2013). Probably, the home gardens confer the humankind's oldest way of farming system and predominant resilient agroforestry land use (Devi and Das, 2010). The home garden is the privately-owned bounded land piece mostly cultivated with a varied combination of annual crops and perennial trees or shrubs around the house (Karyono, 1990). The primary functions of the home gardens are subsistence livelihood and cash income, especially in the rural families (Kumar and Nair, 2004). The rich plant biodiversity ensures the production of a wide variety of multiple-use agricultural and forest/tree products through low costs, labor, planting materials, or other inputs. The diversified and constant production of cereals, horticultural crops, and vegetables through home gardens ensures food and nutritional security during the time of exigency, especially in meeting the micronutrient supplies of the children (Gajaseni and Gajaseni, 1999). Due to its year round production seasonality and large labor engagements, the home gardens are considered to be small-scale ventures for employment and economic security, and the whole production unit is maintained solely by family members, especially the women (Gautam et al., 2008). The home gardens have been regarded as integral traditional farming systems and well-managed complex ecosystems with multiple social, economic, cultural, and ecological functions (Adekuncle, 2013). The multi-storied woody vegetation of trees, palms, bamboos, and shrubs adjoining the homesteads substantially contributes to the ecological sustainability of the village landscapes (Kehlenbeck and Maass, 2004). This forest-like vegetation conserves wild plants, averts soil erosion, provides nutrients/microorganisms for soil development, and improves the microclimate of the habitat (Bhat et al., 2014). The rich plant biodiversity of home gardens enhances biomass production, augments nutrient cycling, and increases biomass and soil carbon. In addition, the home gardens have been recognized as depositories of medicinal, aromatic, and ornamentals plants that play a significant role in traditional remedies, healthcare, and cultural rituals (Bajpai et al., 2013). The significance of home gardens as a safety net in providing substitute livelihood sources during scarcity was emphasized by earlier workers (Schroth and Harvey, 2007). Home gardens, which ensures the production of forest trees, fruit trees, crops, animals, and fish under an integrated farming system, is believed to have higher efficiency than pastures or field crops in terms of production and protection values (Nair et al., 2009). The species composition of home gardens is consistent with the local climatic and biophysical conditions and people's socioeconomic features. The home gardens provide major bulk of the total supply of forest resources in terms of fuelwood, fodder, timber, fruits, vegetable, medicines, and other non-timber forest products (NTFPs) that enable the homestead owners prosperity (Huai and Hamilton, 2009).

Economic valuation studies undertaken in different countries showed that the share of home garden forest income in the total household income was 21.10% in Indonesia and 35.00% in Nicaragua (Mendez et al., 2006). Studies (Motiur et al., 2005, 2006) indicated that homestead owners derived about 15.90% and 11.80% of their annual household income from home garden forests in southwest and northeastern Bangladesh, respectively. In addition to cash income, the home garden forest resources perform the safety net functions by the supply of substitute livelihood alternatives for rural people at the times of crisis, including natural calamities (Kabir and Webb, 2009). Nonetheless, the home garden forest resources substantially contribute to food and nutrition security, annual income and traditional healthcare of rural households besides many socioeconomic and religio-cultural functions, a major concern continues in the economic valuation of these resources as invaluable livelihood component for rural communities. Moreover, the value of home garden forest resources in the household economy is remained unnoticed because of the unsystematic and unorganized system of countrywide data collection in developing nations. The survey of literatures (Islam et al., 2015, 2017; Dar et al., 2018; Bhat et al., 2019) distinctly established the fact that the researches on the production of diverse forest resources in the home gardens and their economic importance in Kashmir valley is lacking where the practice of home gardening is quite old. In light of this, the present research has been undertaken to evaluate the household economic dependence on home garden forest resources and determine the socioeconomic factors influencing the economic contribution in Kashmir Himalaya, India.

14.2   MATERIALS AND METHODS

14.2.1   STUDY AREA

This study was undertaken in Budgam district located between the geographical coordinates of 34°1'12" N and 74°46' 48" E at an altitude of 1610 mamsl in the rugged landscape of Jammu and Kashmir state (Figure 14.1). The district has a total geographic area of 1371 km² and has an area of 477 km² under forest cover. The general topography of the area is both mountainous and plain, and some areas remain inaccessible for quite some time during winter. The district is surrounded in the north by Srinagar and Baramulla, in the south by Pulwama, and in the southwest by Poonch districts. The river Doodhganga, an important offshoot to the Jhelum, runs through the district. The total population of the district is 735753 and the literacy rate is 57.98% (Census of India, 2011). The district is divided into seventeen blocks and has 496 villages, out of which 475 are inhabited and 21 villages are uninhabited. The dominant livelihood strategies among the local people comprise agriculture, horticulture, animal husbandry, forestry interventions, and petty trade. The major land-use types in the district are agricultural fields, non-agricultural lands, forests, fallows, pastures, wastelands, agroforests, and tree groves (Anonymous, 2011). The district is characterized by a temperate monsoon type of climate; its higher regions get heavier snowfalls and experience severe cold in winter. The district has maximum rainfall of 704 mm and temperature ranging between 29.8°C and –1.92°C. The study site covers the landscape of the district situated along the outer stretch of the Special Forest Division in the state.

14.2.2   SAMPLING PROCEDURE

A multistage random sampling method (Kumar, 2012) was administered in the selection of blocks, villages, and households. At the 1st stage, five blocks,

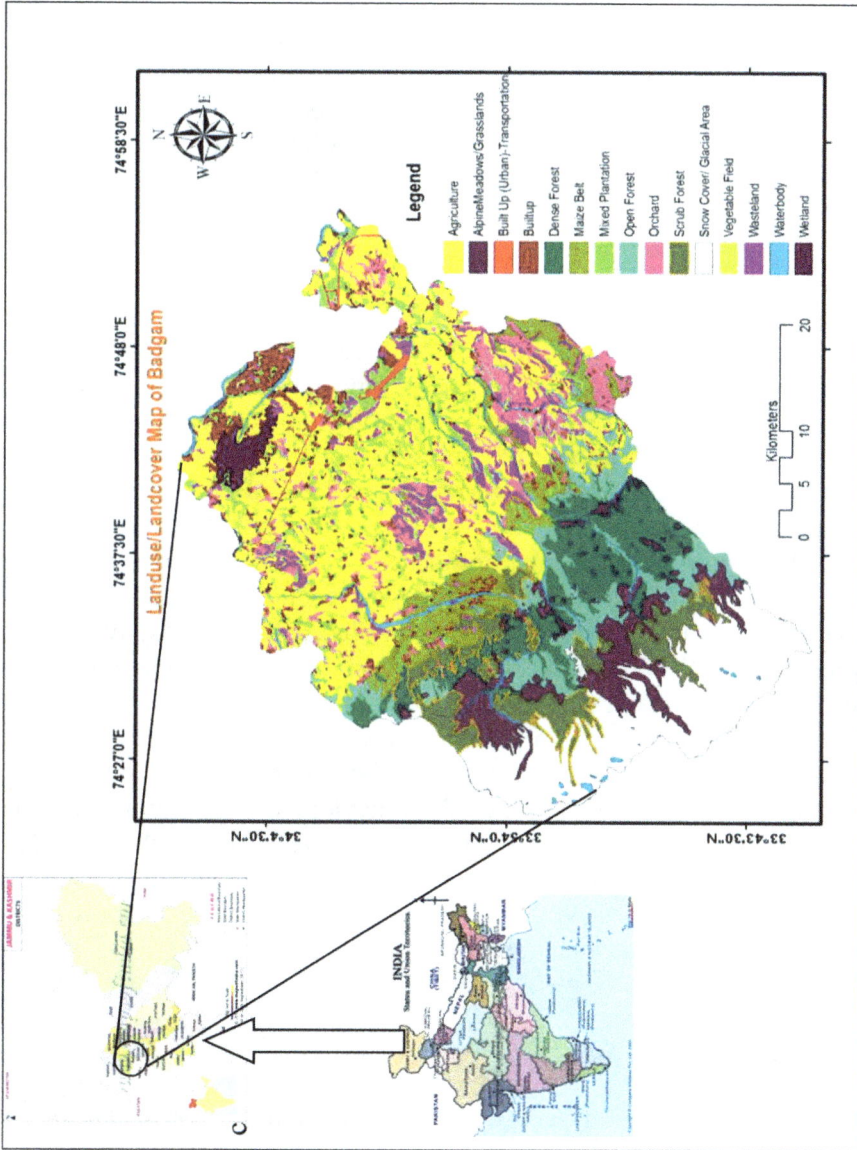

FIGURE 14.1 Map showing the location of the study site.

including Narbal, Pakherpora, Nagam, Ratsun, and Waterhail, were selected randomly in the district. At the 2nd stage, 12 villages, namely, Nowpora, Kawoosa-jagir, and Rakhi-Arath from Narbal block, Karapora, and Shankarpora from Pakherpora block, Wampora, Chowdrigund, and Showpora from Nagam block, Dragger, and Radbugh from Ratsun block and Brail and Udroo from Waterhail block were sampled randomly. At the 3rd stage, 106 households owning home gardens were selected, comprising a sampling intensity of 10% in the selected villages. The interviews were conducted with the family heads or eldest members in the field study.

14.2.3   DATA COLLECTION

Data for this study were gathered at both secondary and primary levels. Secondary information was gathered from all the available sources. The primary data were collected from the structured interviews and non-participant observations administered to the respondents (Tripathi, 1987). The interview schedule was structured based on relevant studies, preliminary survey, and dialog with experts and consultation with knowledgeable people. The interview schedule was comprised of household issues relevant to socioeconomic variables, ecological characteristics of home gardens, forest resources collected/produced, households involvement in production/collection/marketing, collection/annum, consumption/annum, sale of forest resources, rate, and income (₹). The monetary value of the forest resources were estimated by a periodical market survey of the locality. The non-participant observation involved the data collection through personal watching, recording, and inspecting the behaviors in normal situations. The primary data were collected at the household level, whereas the secondary data were collected for blocks, villages, and households/individual level. The scale (Venkataramaiah, 1990) was employed to measure the socioeconomic characteristics of the people (Table 14.1) in the study.

14.2.4   DATA ANALYSIS

The survey data were analyzed by descriptive as well as inferential statistics. Frequency (f), percentage (%), average (x), standard deviation, confidence interval and range were the descriptive statistics (Snedecor and Cochran, 1967) applied for the present analysis. The data were run through the software, including statistical package for social sciences (SPSS) and MS

Excel to test statistical significance ($p < 0.05$), and the results were displayed through tables and graphs.

TABLE 14.1   Explanation of the Household Characteristics

Characteristics (Symbol)	Explanation
Respondent's age(X_1)	Number of years lived by the respondent
Education level (X_2)	6 = graduate and over, 5 = intermediate, 4 = high school, 3 = middle, 2 = primary, 1 = <primary, 0 = illiterate
Social membership (X_3)	4 = public leader, 3 = office bearer, 2 = membership of > 1 organization, 1 = membership of 1 organization, 0 = no membership
Household size (X_4)	2 = >5 members, 1 = ≤ 5 members
Household labor (X_5)	4 = >3 workers, 3 = 3 workers, 2 = 2 workers, 1 = 1 worker
Farm size (X_6)	4 = large (> 4.0 ha), 3 = medium (2.1 to 4.0 ha), 2 = small (1.1 to 2.0 ha), 1 = marginal (up to 1.0 ha), 0 = landless,
Livestock ownership (X_7)	3 = >10 livestock, 2 = 6 to 10 livestock, 1 = ≤5 livestock, 0 = no livestock
Main occupation (X_8)	6 = any other, 5 = service, 4 = business, 3 = cultivation, 2 = caste occupation, 1 = wage labor
Wealth status (X_9)	2 = rich, 1 = medium, 0 = poor
Gross annual income (X_{10})	₹/annum

14.2.5   MODEL SPECIFICATION

Multiple regression analysis (Gujrati and Sangheeta, 2007) was used to determine the socioeconomic factors that influence the home garden forest resources based household income levels. In this case, it was hypothesized that household income accrual by the home garden forest resources is inextricably associated with the household's socioeconomic attributes. In the analysis, home garden forest resources based income was the regress and socioeconomic characteristics were the regressors. The econometric model based on multivariate analysis is stated as follows:

Y   = a + b_1x_1 + b_2x_2 +............. + $b_{10}x_{10}$+ e;
Y   = home garden forest resources based income (₹/annum);
$X_1 - X_{10}$ = household socioeconomic characteristics;
a   = constant or intercept;
$b_1 - b_{10}$ = regression co-efficient;
e   = error term.

14.3   RESULTS AND DISCUSSION

14.3.1   *SOCIOECONOMIC CHARACTERISTICS OF SAMPLE HOUSEHOLDS*

Descriptive statistics (Table 14.2) of household socioeconomic characteristics indicated that the families were headed by middle-aged (41.61), low literate (1.74) people having membership of at least 1 social organization (1.41), with large family size (1.69) and labor force above 3 (3.09). The sample families owned marginal size of farmland (1.09), medium size of herd (2.33) and medium wealth status (1.33). Furthermore, the predominant main occupation was agriculture (3.18) and average gross annual income earned from different livelihood sources was ₹ 95921.73 in the sample households. The dominance of middle-aged family heads in home garden forest resource production is attributable to the fact that these people are relatively more innovative, experienced, enthusiastic, and laborious than the youngers and elderly people. Low literacy among the rural people is because of low socioeconomic status, scarcity of educational amenities, high livelihood concerns, and unawareness towards schooling. The grousing magnitude of social participation of the people is owing to their least interest and willingness towards alliance with social organizations. Considering the child an indispensable asset to a family who can enhance the household earnings and un-acquaintance towards family planning could be the reasons for the prevalence of large-sized families. Possession of large-sized families is unquestionably responsible for high family labor force. The preponderance of marginal farm size is because of neo-local and primary family provisions in rural communities which induced land disintegration from one generation to another and within the married family members. Families with marginal farming land are unable to generate ample earnings for their households; therefore, they depend profoundly upon livestock rearing as the most preferred subsidiary occupation. However, the families possessed numerous and diverse varieties of household wealth items, the general scenario was unsatisfactory, particularly perspective to the modern, improved, and prestigious material assets. Poverty, low socio-economy, illiteracy, unawareness, lack of exposure, etc., are the main reasons for such grousing wealth situations. Agriculture is the backbone of the economy; hence, the largest numbers of rural households are engaged mainly in crop production for cash income, subsistence, and food security. The families occupied in other economic activities, including livestock production, petty trade, cottage industries, horticulture, etc., were also doing agriculture as their secondary occupation. The average annual income among the sampled families was inadequate because the contributions of the core livelihood sources to household annual income are very low. Probable reasons for such low gross annual income might be dominance of marginal farmers or petty businessmen, lack of scientific know-how, low crop production, scarce irrigation avenues, conventional tools and implements, monoculture cropping, inadequate fertility of soil and unpredictable climate that accumulate insignificant earnings to rural households. The findings are inconsistent with earlier workers such as Kabir and Webb (2009), Freedman (2015) and Gbedomon et al. (2015).

TABLE 14.2 Summary of descriptive Information of Household Variables (N = 106)

Factors (Symbol)	Mean	Std. Dev.	Minimum	Maximum	95% CI for Mean	
					Lower	Upper
Respondent's age (X_1)	41.61	10.95	18	70	39.50	43.72
Education level (X_2)	1.74	1.90	0	6	1.37	2.11
Social membership (X_3)	1.41	0.84	0	4	1.25	1.57
Household size (X_4)	1.69	0.46	1	2	1.60	1.78
Household labor (X_5)	3.09	0.95	1	4	2.91	3.27
Farm size (X_6)	1.09	0.29	1	2	1.03	1.15
Livestock ownership(X_7)	2.33	1.12	0	3	2.11	2.54
Main occupation (X_8)	3.33	1.31	1	6	3.07	3.58
Wealth status (X_9)	1.33	0.76	0	2	1.19	1.48
Gross annual income (X_{10})	95921.73	87319.12	35814.00	491814.00	79105.11	112738.35

14.3.2 ECOLOGICAL CHARACTERISTICS OF HOME GARDENS

The survey of 106 sampled home gardens indicated that the average size of home gardens was 0.063 ha with a range of 0.0025–0.50 ha. The total area of home gardens surveyed during the study was 6.68 ha. The number of forest plants per home garden ranged from 43 to 342 with an average of 218.17. In sum, about 77 forest species belonged to 62 genera and 34 families were grown in the surveyed home gardens. Simpson's diversity index varied from 0.11 to 0.32 with an average of 0.21 in the home gardens (Table 14.3). Almost, all the households' posses a home garden adjacent to their home, mostly fenced by wood poles in the study villages. In certain home gardens, live fences of trees including *Salix alba*, *Populus deltoides*,

Robinia pseudoacacia, etc., were utilized for demarcation. The home garden forest plants were distributed in several distinct strata based on the age and size of the species. The layering of vegetation varied from home garden to home garden due to the home garden owner's tree-crop combination, species preference, and management practices for sustainable production. In the Kashmiri home gardens maximum forest species were found in the ground layer. In accordance with the plant habit of the species, of the total 77 forest species recorded in the home garden, the majority (62.34%) were herbs followed by trees (27.27%), shrubs (9.09%) and climber (1.30%) in surveyed home gardens. The home garden forest species were consumed under twelve different use categories among the surveyed population. Corresponding to the use of forest species majority (49) were utilized as fodder followed by fuel (16), vegetable (14), medicines (9), ornamental (9), edible fruit (8), timber (6), fencing materials (7), cottage industry (3), spice (1), toothpick (1) and edible seed (1). According to the plant parts utilized the home garden forest resources were mostly (51) utilized as entire plants followed by leaves (23), branch (15), bole (12), shoot (11), fruit (8), flower (3), seed, and root (1). The outcome indicated that the home garden owners have incredible ecological knowledge of forest species which they accumulated from their ancestors transmitted over generations. The ecological parameters of the Kashmiri home gardens were very similar to the other previously documented home gardens (Rahman et al., 2013; Darcha et al., 2015; Regassa, 2016).

TABLE 14.3 Ecological Characteristics of Home Gardens of the Sample Households (N = 106)

Ecological Characteristics	Statistics	
Home garden size (ha)	Average	0.063
	Range	0.0025–0.50
Total area of home gardens sampled (ha)	6.68	–
Forest plants/home garden	Average	218.17
	Range	43–342
Species	77	–
Genus	62	–
Families	34	–
Simpson's diversity index	Average	0.21
	Range	0.11–0.32

14.3.3 HOUSEHOLD INCOME FROM HOME GARDEN FOREST RESOURCES

The results (Table 14.4) indicated that the extraction and marketing of home garden forest resources fetched a total income of ₹ 1593323.00/annum among the surveyed population @₹ 15031.35/household/annum. Of the average household home garden forest resources based annual income, fruits fetched the maximum share (45.33%) followed by wicker (17.88%), fodder (13.34%), fuelwood (10.93%), timber (7.98%), vegetable (2.66%), and medicine (1.88%). Fuelwood, fodder, timber, and vegetable are the main forest resources produced/collected by more or less every families (91.51 to 100.00%) from home gardens whereas the involvement of households in production/collection of wicker, fruits, and medicines is low (14.15 to 41.51%). However, of the total households interviewed, the proportion of households involved in the sale of home garden forest resources was only 14.15–34.91% (Figure 14.2).

Forest resources collected from home gardens are used mainly for household self-consumption while a small portion is sold for income earnings. Fruits collected from home gardens play a vital role in food and nutritional security; hence, the fruits are collected and sold in considerable quantity. Wicker handicraft is a prominent forest resources based cottage industry which fetched substantial earnings among the sample households. As livestock production is an important subsidiary occupation among the sample households; hence, fodder is an integral forest resource that is collected/produced, consumed, and sold by the local people. The high demand for fuelwood in rural households is merely because of scarcity of inexpensive substitute of energy sources. The small timber collected from home gardens were mainly consumed for building houses and repairing, huts, and fences, domestic furnishings, agricultural tools, scaffolding, ladders, electric/telephone props, etc., and hence, it fetches a handsome return to the sellers after rotational harvesting. Wild vegetables collected from home gardens are the major source of health and nutrition, which play a vital role in food security. Although the home gardens produce a variety of medicines, but they are mostly consumed domestically; only a little fraction is marketed, yielding a paltry income to the home garden owners. Harvesting and selling of home garden forest resources is the main income component in rural families in Kashmir Himalaya. The home garden forest resources provide both subsistence and cash income, which contributes substantially to household food and livelihood security. The studies (Kebebew et al., 2011; Islam et al., 2013; Nath et al., 2014) across the globe have emphasized the potential role of

TABLE 14.4   Household Income by Different Home Garden Forest Resources (N = 106)

Forest Resources	Collection (tons/annum)	Consumption (tons/annum)	Sale (tons/annum)	Total Income (₹/annum)	Mean Income (₹/annum)	Standard Error	Income Share (%)
Fuelwood	125.74	96.73	29.01	174060	1642.07	216.36	10.93
Fodder	123.92	88.51	35.41	212460	2004.34	244.88	13.34
Timber	33.39^Δ	19.08^Δ	14.31^Δ	127200	1200.00	167.82	7.98
Wicker	2.28	0.00	2.28	284900	2687.74	259.12	17.88
Fruits	83.82	47.70	36.12	722303.08	6814.18	666.16	45.33
Vegetable	9.01	7.95	1.06	42400.00	400.00	43.28	2.66
Medicine	0.52	0.28	0.24	30000.00	283.02	29.92	1.88
Total	–	–	–	1593323.00	15031.35	1469.48	100

Note: $\Delta = m^3$.

home garden forest resources in meeting the food, livelihood, and health security besides cash income and safety net functions.

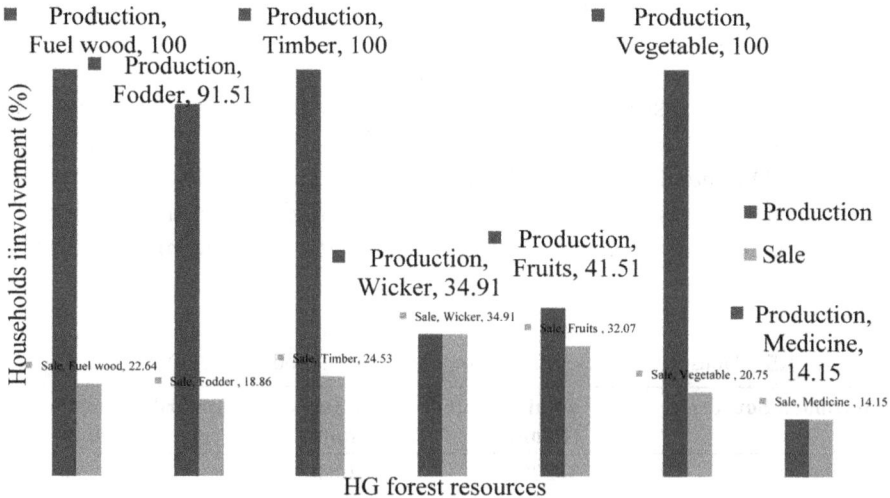

FIGURE 14.2 Household involvements in production and marketing of home garden forest resources.

14.3.4 *LIVELIHOOD SOURCES AND ECONOMIC CONTRIBUTION OF HOME GARDEN FOREST RESOURCES*

The annual gross income including all non-farm and farm income sources was ₹ 95921.73/household/annum in the sample households (Table 14.5). Agricultural income contributed the major share (37.60%) which was followed by livestock (16.09%), business (20.75%), home garden forest resources (15.67%), service (6.90%), wage labor (2.38%), and others (0.61%) (Figure 14.3). Hence, the home garden forest resources are the 4th major contributor of rural household economy.

The finding implies that the home garden forest resources constitute a dominant constituent of rural economy since it accounts significant share in the total households' income. The production/collection of home garden forest resources is the prominent livelihood intervention for survival, currency, and safety net since the alternative sources are lacking in rural Kashmir. In several households, the production/collection of forest resources through home gardens is just their complementary livelihood source on a part-time basis. Likewise, the home garden forest resources are used by all categories of people in the society, whether poor or wealthy, literate or illiterate, rural or urban. However, the income accrual through the home garden forest products is rather little; involvement in the activities is a matter of self-respect, honor, and self-reliance. The income earned from the home garden forest resources are spent to secure domestic basic needs in terms of children's education, healthcare expenditures, wedding expenses, agricultural investments, assets for entrepreneurial activities, savings as safety nets, and others. The researches (Kebebew et al., 2011; Zimik et al., 2011; Nath et al., 2014; Darcha et al., 2015; Neelamegam et al., 2015; Regassa, 2016) on home gardens forest resources across the world advocated that the income from forest resources have substantial contributions to the rural household livelihoods.

TABLE 14.5   Household Annual Income by Major Livelihood Sources (N = 106)

Livelihood Sources	Total Income (₹/annum)	Mean Income (₹/annum)	Standard Error	Income Share (%)
Agriculture	3822554.00	36061.83	2791.21	37.60
Livestock	1635789.00	15431.97	1946.09	16.09
Business	2110103.00	19906.63	1012.82	20.75
Service	701793.10	6620.69	362.10	6.90
Wage labor	241801.90	2281.15	801.14	2.38
Home garden forest resources	1593323.00	15031.35	81.81	15.67
Others	62339.66	588.11	173.37	0.61
Total	**10167704.00**	**95921.73**	**8851.84**	**100.00**

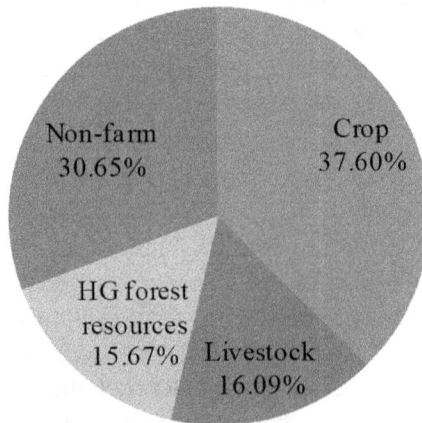

FIGURE 14.3   Income contribution of major livelihood sources (N = 106).

14.3.5 DETERMINANTS OF ECONOMIC DEPENDENCE ON HOME GARDEN FOREST RESOURCES

Multiple linear regression model (Table 14.6) was carried out to identify the household's socioeconomic factors influencing the home gardens forest resources based annual income. The (coefficient of determination) value of the analysis was 0.872, which indicated that the socioeconomic variables altogether had contributed to 87.20% variation in home garden forest resources based annual income. The magnitude of F value (64.48) signified that the coefficient of determination (R^2) is significant ($p<0.05$) which clearly showed that the model is very strong, reliable, and has high predictive ability. The 't' values of regression analysis revealed that among the ten household variables, eight variables namely, education, family size, family labor, farm size, livestock possession, wealth status, main occupation and annual income had substantial contribution in influencing home garden forest resources based income. The explicit form of multiple regression equation fitted for household home garden forest resource-based income is presented as:

$$Y = 12675.77 + 1.486X_1 + 65.066X_2 + 37.480X_3 + 119.507X_4 \\ + 178.172X_5 + 421.673X_6 + 191.761X_7 + 42.261X_8 \\ + 187.924X_9 + 0.001X_{10}$$

where; Y = household home garden income (₹/annum).

$X_1 - X_{10}$ = socioeconomic characteristics.

The education of the household heads has a direct influence in the adoption behavior of occupational activities and subsequently the income accrual. Moreover, the education helps the household heads in understanding the problems and prospects of forest resources production through home gardening for further improvement. Thus, education plays a crucial in awareness creation, motivation, enhancing scientific expertise, production, and management of home garden forest resources. The family size has direct linkage with the quantity of forest resource exploitation. This implies that larger families exploit more home garden forest resources than their counterparts with smaller families. The family labor is the prime input for the home gardens forest resources production/collection, protection, and management, processing, consumption, and marketing. Farm size is the prominent physical asset which has direct linkage with the forest resources production at household level. Livestock keeping provides people with another important source of household food security besides employment and income opportunities at the homestead level. Main occupation, wealth status, and gross annual income are the dominant socioeconomic factors which have a direct effect on home garden forest resources based income. Hence, the variations in the magnitudes of these household variables are directly proportional to the variations in home garden forest resources production, use, and income. There is a multitude of studies (Guuroh et al., 2013; Rahman et al., 2013; Nath et al., 2014; Gbedomon et al., 2015; Regassa, 2016) which emphasized that the household characteristics are vital predictors in home garden forest resources based income.

TABLE 14.6 Regression of Household Home Garden Forest Resource Income against Socioeconomic Characteristics

Socioeconomic Characteristics (Symbol)	Coefficient (b)	SE of 'b'	B	t-Value
Respondent's age (X_1)	1.486	2.346	0.026	0.638
Education level (X_2)	65.066	18.597	0.196	3.499*
Social membership (X_3)	37.480	47.906	0.050	0.782
Household size (X_4)	119.507	59.694	0.087	2.002*
Household labor (X_5)	178.172	35.566	0.267	5.010*
Farm size (X_6)	421.673	131.892	0.195	3.197*
Livestock ownership(X_7)	191.761	25.109	0.341	7.637*
Main occupation (X_8)	42.261	19.907	0.088	2.123*
Wealth status (X_9)	187.924	39.039	0.227	4.814*
Gross annual income (X_{10})	0.001	0.000	0.104	2.107*

a = 12675.77; F = 64.48*; R^2 = 0.872; Multiple R = 0.934; Adjusted R^2 = 0.858.
* = Significant at 5% level of probability

14.4 CONCLUSION

The findings led to conclude that the home gardens play a crucial role in rural livelihoods by providing various resources such as fruits, wicker, fodder, fuelwood, timber, vegetable, and medicine for domestic use and contribute substantially to the gross annual income besides acting as safety net in cases of exigency. Such livelihood contributions of home garden forest resources must be given due recognition in rural developmental schemes and land-use prioritizations to harmonize socioeconomic development, poverty alleviation and livelihood security of the local communities. Further, the potential opportunities for economic diversification through value addition of home garden forest resources, fortunate marketing, and better commercialization should be explored and accordingly, capacity building and skill development of stakeholders on production of valuable forest resources, sustainable harvesting, value addition and commercialization should be strengthened. Likewise, results of the regression analysis indicated that the economic support from home garden forest resources depends on a multitude of household socioeconomic factors like education, family size, family labor, farm size, livestock possession, main occupation, wealth status, and gross annual income. Hence, these factors should be given due consideration during planning, implementation, and execution of specific strategies for improvement and strengthening of home gardens by the scientists, extension providers, and policymakers.

KEYWORDS

- **forest resources**
- **home garden**
- **livelihood**
- **livestock possession**
- **non-timber forest product**
- **wealth status**

REFERENCES

Adekuncle, O. O., (2013). The role of Home gardens in household food security in Eastern Cape: A case study of three villages in Nkonkobe Municipality. *Journal of Agricultural Science, 5*(10), 67–76.

Anonymous, (2011). *Directorate of Economics and Statistics*. District Statistics and Evaluation Office, Budgam, Jammu and Kashmir.

Bajpai, S., Sharma, A. K., & Kamungo, V. K., (2013). Traditional gardens: A preserve of medicinal plants. *International Journal of Herbal Medicine, 1*(2), 152, 161.

Bhat, G. M., Islam, M. A., Malik, A. R., Rather, T. A., Shah, K. F. A., & Mir, A. H., (2019). Productivity and economic evaluation of willow (*Salix alba* L.) based silvopastoral agroforestry system in Kashmir valley. *Journal of Applied and Natural Science, 11*(3), 743–751.

Bhat, S., Bhandary, M. J., & Rajanna, L., (2014). Plant diversity in the home gardens of Karwar, Karnataka, India. *Biodiversitas, 15*(2), 229–235.

Census of India, (2011). *A-5 State Primary Census Abstract*. India.

Dar, M., Qaisar, K. N., Ahmad, S., & Wani, A. A., (2018). Inventory and composition of prevalent agroforestry systems of Kashmir Himalaya. *Advances in Research, 14*(1), 1–9.

Darcha, G., Birhane, E., & Abadi, N., (2015). Woody species diversity in *Oxytenanthera abyssinica* based homestead agroforestry systems of serako, Northern Ethiopia. *Journal of Natural Sciences Research, 5*(9), 18–26.

Devi, N. L., & Das, A. K., (2010). Plant species diversity in the traditional homegardens of Meitei community: A case study from Barak Valley, Assam. Journal of Tropical Agriculture, 48, 40–43.

Devi, N. L., & Das, A. K., (2013). Diversity and utilization of tree species in Meitei home gardens of Barak Valley, Assam. *Journal of Environmental Biology, 34*(1), 211–217.

Freedman, R. L., (2015). Indigenous wild food plants in home gardens: Improving health and income, with the assistance of agricultural extension. *International Journal of Agricultural Extension, 3*(1), 63–71.

Gajaseni, J., & Gajaseni, N., (1999). Ecological rationalities of the traditional home garden system in the chao Phraya basin, Thailand. *Agroforestry Systems, 46,* 13–23.

Gautam, R., Sthapit, B., Subedi, A., Poudel, D., Shrestha, P., & Eyzaguirre, P., (2008). Home gardens management of key species in Nepal: A way to maximize the use of useful diversity for the well-being of poor farmers. *Plant Genetic Resources Characterization and Utilization NIAB,* 1–12.

Gbedomon, C., Fandohan, A. B., Salako, V. K., Idohou, A. F. R., Kakai, R. G., & Assogbadjo, A. E., (2015). Factors affecting home gardens ownership, diversity and structure: A case study from Benin. *Journal of Ethnobiology and Ethnomedicine, 11*(56), 2–15.

Gujarati, D. N., & Sangeetha, (2007). *Basic Econometrics.* Tata McGraw-Hill Publishing Company Limited, New Delhi, India.

Guuroh, R. T., Uibrig, H., & Acheampong, E., (2013). *How Does Home Garden Size Affect Input and Output Per Unit Area?: A Case Study of the Bieha District, Southern Burkina Faso* (pp. 34–42). Conference on International Research on Food Security, Natural Resource Management and Rural Development organized by the University of Hohenheim.

Hui, H. & Hamilton, A. (2009). Characteristics and functions of traditional homegardens: A review. Frontiers of Biology in China, 4, 151–157.

Islam, M. A., Masoodi, T. H., Gangoo, S. A., Sofi, P. A., Bhat, G. M., Wani, A. A., Gatoo, A. A., et al., (2015). Perceptions, attitudes and preferences in agroforestry among rural societies of Kashmir, India. *Journal of Applied and Natural Science, 7*(2), 976–983.

Islam, M. A., Qaisar, K. N., & Bhat, G. M., (2017). Indigenous knowledge in traditional agroforestry systems of Kashmir valley: Current challenges and future opportunities. *International Journal of Forestry and Crop Improvement, 8*(1), 68–77.

Islam, S., Miah, Q., & Habib, M., (2013). Diversity of fruit and timber tree species in the coastal homesteads of southern Bangladesh. *Journal of Asiatic Society Bangladesh, 39*(1), 83–94.

Kabir, M. E., & Webb, E. L., (2009). Household and home garden characteristics in southwestern Bangladesh. *Agroforestry System, 75*(2), 129–145.

Karyono, (1990). Home gardens in java: Their structure and functions. In: Landauer, K., & Brazil, M., (eds.), *Tropical Home Gardens* (pp. 138–146). Tokyo, Japan: United Nations University Press.

Kebebew, Z., Garedew, W., & Debela, A., (2011). Understanding home garden in household food security strategy: Case study around Jimma, Southwestern Ethiopia. *Research Journal of Applied Sciences, 6*(1), 38–43.

Kehlenbeck, K., & Maass, B. L., (2004). Crop diversity and classification of home gardens in Central Sulawesi, Indonesia. *Agroforestry Systems, 63,* 53–62.

Kumar, B. M., & Nair, P. K. R., (2004). The enigma of tropical home gardens. *Agroforestry Systems, 61,* 135–152.

Kumar, R., (2012). *Research Methodology: A Step Guide for Beginners.* Dorling Kindersley India Pvt., Ltd. New Delhi, India.

Mendez, V. E., Lok, R., & Somarriba, E., (2006). Interdisciplinary analysis of home gardens in Nicaragua: Micro-zonation, plant use and socioeconomic importance. *Agroforestry Systems, 51*(2), 85–96.

Motiur, M. R., Furukawa, Y., & Kawata, I., (2005). Homestead forest resources and their role in household economy: A case study in the villages of Gazipur Upazilz of central Bangladesh. *Small-Scale Forest Economics, Management and Policy, 4,* 359–376.

Motiur, R. M., Furukava, Y., Kawata, I., Rahman, M., & Alam, M., (2006). Role of homestead forest in household economy and factors affecting forest production: A case study in southwest Bangladesh. *Journal of Forest Research, 11,* 89–97.

Nair, P. K. R., Kumar, B. M., & Nair, V. D., (2009). Agroforestry as a strategy for carbon sequestration. *Journal of Plant Nutrition and Soil Science, 172*(1), 10–23.

Nath, T. K., Aziz, N., & Inoue, M., (2014). *Contribution of Homestead Forests to Rural Economy and Climate Change Mitigation: A Study from the Ecologically Critical Area of Cox's Bazar-Teknaf Peninsula, Bangladesh* (Vol. 13, No. 2, pp. 66–73). Springer.

Neelamegam, R., Pillai, V. M., Priyanka, A. M. A., & Roselin, S., (2015). Status and composition of home garden plants in rural and urban areas in Kanyakumari District, Tamil Nadu, India. *Scholars Academic Journal of Biosciences, 3*(8), 656–667.

Rahman, S., Baldauf, C., Molle, E., & Al-Pavel, M., (2013). Cultivated plants in the diversified home gardens of local communities in Ganges valley, Bangladesh. *Science Journal of Agricultural Research and Management, 197,* 2276–8572.

Regassa, R., (2016). Useful plant species diversity in home gardens and its contribution to household food security in Hawassa city, Ethiopia. *African Journal of Plant Science, 10*(10), 211–233.

Schroth, G. F., & Harvey, C. A., (2007). Biodiversity conservation in cocoa production landscapes: An overview. *Biodiversity Conservation, 16,* 2237–2244.

Snedecor, G. W., & Cochran, W. G., (1967). *Statistical Methods.* Iowa State University Press, Ames, Iowa-50010.

Tripathi, P. C., (1987). *A Text Book of Research Methodology in Social Sciences*, Sultan Chand and Sons, 216 pages.

Tynsong, H., & Tiwari, B. K., (2010). Plant diversity in the home gardens and their significance in the livelihoods of *War Khasi* community of Meghalaya, North-east India. *Journal of Biodiversity, 1*(1), 1–11.

Venkataramaiah, P., (1990). *Development of Socioeconomic Status Scale.* PhD thesis, Department of Agricultural Extension, University of Agriculture Sciences, Bangalore.

Zimik, L., Saikia, P., & Khan, M. L., (2012). Comparative study on home gardens of Assam and Arunachal Pradesh in terms of species diversity and plant utilization pattern. *Research Journal of Agricultural Sciences, 3*(3), 611–618.

Index

For Product Safety Concerns and Information please contact our EU
representative GPSR@taylorandfrancis.com
Taylor & Francis Verlag GmbH, Kaufingerstraße 24, 80331 München, Germany